Functional Analysis

Functional Analysis

Second Edition

PAWAN K JAIN

Ph.D., F.N.A.Sc.

Emeritus Fellow (U.G.C.)
Formerly, Professor and Head
Department of Mathematics, University of Delhi
Delhi, India

OM P AHUJA

Ph.D.

Associate Professor of Mathematics
Kent State University, Ohio, USA

New Age Science Limited

The Control Centre, 11 A Little Mount Sion
Tunbridge Wells, Kent TN1 1YS, UK
www.newagescience.co.uk • e-mail: info@newagescience.co.uk

ISBN : 978 1 906574 67 3

British Library Cataloguing in Publication Data
A Catalogue record for this book is available from the British Library

Every effort has been made to make the book error free. However, the author and publisher have no warranty of any kind, expressed or implied, with regard to the documentation contained in this book.

Printed and bound in India by Replika Press Pvt. Ltd.

PREFACE

Functional Analysis, which provides a unifying framework for many areas of mathematics, e.g. Function Theory, Measure and Integration Theory, Approximation Theory, Distribution Theory, Analytic Number Theory, Fourier Analysis, Sobolev Space Theory, etc. plays a significant role in pure and applied sciences. It is the key to solve different types of problems in mathematics, physics, mechanics, structural engineering, operations research, system analysis, economics, biotechnology, military sciences, medical sciences as well as other disciplines. In this way, the knowledge of functional analysis is quite essential to pursue mathematics, especially applied mathematics and applied sciences.

This book, *Functional Analysis*, based on our lecture notes prepared for teaching the subjects in various countries, has been designed to cater to the need of students who are yet to be exposed to the subject as well as senior undergraduate and graduate level students in various universities the world over. The first edition of the book was published in 1995. Since then many developments have taken place in the subject. We too had to revise our lecture notes based both on developments of the subject as well as the need for providing clarity to our students so as to present it within the comprehension of the subject. As a result of all this, the new edition, written in a simple and lucid language, and illustrated with familiar examples, it should be an ideal textbook designed for easy comprehension of the subject. It should be useful to students of applied mathematics, statistics, engineering and theoretical physics in addition to those pursuing the subject at undergraduate and graduate level.

The text has been arranged in sections, spread out in seven chapters and one appendix, beginning with a chapter on preliminaries and providing a sequence of definitions and theorems. This chapter is intended to establish a uniform notation and covers the background material in real analysis, linear algebra and metric spaces. It is followed by vital information on functional analysis; namely, normed and Banach spaces, bounded linear operators, bounded linear functionals, etc. An appendix at the end of the book provides an introduction to Schauder Bases. It has its own importance as it has several interesting properties which are pointed out in various chapters of the book.

Each chapter of the book starts with a brief introduction intended to invoke interest of the students in what follows in the text. The book is profusely illustrated with familiar examples to bring home to the students the underlying concept of the subject and in solving concrete problems. Most of the theoretical portions have also been introduced to the students by citing examples. This, we feel, should help in easy comprehension of the subject. Wherever essential, students are provided with hints to solve the problems. The problems graded in a proper way are presented throughout each chapter.

Without claiming originality of results we do claim simplicity and lucidity of presentation coupled with comprehensiveness of the material. The various sources that have inspired the authors are listed in the bibliography. Yet the works of Kreyszig and Simmons have made contribution in making the book useful for the students. We have not tried to attribute the various theorems and proofs to their original discoverers. However, we take it as our duty to place on record our gratitude to all of them. In fact, many distinguished authors have contributed immensely to the subject. Moreover, we are thankful to the generation of students who have made valuable contributions in injecting simplicity in presentation of the material so as to the intelligible to the students' community in general.

We had an opportunity to show and discuss the text and its presentation with many of our colleagues and are indeed grateful to them for their valuable suggestions. All this has been taken into account while bringing out the Second Edition of the book. Our approach in presentation of the subject has been to make it as simple as possible while ensuring that the interest of students is sustained in the subject through logical instances.

The authors are grateful to the readers who made valuable suggestions for improvement of the book, especially to Dr. Pankaj Jain and Dr. Arun Pal Singh who read the book thoroughly and, besides, pointing out several printing mistakes, gave various ideas and suggestions for the benefit of the readers. In future also, the authors will be happy to receive suggestions and comments for further improvement.

We owe our gratefulness to our respective families for their moral support they have been extending to us to concentrate on this project.

P. K. Jain

O. P. Ahuja

LIST OF SYMBOLS

$\in$	belongs to
$\notin$	does not belong to
$\subset$	subset
$\subsetneq$	proper subset
$\supset$	contains or super set
$\cup$	union
$\cap$	intersection
$\forall$	for all
$\exists$	there exists
$\blacksquare$	Q.E.D.
$\Rightarrow$	implies
$\Leftrightarrow$	implies and is implied by (or if and only if)
$\sim$	equivalent
$\leq$	partial order
ϕ	empty set
$A - B$	set of all points in A which are not in B
$A \times B$	cartesian product of A and B
$A \sim B$	A is equivalent to B
A^c	complement of A
A°	interior of A
A'	set of all limit points of A
$\overline{A}$	closure of A
$\mathcal{P}(A)$	power set of A
$\mathbb{N}$	set of natural numbers
$\mathbb{Z}$	set of integers

$\mathbb{Q}$	set of rational numbers
$\mathbb{R}$	set of real numbers
$\mathbb{C}$	set of complex numbers
C	Cantor set
$[a, b]$	closed interval
$]a, b[$	open interval
$[a, b[$	interval closed at a and open at b
$]a, b]$	interval open at a and closed at b
$\overline{\overline{A}}$	cardinal number of A
$\aleph_0$	aleph nought, the cardinal number of $\mathbb{N}$
c	cardinal number of the continuum
$f: A \to B$	function from A to B
I	identify function
gof	composite function
$f\vert_M$	restriction of f to the set M
$d(x, y)$	distance between x and y
$d(x, A)$	distance of x from the set A
$d(A, B)$	distance between sets A and B
$d(A)$	diameter of A
$\inf A$	infimum (or greatest lower bound) of A
$\sup A$	supremum (or least upper bound) of A
$S(x ; r)$	open sphere with radius r and centre at x
$S[x ; r]$	closed sphere with radius r and centre at x.
$\mathbb{R}^n$	n-dimensional Euclidean space
$\mathbb{C}^n$	n-dimensional unitary space
ω	space of all sequences
c	space of convergent sequences
c_0	space of sequences converging to 0
l^P	a sequence space
l^∞	space of bounded sequences
bv	space of all sequences of bounded variation
bts	space of all sequences of bounded partial sums
$P[a, b]$	space of all polynomials defined on $[a, b]$

$P_n\,[a,\,b]$	space of all polynomials defined on $[a,\,b]$ of degree $\leq n$
$B\,[a,\,b]$	space of all bounded functions on $[a,\,b]$
$BV\,[a,\,b]$	space of all functions of bounded variation on $[a,\,b]$
$C\,[a,\,b]$	space of all continuous functions on $[a,\,b]$
$C^n\,[a,\,b]$	space of all n-times continuously differentiable functions on $[a,\,b]$
$C_0\,(\mathbb{R})$	space of continuous functions vanishing at ∞
$L^p\,[a,\,b]$	space of p-integrable functions on $[a,\,b]$
$L^\infty\,[a,\,b]$	space of essentially bounded measurable functions on $[a,\,b]$
X/M	quotient space
$\dim X$	dimension of X
$\operatorname{codim}_X Y$	codimension of Y in X (dimension of X/Y)
$\mathscr{D}\,(T)$	domain of T
$\mathscr{R}\,(T)$	range of T
$\mathscr{N}(T)$	null space of T
$\ker T$	kernel of T
$\|\ \|_X$	norm on X
$L\,(X,\,Y)$	space of all linear operators from X into Y
$B\,(X,\,Y)$	space of all bounded linear operators from X into Y
$B\,(X)$	$B\,(X,\,X)$
$\operatorname{span} A$	set of all linear combinations of elements in A
$\overline{\operatorname{span} A}$	closure of span A
$M^\perp$	set, orthogonal to M
$M^{\perp\perp}$	set, orthogonal to $M^\perp$
$\mathscr{S}\,(H)$	space of all self-adjoint operators on H
$\mathscr{S}^+\,(H)$	space of all positive operators on H
$\mathscr{N}\,(H)$	space of all normal operators on H
$\mathscr{U}(H)$	space of all unitary operators on H
$T^\times$	adjoint operator of T
T^*	Hilbert-adjoint operator of T
$X^{\#}$	algebraic dual of X
$X^{\#\#}$	algebraic dual of $X^{\#}$ (or algebraic bidual of X)
X^*	topological dual of X
X^{**}	topological dual of X^* (or topological bidual of X)

$\pi : X \to X^{**}$ canonical imbedding

$\rho(T)$ resolvent set of T

$R_\lambda(T)$ resolvent operator of T

$\sigma(T)$ spectrum of T

$\sigma_c(T)$ continuous spectrum of T

$\sigma_p(T)$ point specturm of T

$\sigma_r(T)$ residual spectrum of T

$r_s(T)$ spectral radius of T

$\det A$ determinant of matrix A

SCHEME OF NUMBERING

- Chapters are numbered by Arabic numerals.
- Sections are numbered in two digits. The first digit indicates the chapter while the second one the section of that chapter.
- Within each section, all items such as Definitions, Examples, Lemmas, Theorems and Corollaries, irrespective of the nature of the item, are numbered by using a period after the section number.
- Equations or display are numbered serially, within small parenthesis, for each chapter.
- Diagrams are numbered serially in two digits for each chapter. The first digit indicates the chapter while the second one the serial number of the diagram of that chapter.
- Problems are numbered serially for each chapter.

CROSS-REFERENCING

- A reference to Problem 5 of Chapter 3 is made as Problem 5 within Chapter 3 and as Problem 3.5 in other chapters.
- A reference to Example 4 of Examples 5.1.6 is made as Example 4 within the Examples 5.1.6 and Example 5.1.6. (4) elsewhere.

CONTENTS

Preface   *v*

List of Symbols   *vii*

Scheme of Numbering   *xi*

1 Preliminaries   1–60

1.1   Sets   1

1.2   Relations   4

1.3   Functions   6

1.4   Countable Sets, Uncountable Sets and Cardinal Numbers   12

1.5   Inequalities   14

1.6   Linear Spaces   16

1.7   Metric Spaces   37

2 Normed and Banach Spaces   61–99

2.1   Definitions and Elementary Properties   61

2.2   Some Concrete Normed and Banach Spaces   68

2.3   Subspaces   89

2.4   Quotient Spaces   92

2.5   Completion of Normed Spaces   96

3 Bounded Linear Operators   100–150

3.1   Definitions, Examples and Basic Properties   100

3.2   Spaces of Bounded Linear Operators   111

3.3   Equivalent Norms   116

3.4 Finite Dimensional Normed Spaces and Compactness 123

3.5 Open Mapping Theorem and its Consequences 131

3.6 Closed Graph Theorem and its Consequences 138

3.7 Uniform Boundedness Principle 145

4 Bounded Linear Functionals 151–232

4.1 Definitions, Examples and Basic Properties 152

4.2 The Form of Some Dual Spaces 164

4.3 Hahn-Banach Theorem and its Consequences 179

4.4 Embedding and Reflexivity of Normed Spaces 195

4.5 Adjoint of Bounded Linear Operators 204

4.6 Weak Convergence 212

4.7 Weak* Convergence 222

5 The Concept and Specific Geometry of Hilbert Spaces 233–286

5.1 Definitions and Basic Properties of Inner Product Spaces and Hilbert Spaces 234

5.2 Completion of Inner Product Spaces 248

5.3 Orthogonality of Vectors 250

5.4 Orthogonal Complements and Projection Theorem 254

5.5 Orthonormal Sets and Fourier Analysis 262

5.6 Complete Orthonormal Sets 279

6 Functionals and Operators on Hilbert Spaces 287–333

6.1 Bounded Linear Functionals 287

6.2 Hilbert-Adjoint Operators 295

6.3 Self-Adjoint Operators 308

6.4 Normal Operators 320

6.5 Unitary Operators 323

6.6 Orthogonal Projection Operators 327

7 Introduction to Spectral Theory

7.1 Eigenvalues of a Linear Operator 334

7.2 The Spectrum of a Bounded Linear Operator 338

7.3 Spectral Properties of Bounded Linear Operators 344

7.4 Complex Analysis and Spectral Theory 349

Appendix—Schauder Bases   356–366

Bibliography   367–369

Index   371–375

Preliminaries

1

CHAPTER

This chapter is to help the reader in reviewing the preliminaries needed subsequently in this book. It is presumed that the reader has pursued an elementary course is real analysis, linear algebra and topology. The approach adopted in this chapter is somewhat different from that used in other chapters. It is descriptive and the arguments given are directly towards plausibility and understanding rather than rigorous proofs. The preliminaries are divided into seven sections.

1.1 Sets

A *set* is a well-defined collection or system of objects. Other words such as *collection, class* and *aggregate* are used synonymously for the term set. 'Well-defined' means that it is possible to determine readily whether an object is a member of a given set or not. The objects that belong to a set are called its *elements* (or *points* or *members*). If A is a set, then $a \in A$ denotes that a is an element of A and the notation $a \notin A$ denotes the negation of $a \in A$. For any element a and a set A, either $a \in A$ or $a \notin A$.

Two methods used frequently to describe sets are the '*tabulation method*' and the '*defining-property method*'.

The first, the tabulation method, enumerates or lists the individual elements separated by commas and enclosed in braces. By this method, the set of vowels of English alphabets is written as $\{a, e, i, o, u\}$. Sets which are difficult to describe by an enumeration are described by the second method, the defining-property method. In fact, this method is often more compact and convenient. A defining property of a set is the property which is satisfied by each element of that set and by nothing else. The standard notation for a set so described is $\{x \mid \}$ or $\{x : \}$. Here x is a dummy symbol and the space between : and } is filled by a defining property. The above set by this method is described as $\{x : x$ is a vowel of English alphabets$\}$.

Given two sets A and B, if the relation $a \in A$ implies $a \in B$ for all a, we say that A is a *subset* of B (or B is a *superset* of A, or A is contained in B or B contains A). In symbol, it is written as $A \subset B$ or $B \supset A$.

Two sets A and B are equal if $A \subset B$ and $B \subset A$. Generally, a set is completely determined by its elements but there is a set which has no elements, and we call it as the *empty* (or *void* or *null*) *set* and denote it by ϕ (phi). If A is any set, each element of ϕ (there is none) is an element of A, and so $\phi \subset A$. Thus, the empty set is a subset of every set. Further, if A is a subset of B with $A \neq \phi$ and $A \neq B$, the A is a *proper subset* of B (or B properly contains A). In other words, set A is a proper subset of B if and only if $a \in A$ implies $a \in B$, and there exists atleast one $b \in B$ such that $b \notin A$ and $A \neq \phi$.

Let A be a set. Then, the collection of all subsets of A is called the *power set* of A and is denoted by $\mathscr{P}(A)$. For instance, if A is a set containing a, b and c as its elements, there are eight subsets of A. Hence, the power set $\mathscr{P}(A)$ would contain eight elements, each being a subset of A. It is obvious that the sets ϕ and A are always members of $\mathscr{P}(A)$. In particular, $\mathscr{P}(A)$ is always a non-empty set. If A is a finite set containing n (distinct) elements, $\mathscr{P}(A)$ has 2^n elements and this is reason for the name 'power set'.

The following sets are of particular interest to us throughout the book:

$\mathbb{N}$: Set of all natural numbers.

$\mathbb{Z}$: Set of all integers.

$\mathbb{Q}$: Set of all rational numbers (considered as an ordered field).

$\mathbb{R}$: Set of all real numbers (considered as a complete ordered field).

$\mathbb{C}$: Set of all complex numbers (considered as a field).

C : Cantor set.

Operations on sets

Let A and B be two sets. Using certain operations on A and B, we can obtain four other sets. One of these is called the *union* of the two sets; written $A \cup B$ (sometimes, called the *sum* and is written as $A + B$); it consists of all elements that are in A or in B or in both (an element that is in both is counted once). The second is called the *intersection* of two sets, written $A \cap B$ (sometimes, called the *product* and written as $A \cdot B$); it consists of all elements in A as well as in B. The third one is called the *difference* of the two sets, written $A - B$; it consists of all those elements of A which are not elements of B. The fourth one is called the *cartesian product* of the two sets, written as $A \times B$; it consists of all ordered pairs (a, b), where $a \in A$ and $b \in B$. Two sets A and B are said to be *disjoint* if $A \cap B = \phi$, otherwise A intersects B. If $B \subset A$, $A - B$ is called the *complement* of B with respect to A. In case A is taken as a universal set, $A - B$ is written as B^c (or $\sim B$) and simply read as complement of B. If $A = \mathbb{R}$, the set of all real numbers, then $\mathbb{Q}^c$ (complement of $\mathbb{Q}$, the set of all rational numbers) is the set of all irrational numbers.

Let A, B and C be any three non-empty sets. Then, the following laws hold good:

1. Commutative Laws

$$A \cup B = B \cup A \text{ and } A \cap B = B \cap A.$$

2. Associative Laws

$$A \cup (B \cup C) = (A \cup B) \cup C \text{ and } A \cap (B \cap C) = (A \cap B) \cap C.$$

3. Distributive Laws

$$A \cap (B \cup C) = (A \cap B) \cup (A \cap C),$$
$$A \cup (B \cap C) = (A \cup B) \cap (A \cup C).$$

4. de Morgan's Laws

$$(A \cup B)^c = A^c \cap B^c \text{ and } (A \cap B)^c = A^c \cup B^c.$$

A set whose elements are used as names is called an *index set*. An index set may be finite or infinite. Suppose for each member α of a fixed set Λ, we have a set A_α. Then, Λ is the index set and the sets A_α are called the *indexed sets*, and the subscript α of A_α, i.e., each $\alpha \in \Lambda$, is called an *index*. The collection of sets A_α is called a *family of indexed sets* and is denoted by $\{A_\alpha\}_{\alpha \in \Lambda}$. We shall be using the symbol Λ for index set throughout the book. Let X be a set and Λ an index set. Let $\{A_\alpha\}_{\alpha \in \Lambda}$ be an indexed family of subsets of X. Then, the union of all the sets A_α is the set $\{x \in X : x \in A_\alpha, \alpha \in \Lambda\}$. We denote it by $\bigcup_{\alpha \in \Lambda} A_\alpha$. We may define $\bigcap_{\alpha \in \Lambda} A_\alpha$ similarly. De Morgan's laws hold good in an indexed family of sets. If $\{A_\alpha\}_{\alpha \in \Lambda}$ is an indexed family of subsets of X, then

$$\left(\bigcup_{\alpha \in \Lambda} A_\alpha \right)^c = \bigcap_{\alpha \in \Lambda} A_\alpha^c \quad \text{and} \quad \left(\bigcap_{\alpha \in \Lambda} A_\alpha \right)^c = \bigcup_{\alpha \in \Lambda} A_\alpha^c.$$

1.2 Relations

Let A and B be any two non-empty sets. *A relation R* from A to B is a subset of $A \times B$. We write $x\,R\,y$, $x \in A$ and $y \in B$. The set of first (second) elements of the ordered pairs in R is called the *domain* (resp., *range*) of the relation R. The whole set B is called the *co-domain* of R. In case $B = A$, R is said to be a relation on the set A.

A relation R on a set A is said to be an *equivalence relation* if

  (a) $x\,R\,x$, for every $x \in A$ (Reflexivity)

  (b) $x\,R\,y \;\Rightarrow\; y\,R\,x$ (Symmetry)

  (c) $x\,R\,y$ and $y\,R\,z \;\Rightarrow\; x\,R\,z$ (Transitivity)

The symbol '~' is frequently used for an equivalence relation.

Given an equivalence relation '~' on a set A and an element $x \in A$, we consider the set

$$E = \{\, y : y \sim x \,\}.$$

The set E is called an *equivalence class* determined by x for the relation R.

1.2.1 Lemma. Two equivalence classes E and E' are either disjoint or equal.

1.2.2 Definition. A *partition* of a set A is a collection of disjoint non-empty subsets of A whose union equals A.

1.2.3 Lemma. Given a partition $\mathcal{P}$ of a set A, there is exactly one equivalence relation on A from which it is derived, and conversely.

1.2.4 Definition. The collections of equivalence classes, of a set A determined by the relation R, which is also a partition of A, is called the *quotient set* of A with respect to R.

1.2.5 Definition. A relation R on a set A is said to be an *order relation* (or, a *simple order* or a *linear order*) if the following properties are satisfied:

  (i) $x, y \in A$, $x \neq y \;\Rightarrow\;$ Either $x\,R\,y$ or $y\,R\,x$ (Comparability)

  (ii) $x \in A \;\Rightarrow\; x\,R\,x$ does not hold   (Non-reflexivity)

  (iii) $x\,R\,y$ and $y\,R\,z \;\Rightarrow\; x\,R\,z$ (Transitivity)

The less than symbol '<' is commonly used to denote an order relation. Also, we shall use the notation $x \leq y$ for the statement "Either $x < y$ or $x = y$"; the notation $y > x$ for the statement "$x < y$" and similarly $y \geq x$ for "Either $x < y$ or $x = y$". Finally, the notation $x < y < z$ means "$x < y$ and $y < z$".

Supremum and infimum

A set $A \subset \mathbb{R}$ is said to be *bounded below* if there is a real number m such that $x \geq m$ for all $x \in A$. The real number m in this case is called a *lower bound* of A. It is easily seen that if m is a lower bound of A, then any number $m' \leq m$ is also a lower bound of A. On similar lines, the set A is said to be *bounded above*, if there is a real number M such that $x \leq M$ for all $x \in A$; the number M, in this case, is called an upper bound of A; again, it is easily seen that if M is an upper bound of A, then any number $M' \geq M$ is also an upper bound of A. The set A is *bounded* if it is both bounded below and bounded above; in other words, the set A is bounded if $\exists$ a real number $K > 0$ such that $|x| \leq K$ for all $x \in A$; otherwise, A is said to be *unbounded*. Note that an unbounded set may be unbounded above, unbounded below or both.

A real number M is called the *least upper bound* (or *supremum*) of a non-empty set $A \subset \mathbb{R}$ if (i) $x \leq M$ for all $x \in A$, and (ii) given $\varepsilon > 0$, however small, $\exists$ and element $x_\varepsilon \in A$ such that $x_\varepsilon > M - \varepsilon$; in other words, M is an upper bound of A and no other real number smaller than M is an upper bound of A. In fact, the least upper bound of A is the smallest of all the upper bounds of A. We denote it by lub A or sup A or $\sup_{x \in A} x$. Similarly, a real number m is called the *greatest lower bound* (or *infimum*) of a non-empty set $A \subset \mathbb{R}$ if (i) $x \geq m$ for all $x \in A$, and (ii) given $\varepsilon > 0$, however small, $\exists$ an element $x_\varepsilon \in A$ such that $x_\varepsilon < m + \varepsilon$. We denote it by glb A or inf A or $\inf_{x \in A} x$.

It is known that every non-empty set of real numbers bounded above possesses the supremum while the one bounded below possesses the infimum.

It is obvious that the supremum and infimum, if they exist, are determined uniquely. The supremum and the infimum of a set may or may not belong to the set. For finite sets, the supremum coincides with the greatest member of the set and the infimum with the smallest member of the set. For an unbounded set A having no upper bound, we write sup $(A) = + \infty$, and for a set A having no lower bound, inf $(A) = -\infty$. The following equalities of supremum and infimum are obvious.

$$\inf (A) = \inf_{x \in A} x = - \sup_{x \in A} (-x) = - \sup (-A).$$

Since the least upper bound of a set A is a special upper bound of A, it is clear that only sets bounded above can have the least upper bound. However, the empty set ϕ has no least upper bound even though it is bounded above by any real number. One can discuss similarly about the greatest lower bound of A.

Intervals

The set $\mathbb{R}$ of real numbers has an order relation '$<$'. Given $a, b \in \mathbb{R}$ such that $a < b$, we have the following four subsets of $\mathbb{R}$ determined by a and b, called the intervals:

$$]a, b[= \{x : a < x < b\} \qquad \text{(Open interval)}$$

$$]a, b[= \{x : a < x \le b\} \qquad \text{(Semi-open interval, closed at right)}$$

$$[a, b[= \{x : a \le x < b\} \qquad \text{(Semi-open interval, closed at left)}$$

$$[a, b] = \{x : a \le x \le b\} \qquad \text{(Closed interval)}$$

In case $a = b$, $]a, b[= \phi$, $[a, b] = \{a\}$ and the remaining two are not defined. All the four intervals defined above are bounded.

The set $\mathbb{R} \cup \{-\infty, \infty\}$, denoted by $\mathbb{R}^*$, is called the extended real number system. For $a \in \mathbb{R}$, the sets $\{x \in \mathbb{R} : x > a\}$, denoted by $]a, \infty[$, and $\{x \in R : x \ge a\}$, denoted by $[a, \infty[$ are called, respectively, the open interval and the closed interval from a to ∞. It may be noted that ∞ and $-\infty$ are not real numbers. By convention, $]-\infty, \infty[$ is the entire set $\mathbb{R}$. An interval which has at least one end point as ∞ or $-\infty$ is called an unbounded interval.

1.3 Functions

Let A and B be arbitrary given sets. By a *function* (or *mapping*) $f : A \to B$, we mean a rule which assigns to each element a of A, a unique element b of B. If $a \in A$, the corresponding element $b \in B$ is called the *f-image* of a and is denoted by $f(a)$, *i.e.*, $b = f(a)$. In this case, a is called the *pre-image* of b. The set A is called the *domain* of the function f, and B the *co-domain* of f. The set $B_1 \subset B$ consisting of all *f*-images of elements of A is called the *range* of f, denoted by $f(a)$. A function f whose co-domain is $\mathbb{R}$ is called a *real valued function*.

If f and g are two functions defined on the same domain A and if $f(a) = g(a)$ for every $a \in A$, the functions f and g are said to be equal and we write $f = g$. Let f be a function of A into B. Then, $f(A) \subset B$. If $f(A) = B$, f is a function of A *onto* B, or $f : A \to B$ is an *onto* (*surjective*) function. The function $f : A \to B$ is one-to-one (injective) if for any two elements a_1 and a_2 of A, $a_1 \ne a_2$ implies $f(a_1) \ne f(a_2)$. A function which is both injective and surjective is called *bijective*.

Let A be any set. Then, $f : A \to A$ defined by $f(a) = a$, $a \in A$, is called the *identity function*, denoted by I_A. An identity function is bijective. A function

f is called a *constant function* if its range consists of only one element. Let $f : A \rightarrow B$ and $g : B \rightarrow C$ be two functions such that $f(a) = b$, $a \in A$ and $b \in B$; and $g(b) = c$, where $c \in C$. Then, a function $h : A \rightarrow C$ defined by

$$h(a) = c = g(b) = g(f(a)), \qquad a \in A$$

is called the *composite function* of two functions f and g, denoted by *gof*. If $f : A \rightarrow B$, then $I_B o f = f o I_A$. Let $f : A \rightarrow B$ be a function and $E \subset A$ the function $f o I_E$ is called the *restriction* of f to the set E, denoted by $f\big|_E$; dually, *the function f is referred to as the extension* of $f\big|_E$ to the set A.

Let $f : A \rightarrow B$ and $b \in B$. Then, the f-inverse of b, denoted by $f^{-1}(b)$, consists of those elements of A which are mapped onto b by f, *i.e.*, those elements of A which have b as their f-image. More precisely, if $f : A \rightarrow B$, then

$$f^{-1}(b) = \{x \in A : f(x) = b\}.$$

It is obvious that $f^{-1}(b)$ is a subset of A. We read f^{-1} as *f-inverse*. It is easy to verify that a function $f : A \rightarrow B$ is injective if and only if for each $b \in B$, $f^{-1}(b)$ is either empty or singleton (set consisting of only one element). Let $f : A \rightarrow B$ and B_1 be a subset of B. Then, the inverse image of B_1 under the function f, denoted by $f^{-1}(B_1)$, consists of those elements of A which are mapped by f onto an element in B_1, in other words

$$f^{-1}(B_1) = \{x \in A : f(x) \in B_1\}.$$

It is easy to prove that a function $f : A \rightarrow B$ is onto if and only if for every non-empty subset $B_1 \subset B$, $f^{-1}(B_1)$ is a non-empty set. For a function $f : A \rightarrow B$ which is bijective, we note that

$$f^{-1} o f = I_A \text{ and } f o f^{-1} = I_B.$$

It may also be seen that $I_A^{-1} = I_A$.

1.3.1 Let X and Y be two non-empty sets and $f : X \rightarrow Y$ be a function. Let $\{A_\alpha\}_{\alpha \in \Lambda}$ be a family of subsets of X and $\{B_\alpha\}_{\alpha \in \Lambda}$ be a family of subsets of Y. Then:

(*i*) $A_\alpha \subset A_\beta \;\Rightarrow\; f(A_\alpha) \subset f(A_\beta), \qquad \alpha, \beta \in \Lambda.$

(*ii*) $B_\alpha \subset B_\beta \;\Rightarrow\; f^{-1}(B_\alpha) \subset f^{-1}(B_\beta), \qquad \alpha, \beta \in \Lambda.$

(*iii*) $f\left(\bigcup_{\alpha \in \Lambda} A_\alpha\right) = \bigcup_{\alpha \in \Lambda} f(A_\alpha).$

(*iv*) $f\left(\bigcap_{\alpha \in \Lambda} A_\alpha\right) = \bigcap_{\alpha \in \Lambda} f(A_\alpha).$

$$(v)\, f^{-1}\left(\bigcup_{\alpha \in \Lambda} B_\alpha\right) = \bigcup_{\alpha \in \Lambda} f^{-1}(B_\alpha)\,.$$

$$(vi)\, f^{-1}\left(\bigcap_{\alpha \in \Lambda} B_\alpha\right) = \bigcap_{\alpha \in \Lambda} f^{-1}(B_\alpha)\,.$$

$(vii)\, A_\alpha \subset f^{-1}(f(A_\alpha)),$

and if f is injective, then

$$A_\alpha = f^{-1}(f(A_\alpha)), \qquad\qquad \alpha \in \Lambda.$$

$(viii)\, f(f^{-1}(B_\alpha)) \subset B_\alpha,$

and if f is surjective, then

$$f(f^{-1}(B_\alpha)) = B_\alpha, \qquad\qquad \alpha \in \Lambda.$$

$$(ix)\, f^{-1}(Y - B_\alpha) = X - f^{-1}(B_\alpha), \qquad \alpha \in \Lambda.$$

Sequences

A function $f : \mathbb{N} \to X$, where X is any set, is called a *sequence* in X. Since a sequence is uniquely and completely determined by the values $x_n\ (= f(n))$ for $n \in \mathbb{N}$, a sequence is usually denoted by $\{x_n\}$ without explicit reference to f. The value x_n is called the value of the nth term of the sequence $\{x_n\}$.

In the particular case, when $X = \mathbb{R}$ or $\mathbb{C}$, the sequence is, respectively, called the real or complex. Further, a real sequence $\{x_n\}$ is said to be an *increasing sequence* if $x_{n+1} \geq x_n$ for all n. In case $x_{n+1} > x_n$ for all n, it is said to be *strictly increasing*. The *decreasing* and *strictly decreasing sequences* are defined in a similar way. Sequences, which are either increasing or decreasing, are called *monotone (monotonic)* sequences.

If $\{x_n\}$ is a given sequence in X and $\{n_k\}$ is an strictly increasing sequence of positive integers, then $\{x_{n_k}\}$ is called a *subsequence* of $\{x_n\}$. Clearly:

(i) Every sequence is a subsequence of itself.

(ii) If $\{z_n\}$ is a subsequence of $\{y_n\}$ and $\{y_n\}$ is a subsequence of $\{x_n\}$, then $\{z_n\}$ is a subsequence of $\{x_n\}$.

A sequence $\{x_n\}$, real or complex, is said to converge to a *limit l* if for each $\varepsilon > 0$, $\exists$ a positive integer N (depending on ε) such the $\left| x_n - l \right| < \varepsilon$ for all $n \geq N$.

We write $\lim_{n \to \infty} x_n = l$. Such a sequence is said to be *convergent*.

It may be noted that a convergent sequence has a unique limit. Also, every convergent sequence is bounded but the converse is not true; for instance, the sequence $\{(-1)^n\}$ is bounded but not convergent. A sequence $\{x_n\}$ is said to be a *Cauchy sequence* if for each $\varepsilon > 0$, $\exists$ a positive integer N such that $|x_n - x_m| < \varepsilon$, $\forall\, n, m \geq N$. One has the following:

1.3.2 A real (complex) sequence is convergent if and only if it is a Cauchy sequence.

1.3.3 A sequence $\{x_n\}$, real or complex, converges to l if and only if each of its subsequences converges to l.

1.3.4 Let $\{x_n\}$ and $\{y_n\}$ be two sequences, real or complex, such that

$$\lim_{n \to \infty} x_n = l_1 \text{ and } \lim_{n \to \infty} y_n = l_2. \text{ Then:}$$

(a) $\displaystyle \lim_{n \to \infty} (x_n \pm y_n) = l_1 \pm l_2.$

(b) $\displaystyle \lim_{n \to \infty} x_n y_n = l_1 \, l_2.$

(c) $\displaystyle \lim_{n \to \infty} \frac{x_n}{y_n} = \frac{l_1}{l_2}$, provided $l_2 \neq 0.$

1.3.5 A monotonic sequence, of course real, either converges or diverges to $\pm \infty$. More precisely, a monotonic sequence bounded below (above) converges to its infimum (supremum).

There are many bounded (real) sequences which are not convergent. For such sequences, the concepts of *limit superior* and *limit inferior* are introduced.

Let $\{x_n\}$ be a bounded real sequence and let $M_k = \sup_{n \geq k} x_n$. It is evident that

$M_{k+1} \leq M_k$. Thus, the sequence $\{M_k\}$ has a limit; namely, the infimum of the M_k's. This infimum is defined to be the *limit superior* (or the *upper limit*) of

$\{x_n\}$. It is denoted by $\displaystyle \lim_{n \to \infty} \sup x_n$ or $\displaystyle \overline{\lim_{n \to \infty}} \, x_n$ or simply $\overline{\lim} x_n$. Thus

$$\lim_{n \to \infty} \sup x_n = \inf_{k \geq 1} \, \sup_{n \geq k} x_n.$$

The *limit inferior* or the *lower limit* of the sequence $\{x_n\}$ is defined in a similar way. It is denoted by $\lim\limits_{n \to \infty} \inf x_n$ or $\varliminf\limits_{n \to \infty} x_n$ or simply $\varliminf x_n$. It may be noted that

$$\lim_{n \to \infty} \inf x_n = \sup_{k \geq 1} \inf_{n \geq k} x_n.$$

From the definition, it is obvious that $\lim\limits_{n \to \infty} \inf x_n \leq \lim\limits_{n \to \infty} \sup x_n$ and a sequence $\{x_n\}$ converges to l if and only if

$$\lim_{n \to \infty} \sup x_n = l = \lim_{n \to \infty} \inf x_n.$$

The idea of upper and lower limits can also be generalised to include the case of unbounded sequences. If the sequence $\{x_n\}$ is unbounded above, $\lim\limits_{n \to \infty} \sup x_n = \infty$; and, in case it is unbounded below, $\lim\limits_{n \to \infty} \inf x_n = -\infty$.

Series

An expression $\sum\limits_{n=1}^{\infty} x_n$ (or simply $\Sigma\, x_n$), where x_n are real or complex numbers and depend on the index $n = 1, 2,\ldots\ldots$, is called a *series*. For each $n \in \mathbb{N}$, let $S_n = x_1 + x_2 + \ldots + x_n$. The sequence $\{S_n\}$ is called the *sequence of partial sums* of the series $\Sigma\, x_n$ and the number S_n is called the n-th partial sum of the series $\Sigma\, x_n$. The series $\Sigma\, x_n$ is convergent if the $\lim\limits_{n \to \infty} S_n$ exists, say, s. In this case, we write

$$s = x_1 + x_2 + \ldots = \sum_{n=1}^{\infty} x_n$$

and call s the sum of the series. A series $\Sigma\, x_n$ is *absolutely convergent* if the series $\Sigma\, |x_n|$ is convergent. Every absolutely convergent series is convergent and a convergent series with terms as non-negative real numbers is absolutely convergent. A series with arbitrary terms though being convergent may not be absolutely convergent.

A sequence $\{x_n\}$ of real or complex numbers is said to be *summable* (*absolutely summable*) if the series $\Sigma\, x_n$ is convergent (absolutely convergent).

Finally, in the section, we would like to discuss some very significant principles of the set theory on which the foundation of modern mathematics is lying.

A relation defined on a non-empty set X is said to be a *partial order*, denoted by $\lesssim x$, on X if it is reflexive, antisymmetric and transitive. The ordered pair $(X, \lesssim)$ is called a *partially ordered set*. The elements $x, y \in X$ are said to be *comparable* if either $x \lesssim y$ or $y \lesssim x$. In case every pair of elements in X are comparable with respect to $\lesssim$, then $\lesssim$ is said to be a *total order relation* and $(X, \lesssim)$ is called a *totally ordered set* (or *chain* or a *linearly ordered set*). In general, every pair of elements in X need not be comparable with respect to $\lesssim$ which, infact, justifies the name partial ordering. Note that $\mathscr{P}(X)$ with inclusion relation is not a totally ordered set.

1.3.6 Definition. Let $(X, \lesssim)$ be a partially ordered set and $A \subset X$.

(a) An element $a \in X$ is said to be an *upper bound* for A if $x \lesssim a, \forall\, x \in A$. An upper bound a^* of A is said to be a *least upper bound* (*supremum*) of A if $a^* \lesssim a$ for every upper bound a of A.

(b) *Lower bound* and *greatest lower bound* (*infimum*) for A can be defined similarly.

(c) An element $M \in A$ is said to be a *maximal element* of A if

$$M \lesssim x \;\Rightarrow\; M = x, \qquad \forall\, x \in A,$$

and, similarly, an element $m \in A$ is said to be a *minimal element* of A if

$$x \lesssim m \;\Rightarrow\; m = x, \qquad \forall\, x \in A.$$

1.3.7 Axiom of Choice. *Let $\mathcal{C}$ be a non-empty collection of non-empty sets. Then, $\exists$ a function F defined on $\mathcal{C}$ which assigns to a set A in $\mathcal{C}$ an element $F(A)$ in A.*

The axiom of choice is equivalent to the following postulate:

1.3.8 Zermelo's Postulate. *Let $\{A_\alpha\}_{\alpha \in \Lambda}$ be any non-empty family of disjoint non-empty sets. Then, $\exists$ a subset $B \subset \bigcup\limits_{\alpha \in \Lambda} A_\alpha$ such that, for each $\alpha \in \Lambda$, $B \cap A_\alpha$ consists of exactly one element.*

The following theorem is proved directly from axiom of choice.

1.3.9 Well Ordering Theorem. *Every set can be well ordered.*

The following is an existence theorem which asserts the existence of a maximal element in a partially ordered set.

1.3.10 Zorn's Lemma. *Let $X \neq \phi$ be a partially ordered set in which every totally ordered subset has an upper bound in X. Then, X has at least one maximal element.*

Now, we state a result which is very basic in the set theory.

1.3.11 Theorem. *The following are equivalent:*

(a) *Axiom of Choice.*

(b) *Well Ordering Theorem.*

(c) *Zorn's Lemma.*

1.4 Countable Sets, Uncountable Sets and Cardinal Numbers

Two sets A and B are said to be *equivalent* if there exists a bijective mapping $f : A \rightarrow B$. A set is said to be *finite*, if it is equivalent to the set $\{1, 2, ..., n\}$ for some $n \in \mathbb{N}$, where $\mathbb{N}$ is the set of all natural numbers; otherwise the set is said to be *infinite*.

A set is said to be *denumerable* if it is equivalent to $\mathbb{N}$. A set which is finite or denumerable is called a *countable* set. Sometimes, denumerable sets are also referred to as *countably infinite* sets. There is no general agreement in English usage governing the meaning attached to the words 'denumerable' and 'countable'. Both of these terms are used by various authors to mean finite or equivalent to $\mathbb{N}$. However, we will follow the convention that 'denumerable' means only equivalent to $\mathbb{N}$ and that 'countable' means finite or equivalent to $\mathbb{N}$.

A set which is not countable is called *uncountable*. Since an uncountable set is necessarily an infinite set, we may also call that as a *non-denumerable* set.

We, now, give (without proof) quite a few results and examples in regard to the above concepts which are useful in the subsequent work.

1.4.1 *A set is infinite if and only if it contains a denumerable subset.*

1.4.2 *An infinite set is equivalent to a proper subset of itself.*

1.4.3 *A subset of a countable set is countable.*

1.4.4 *If A and B are countable sets, then A × B is countable.*

1.4.5 *The union of a countable collection of countable sets is countable.*

1.4.6 *Each of the following sets is countable:*

  (i) The set $\mathbb{Z}$ of all integers.

  (ii) The set $\mathbb{Q}$ of all rational numbers.

  (iii) The set $\mathbb{P}$ of all polynomials

$$P(x) = a_0 + a_1 x + a_2 x^2 + \ldots + a_n x^n$$

  with integral coefficients.

  (iv) The set $\mathbb{A}$ of all algebraic numbers.

  (v) The set of all polynomials with rational coefficients.

  (vi) The set of all complex numbers which are algebraic over the field of rational numbers.

  (vii) The set of all finite sequences whose terms are algebraic numbers.

  (viii) The set of all straight-lines in a plane each of which passes through (at least) two different points with rational coordinates.

  (ix) The set of all rational points in $\mathbb{R}^n$.

1.4.7 *The family of all finite subsets of a countable set is countable.*

1.4.8 *Each of the following is an uncountable set:*

  (i) An open interval]a, b[, a closed interval [a, b], where $a \neq b$; more generally, any interval I which does not degenerate to a single point.

  (ii) The set of all irrational numbers.

  (iii) The set $\mathbb{R}$ of all real numbers.

  (iv) The set of all transcendental numbers.

  (v) The set of all sequences of natural numbers.

  (vi) The set of all points of a plane.

  (vii) The family of all subsets of a denumerable set.

Cardinal numbers

Let all the sets be divided into families such that two sets fall into one family if and only if they are equivalent. This is possible since the relation '~' between the sets is an equivalence relation. To every such family of sets, we assign some arbitrary symbol and call the *cardinal number* (or the *power*) of each set of the given family. If the cardinal number of a set A is α, we write $\overline{\overline{A}} = \alpha$ or card $(A) = \alpha$. The cardinal number of the empty set is defined to be 0 (zero). We designate the number of elements of a non-empty finite set as the cardinal number of the finite set. We assign $\aleph_0$ (read as aleph nought, being the first letter of the Hebrew alphabet) to the class of all denumerable sets and as such $\aleph_0$ is the cardinal number of any denumerable set. We denote by c, the first letter of the word continuum, the cardinal number of the set $[0, 1]$. The cardinal number of a finite set is called *finite cardinal number* and that of an infinite set is called *transfinite cardinal number.*

Let α, β be any two cardinal numbers and A, B be sets such that $\overline{\overline{A}} = \alpha$, $\overline{\overline{B}} = \beta$. We say that $\alpha \leq \beta$ (or $\beta \geq \alpha$) if the set A is equivalent to a subset of B. Further , we write $\alpha < \beta$ if $\alpha \leq \beta$ and $\alpha \neq \beta$. In case $\alpha \neq 0$ and $\beta \neq 0$, we define α^β as the cardinal number of the set A^B, the set of all functions defined on B with range in A. In this regard, we have the following:

1.4.9 $\quad n < \aleph_0 < c, \ \forall\, n \in \mathbb{N}.$

1.4.10 *Let A be any set. Then*

$$\overline{\overline{A}} < \overline{\overline{\mathscr{P}(A)}}.$$

1.4.11 *If α is the cardinal number of a set A, then*

$$\overline{\overline{\mathscr{P}(A)}} = 2^\alpha.$$

1.4.12 $\quad 2^{\aleph_0} = c.$

1.4.13 **(Cantor's Continuum Hypothesis).** *There is no cardinal number μ such that*

$$\aleph_0 < \mu < 2^{\aleph_0} \ (= c).$$

More generally, no matter what the infinite set A is, there is no set with cardinal number μ such that

$$\overline{\overline{A}} < \mu < \overline{\overline{\mathscr{P}(A)}}.$$

The later assumption is known as *generalised continuum hypothesis (g.c.h.).* For more details on the topic see Chapter 2 in [24].

1.5 Inequalities

We give some inequalities, which will be freely used in the subsequent work. Throughout, denote by $\mathbb{K}$, the field $\mathbb{R}$ (or $\mathbb{C}$) of real (or complex) numbers.

1.5.1 **Triangle Inequality.** Let α, $\beta \in \mathbb{K}$. Then

$$|\alpha + \beta| \leq |\alpha| + |\beta|.$$

1.5.2 Let $\alpha, \beta \in \mathbb{K}$. Then

$$\frac{|\alpha + \beta|}{1 + |\alpha + \beta|} \le \frac{|\alpha|}{1 + |\alpha|} + \frac{|\beta|}{1 + |\beta|}.$$

1.5.3 Let $0 < \lambda < 1$. Then

$$\alpha^{\lambda}\,\beta^{1-\lambda} \le \lambda\,\alpha + (1 - \lambda)\,\beta$$

holds good for every pair of non-negative real numbers α and β with equality only if $\alpha = \beta$.

1.5.4 Hölder Inequality (Finite form)

Let $1 < p, q < \infty$ and let p, q be conjugate exponents, *i.e.*, $\dfrac{1}{p} + \dfrac{1}{q} = 1$. If $\alpha_i, \beta_i \in \mathbb{K}$

$(i = 1, 2, ..., n)$, then

$$\sum_{i=1}^{n} |a_i \beta_i| \le \left(\sum_{i=1}^{n} |\alpha_i|^p \right)^{\frac{1}{p}} \left(\sum_{i=1}^{n} |\beta_i|^q \right)^{\frac{1}{q}}.$$

Also

$$\sum_{i=1}^{n} |a_i \beta_i| \le \left(\sum_{i=1}^{n} |\alpha_i| \right) \max_{1 \le i \le n} |\beta_i|.$$

1.5.5 Cauchy-Schwartz Inequality (Finite form)

Note the inequality in 1.5.4 for the case when $p = q = 2$:

$$\sum_{i=1}^{n} |a_i \beta_i| \le \left(\sum_{i=1}^{n} |\alpha_i|^2 \right)^{\frac{1}{2}} \left(\sum_{i=1}^{n} |\beta_i|^2 \right)^{\frac{1}{2}}.$$

1.5.6 Minkowski Inequality (Finite form)

Let $1 \le p < \infty$. If $\alpha_i, \beta_i \in \mathbb{K}$ $(i = 1, 2, ... n)$, then

$$\left(\sum_{i=1}^{n} |\alpha_i + \beta_i|^p \right)^{\frac{1}{p}} \le \left(\sum_{i=1}^{n} |\alpha_i|^p \right)^{\frac{1}{p}} + \left(\sum_{i=1}^{n} |\beta_i|^p \right)^{\frac{1}{p}}.$$

1.5.7   Let $0 < p \leq 1$. If $\alpha_i, \beta_i \in \mathbb{K}$ $(i = 1, 2, ..., n)$, then

$$\sum_{i=1}^{n} |\alpha_i + \beta_i|^p \leq \sum_{i=1}^{n} |\alpha_i|^p + \sum_{i=1}^{p} |\beta_i|^p .$$

1.5.8   Hölder Inequality (Infinite form)

Let $1 < p < \infty$ and q be conjugate to p. If $(\alpha_1, \alpha_2 ...) \in l^p$ and $(\beta_1, \beta_2 ...) \in l^q$

i.e.,
$$\sum_{i=1}^{\infty} |\alpha_i|^p < \infty \text{ and } \sum_{i=1}^{\infty} |\beta_i|^q < \infty,$$

then
$$\sum_{i=1}^{\infty} |\alpha_i \beta_i| \leq \left(\sum_{i=1}^{\infty} |\alpha_i|^p \right)^{\frac{1}{p}} \left(\sum_{i=1}^{\infty} |\beta_i|^q \right)^{\frac{1}{q}} .$$

1.5.9   Cauchy-Schwartz Inequality (Infinite form)

Note the inequality in 1.5.8 for the case $p = q = 2$:

$$\sum_{i=1}^{\infty} |\alpha_i \beta_i| \leq \left(\sum_{i=1}^{\infty} |\alpha_i|^2 \right)^{\frac{1}{2}} \left(\sum_{i=1}^{\infty} |\beta_i|^2 \right)^{\frac{1}{2}} .$$

1.5.10   Minkowski Inequality (Infinite form)

Let $1 \leq p < \infty$, If $(\alpha_1, \alpha_2 ,....)$ and $(\beta_1, \beta_2 ,...) \in l^p$, *i.e.,*

$$\sum_{i=1}^{\infty} |\alpha_i|^p < \infty \text{ and } \sum_{i=1}^{\infty} |\beta_i|^p < \infty,$$

then
$$\left(\sum_{i=1}^{\infty} |\alpha_i + \beta_i|^p \right)^{\frac{1}{p}} \leq \left(\sum_{i=1}^{\infty} |\alpha_i|^p \right)^{\frac{1}{p}} + \left(\sum_{i=1}^{\infty} |\beta_i|^p \right)^{\frac{1}{p}} .$$

1.6 Linear Spaces

Underlying every space in functional analysis, there is a linear space. In this section, we shall give a review and preparatory material on linear spaces.

The definition of a linear space involves a general field $\mathbb{K}$. But, in functional analysis, $\mathbb{K}$ will be the field of real or complex numbers, and so in this book, $\mathbb{K}$ will denote either $\mathbb{R}$ or $\mathbb{C}$. The elements of $\mathbb{K}$ are called scalars.

1.6.1 Definition. Let V be an arbitrary non-empty set and let $\mathbb{K}$ be the field of scalars (real or complex). Define two operations called 'vector addition' and 'scalar multiplication' in V. By addition we mean a rule for associating with each pair of elements x and y in V an element $x + y$, called the sum of x and y; and by scalar multiplication we mean a rule for associating with each scalar $\alpha \in \mathbb{K}$ and each element $x \in V$ an element αx, called the scalar multiple of x by α. The set V together with addition and scalar multiplication defined on V is said to be a *linear space* over $\mathbb{K}$ if the following conditions are satisfied:

(i) $x \in V$ and $y \in V$ $\Rightarrow$ $x + y \in V$ (Closure property)

(ii) $x + y = y + x$, $\forall\, x, y \in V$ (Commutativity)

(iii) $(x + y) + z = x + (y + z)$, $\forall\, x, y, z \in V$ (Associativity)

(iv) $\exists$ unique element $\theta \in V$ such that

$$x + \theta = x, \qquad \forall\, x \in V \qquad \text{(Identity element)}$$

(v) For each $x \in V$, $\exists$ a unique element $w \in V$ such that

$$x + w = \theta \qquad \text{(Additive inverse)}$$

We write $w = -x$

(vi) $\alpha \in \mathbb{K}$ and $x \in V$ $\Rightarrow$ $\alpha x \in V$

(vii) $\alpha(\beta x) = (\alpha \beta)x$, $\forall\, \alpha, \beta \in \mathbb{K}$ and $\forall\, x \in V$

$(viii)$ $\alpha(x + y) = \alpha x + \alpha y$, $\forall\, \alpha \in \mathbb{K}$ and $\forall\, x, y \in V$ (Distributive)

(ix) $(\alpha + \beta) x = \alpha x + \beta x$, $\forall\, \alpha, \beta \in \mathbb{K}$ and $\forall\, x \in V$ (Distributive)

(x) $1\, x = x$, $\forall\, x \in V$.

The elements of a linear space are called *vectors*. V is called a *real linear space* or a *complex linear space* according as $\mathbb{K}$ is $\mathbb{R}$ or $\mathbb{C}$, respectively. A linear space over $\mathbb{K}$ is also called a *vector space* over $\mathbb{K}$.

Notes

1. The vector addition is, in fact, a mapping '+' from $V \times V$ into V. Further, the scalar multiplication as defined above is a mapping from $\mathbb{K} \times V$ into V. Thus, a linear space is a quadruple $(V, \mathbb{K}, +, .)$ which satisfies the above conditions. We shall usually speak of the linear space V over the field $\mathbb{K}$, or simply V, instead of the linear space $(V, \mathbb{K}, +, .)$.

2. It may be observed that in the definition of a linear space, neither the nature of the vectors nor the operations are specified. Any kind of objects

whatsoever can serve as vectors; all that is required is that the linear space axioms are satisfied.

3. Let V be a complex linear space. Obviously, by restricting the mapping $(\alpha, x) \to \alpha x$ of $C \times V \to V$ to $\mathbb{R} \times V \to V$, we obtain a real linear space V. Thus, every complex linear space is also a real linear space. The real linear space $(V, \mathbb{R}, +, .)$ is called the real linear space associated with the complex linear space $(V, C, +, .)$ and is denoted by $V_{\mathbb{R}}$.

1.6.2 Theorem. *Let V be a linear space over the field $\mathbb{K}$. Then:*

(a) $0\, x = \theta, \quad \forall\, x \in V$

   where 0 is the scalar zero and θ is the zero vector.

(b) $\alpha\, \theta = \theta, \quad \forall\, \alpha \in \mathbb{K}.$

(c) $(-1)\, x = -x, \quad \forall\, x \in V.$

(d) $\alpha\, x = \theta \quad \Rightarrow \quad \alpha = 0 \text{ or } x = \theta.$

1.6.3 Examples

1. Let $V = \{\theta\}$ be a singleton set. Define $\theta + \theta = \theta$ and $\alpha\, \theta = \theta, \forall\, \alpha \in \mathbb{K}$. Then V is a linear space, called the *zero space* over the field $\mathbb{K}$.

2. $\mathbb{R}$, with ordinary addition and multiplication taken as linear operations, is a real linear space.

3. $\mathbb{C}$, with addition and multiplication of complex number taken as the linear operations, is a linear space over $\mathbb{K}$ ($\mathbb{R}$ or $\mathbb{C}$).

4. $\mathbb{R}^n$, the set of all n-tuples of real numbers, is a real linear space with the operations defined by

$$x + y = (\xi_1 + \eta_1, \xi_2 + \eta_2, ..., \xi_n + \eta_n)$$

$$\alpha\, x = (\alpha\, \xi_1, \alpha\, \xi_2, ..., \alpha\, \xi_n)$$

   for $x = (\xi_1, \xi_2, ..., \xi_n), y = (\eta_1, \eta_2, ..., \eta_n)$ in $\mathbb{R}^n$ and $\alpha \in \mathbb{R}$.

5. $\mathbb{C}^n$, the set of all n-tuples of complex numbers, is a linear space with algebraic operations defined as in Example 4.

6. Let p be a real number such that $1 \le p < \infty$ and suppose l^p denote the set of all sequences $x = \{\xi_i\}$ in $\mathbb{K}$ ($\mathbb{R}$ or $\mathbb{C}$) such that the infinite series $\displaystyle\sum_{i=1}^{\infty} |\xi_i|^p$

   converges, *i.e.,* $\displaystyle\sum_{i=1}^{\infty} |\xi_i|^p < \infty$. Define the linear operations

$$x + y = (\xi_1 + \eta_1, \xi_2 + \eta_2, ...)$$

$$\alpha\, x = (\alpha\, \xi_1, \alpha\, \xi_2,...,)$$

where $x = (\xi_1, \xi_2,...)$ and $y = (\eta_1, \eta_2,...)$ are in l^p and $\alpha \in \mathbb{K}$.

Since

$$|\, \xi_i + \eta_i \,|^p \le 2^p \max\, \{|\, \xi_i \,|^p, |\, \eta_i \,|^p\}$$

$$\le 2^p\, (|\xi_i|^p + |\, \eta_i \,|^p),$$

if follows that $\displaystyle\sum_{i=1}^{\infty} |\, \xi_i + \eta_i \,|^p < \infty$. Therefore, $x + y \in l^p$. Now, it is easy to verify that l^p is a linear space with respect to the algebraic operations defined above.

7. Each of the following is a linear space over $\mathbb{K}$ ($\mathbb{R}$ or $\mathbb{C}$) with respect to coordinatewise linear operations defined in Example 6.

   (a) ω, the set of all sequences (convergent or not) in $\mathbb{K}$.

   (b) c, the set of all convergent sequences in $\mathbb{K}$.

   (c) c_0, the set of all sequences in $\mathbb{K}$ converging to 0.

   (d) l^∞, the set of all bounded sequences in $\mathbb{K}$.

8. Let $P[a, b]$ be the set of all polynomials with real coefficients defined on the closed interval $[a, b]$. All non-zero constant polynomials (polynomials of degree zero) and the polynomial which is identically zero (zero polynomial) are considered to be contained in $P[a, b]$. Consider the linear operations to be the usual addition of two polynomials and the multiplication by a real number. Then, $P[a, b]$ is a real linear space.

9. For a given positive integer n, let $P_n[a, b]$ be the subset of $P[a, b]$ consisting of the zero polynomial and all non-zero polynomials of degree $\le n$. Then, $P_n[a, b]$ is a real linear space with linear operations defined as in $P[a, b]$.

10. Let $C[a, b]$ be the set of all real functions defined and continuous on the closed interval $[a, b]$. If x and y are any two functions in $C[a, b]$ and α is any real number, define the sum $x + y$ and the scalar multiplication $\alpha\, x$ by

$$(x + y)\, (t) = x\, (t) + y(t)$$

and $\qquad\qquad\qquad (\alpha\, x)\, (t) = \alpha\, x\, (t),\ \forall\ t \in [a, b].$

Note that the value of $x + y$ at t is obtained by adding together the values of x and y at t and the value of $\alpha\, x$ at t is α times the values of x at t. These operations are called pointwise addition and scalar multiplication of functions. Then, the set $C[a, b]$ is a real linear space.

11. Let f be a Lebesgue measurable function defined on $[a, b]$ and $0 < p < \infty$. Designate by $L^p[a, b]$, the class of all (Lebesgue) p-integrable functions over $[a, b]$, *i.e.*,

$$L^p[a, b] = \left\{ f : [a, b] \to \mathbb{R} : \int_a^b |f|^p < \infty \right\}.$$

Then, $L^p[a, b]$ is a linear space over $\mathbb{R}$. Indeed, we observe that:

(a) $f, g \in L^p[a, b] \implies f + g \in L^p[a, b]$

since

$$|f + g|^p \leq 2^p \max \{|f|^p, |g|^p\}$$

$$\leq 2^p (|f|^p + |g|^p)$$

(b) $f \in L^p[a, b], \alpha \in \mathbb{R} \implies \alpha f \in L^p[a, b]$.

12. Let f be a real valued and measurable function defined on $[a, b]$. If

$$|f(x)| \leq M \ a.e. \text{ on } [a, b]$$

then f is said to be *essentially bounded* on $[a, b]$ and M is an essential bound of it. The *essential supremum* of f on $[a, b]$ is defined by

$$\text{ess sup } |f(x)| = \inf \{M : |f(x)| \leq M \quad a.e. \text{ on } [a, b]\}.$$

If f does not have any essential bound, then its essential supremum is defined to be ∞. Let us denote by $L^\infty[a, b]$ the class of all those (Lebesgue) measurable functions defined on $[a, b]$ which are essentially bounded on $[a, b]$, *i.e.*,

$$L^\infty[a, b] = \{f : [a, b] \to \mathbb{R} : \text{ess sup } |f| < \infty\}.$$

It is easy to verify that $L^\infty[a, b]$ is a linear space over $\mathbb{R}$.

1.6.4 Definition. A non-empty subset W of a linear space V over $\mathbb{K}$ is said to be a *linear subspace* (or, simply, a *subspace*) if the following two conditions are satisfied:

(i) $x + y \in W, \quad \forall \, x, y \in W.$

(ii) $\alpha x \in W, \quad \forall \, \alpha \in \mathbb{K} \text{ and } \quad \forall \, x \in W.$

Clearly, a non-empty subset W of a linear space V is a subspace of V if and only if $\alpha x + \beta y \in W, \forall \, x, y \in W \text{ and } \forall \, \alpha, \beta \in \mathbb{K}$. The space V itself and $\{\theta\}$

are the trivial subspaces of V. One can also verify that any intersection (finite or infinite) of subspaces is again a subspace.

A vector of the form

$$x = \alpha_1 x_1 + \alpha_2 x_2 + \dots + \alpha_n x_n$$

is called a *linear combination* of vectors $x_1, x_2, \dots, x_n$ in V, where $\alpha_1, \alpha_2, \dots, \alpha_n$ are scalars in $\mathbb{K}$. If S is any subset of V, then the set of all finite linear combinations of vectors in S forms a subspace of V (Verify !). The subspace so obtained is called the *subspace* spanned (or generated) by S and is denoted by *span* S. The *span* S is, in fact, the smallest subspace of V containing S which can also be obtained by taking the intersection of all subspace of V containing S.

We, now, ask whether some members of the set S could be removed without changing the subspace generated by S. We begin with the concept of linear independence of the set of vectors.

1.6.5 Definition. A finite set of vectors $\{x_1, x_2, \dots, x_n\}$ in V is said to be *linearly independent* if for any scalars $\alpha_1, \alpha_2, \dots, \alpha_n$ in $\mathbb{K}$, we have

$$\alpha_1 x_1 + \alpha_2 x_2 + \dots + \alpha_n x_n = \theta \quad \Rightarrow \quad \alpha_1 = 0 = \alpha_2 = \dots = \alpha_n.$$

A subset S (finite or infinite) of V is *linearly independent* if every non-empty finite subset of S is linearly independent and S is said to be *linearly dependent* if it is not linearly independent. As a convention, we shall regard the empty set to the linearly independent.

Hamel basis and dimension

1.6.6 Definition. A subset S of a linear space V is said to form a *basis* (or *Hamel basis*) for V if

(i) S is a linearly independent set, and

(ii) S spans the whole space V, *i.e.*, $V = span\ S$

In this case, every non-zero vector $x \in V$ has a unique representation as a linear combination of finitely many vectors from S with non-zero scalars as coefficients.

Clearly, any maximal linearly independent set (a set to which no new vector can be adjoined without destroying the linear independence) is a basis for V, and any minimal set spanning V is also a basis for V.

1.6.7 Theorem. *Every linear space $V \neq \{\theta\}$ has a Hamel basis.*

Proof. Let $\mathbb{M}$ be the set of all linearly independent subsets of V. Since $V \neq \{\theta\}$, it has an element $x \neq \theta$ and $\{x\} \in \mathbb{M}$. Therefore, $\mathbb{M} \neq \phi$. Note that 'set inclusion' defines a partial ordering in $\mathbb{M}$. Let $\mathcal{C} = \{M_i\}$ be an arbitrary chain in $\mathbb{M}$. We, now, claim that $\mathcal{C}$ has an upper bound, namely, $M \, (= \bigcup_i M_i)$ is in $\mathbb{M}$.

Let $\{x_i\}_{i=1}^n$ be an arbitrary finite sequence of vectors in M. Then

$$\{x_i\}_{i=1}^n \subset M_j, \quad \text{for some } j$$

$$\Rightarrow \qquad \{x_i\}_{i=1}^n \text{ is linearly independent.}$$

Hence, $M \in \mathbb{M}$. Thus, the condition of Zorn's Lemma 1.3.10 is satisfied and, consequently, $\mathbb{M}$ has a maximal element (say) S. It is, now, enough to show that *span* $S = V$. Let, if possible, *span* $S \neq V$. Then, $\exists \, z \in V - span \, S$. Clearly, $S \cup \{z\}$ is a linearly independent set of V containing S properly. Thus, $S \cup \{z\} \in \mathbb{M}$ and $S \subsetneq S \cup \{z\}$. This contradicts the maximality of S. Hence, *span* $S = V$.

This proves that S is a Hamel basis of V. ∎

1.6.8 Definition. A linear space V is said to be *finite dimensional* if it has a finite basis, otherwise V is said to be *infinite dimensional*.

1.6.9 Examples

1. Let $V = \{\theta\}$ be a trivial linear space. By convention, ϕ is linearly independent set. Note that *span* ϕ is the intersection of all subspaces of V containing ϕ. But θ belongs to every subspace of V. Hence, it follows that *span* $\phi = \{\theta\}$. We thus conclude that ϕ is a basis of V.

2. Consider the real linear space $\mathbb{R}$. Then, any singleton set consisting of a non-zero element from $\mathbb{R}$ forms a basis of $\mathbb{R}$. Thus, $\mathbb{R}$ is finite dimensional.

3. The complex linear space $\mathbb{C}$ is finite dimensional since any singleton set consisting of a non-zero element from $\mathbb{C}$ forms a basis of $\mathbb{C}$.

4. Consider the real linear space $\mathbb{C}_\mathbb{R}$ associated with the complex linear space $\mathbb{C}$. Let $B = \{1, i\}$. Clearly, B is a linearly independent set of vectors in $\mathbb{C}_\mathbb{R}$ and it spans $\mathbb{C}_\mathbb{R}$. Hence, $\mathbb{C}_\mathbb{R}$ is a finite dimensional real linear space.

5. Consider the real linear space $\mathbb{R}^n$ and let

$$e_1 = (1, 0, 0, ..., 0), \; e_2 = (0, 1, 0,.., 0),...., e_n = (0, 0, 0,..., 1)$$

be the unit vectors in $\mathbb{R}^n$. Then, the set $B = \{e_1, e_2,..., e_n\}$ is linearly independent, since

$$\alpha_1 e_1 + \alpha_2 e_2 + ... + \alpha_n e_n = (\alpha_1, \alpha_2,..., \alpha_n) = \theta$$

$$\Leftrightarrow (\alpha_1, \alpha_2,..., \alpha_n) = (0, 0,...., 0)$$

$$\Leftrightarrow \alpha_1 = \alpha_2 = ... = \alpha_n = 0.$$

Next, let $x = (\xi_1, \xi_2,..., \xi_n) \in \mathbb{R}^n$. Then, we can write

$$x = \xi_1 e_1 + \xi_2 e_2 + ... + \xi_n e_n.$$

Thus, any vector $x \in \mathbb{R}^n$ is a linear combination of elements of B. Therefore, $\mathbb{R}^n = span\, B$. Hence, B is a basis for $\mathbb{R}^n$. Furthermore, $\mathbb{R}^n$ is finite dimensional.

6. The complex linear space $\mathbb{C}^n$ is finite dimensional and $B = \{e_1, e_2,...., e_n\}$ forms a basis for it.

Problem 1. Obtain a basis for the linear space $\mathbb{C}_{\mathbb{R}}^n$ and show that this space is finite dimensional.

1.6.10 Theorem. *Let V be a finite dimensional linear space. Then, all the bases for V have the same number of elements.*

Proof. Let S and T be two Hamel bases for V such that S is finite. Suppose S has n elements $x_1, x_2,..., x_n$. Let $y_1, y_2 ..., y_{n+1}$ be $n + 1$ elements in T. Then, we can write

$$y_i = \sum_{j=1}^{n} a_{ij} x_j , \quad i = 1, 2,..., n.$$

Since $y_1, y_2,...., y_n$ are linearly independent (these being the elements of a Hamel basis), the matrix $[a_{ij}]$ has rank n. Also, we have

$$y_{n+1} = \sum_{n=1}^{n} a_{kj} \; x_j$$

and the vector $(a_{k1}, a_{k2},...., a_{kn})$ with $k = n + 1$ is a linear combination of the row vectors of the matrix $[a_{ij}]$. Thus, the vectors $y_1, y_2,..., y_n, y_{n+1}$ are not linearly independent. This proves that any set of $n + 1$ vectors in T is linearly dependent. But, T being a Hamel basis, consists of linearly independent vectors. As such T has m ($\leq n$) vectors. Similarly, interchaning the roles of S and T, it can be shown that $n \leq m$. Hence, the number of elements in S and T is the same. ∎

This enables us to define the dimension of a finite dimensional space as the number of elements in any basis.

1.6.11 Definition. Let V be a finite dimensional linear space. The *dimension* of V, written as dim V, is defined as the number of elements in any of its bases.

1.6.12 Examples

1. dim $\mathbb{R} = 1 = $ dim $\mathbb{C}$.

2. dim $\mathbb{C}_{\mathbb{R}} = 2$.

3. dim $\mathbb{R}^n = n = $ dim $\mathbb{C}^n$

For an infinite dimensional linear space, it can be shown that all Hamel bases are equivalent sets.

1.6.13 Theorem. *Any two Hamel bases of a linear space V have the same cardinal number.*

Proof. Suppose S and T be two Hamel bases of V and let S be infinite. Let $\overline{\overline{S}} = \alpha$ and $\overline{\overline{T}} = \beta$. Since S is a Hamel basis of V, every $y \in V$, in particular, every $y \in T$ is a linear combination, with non-zero coefficients, of a finite number of vectors $x_1, x_2, ..., x_n$ in S and there are only n vectors in T which could be associated in this way with the same set $x_1, x_2, ..., x_n$ or some subset of it. Since the cardinal number of the set of all finite subsets of S is the same as that of S itself, it follows that $\beta \geq \aleph_0 . \alpha = \alpha$. Similarly, by interchanging the roles of S and T, we can prove that $\alpha \leq \beta$. Hence, $\alpha = \beta$. ∎

This theorem motivates to define the dimension of an infinite dimensional linear space.

1.6.14 Definition. Let V be a linear space. The cardinal number of a Hamel basis of V is said to be the *dimension* of V.

In fact, Hamel basis is an extension of the concept of basis (in algebraic sense) for spaces of infinite dimensions. We may sometimes refer to it as an algebraic basis for V.

Direct sums

Let U and W be two subspaces of a linear space V. Define

$$U + W = \{u + w : u \in U, w \in W\}.$$

One can easily verify that the sum $U + W$ is a subspace of V. If the subspaces U and W are such that $V = U + W$, then we say that V is the sum of U and W. If further $U \cap W = \{\theta\}$, then $v \in V$ can be uniquely expressed in the form $v = u + w$ with $u \in U$, $w \in W$. In this case, we say that V is the direct sum of U and W, and we write it as $V = U \oplus W$. More precisely.

1.6.15 Definition. A linear space V is said to be the *direct sum* of two subspaces U and W of V, written

$$V = U \oplus W,$$

if $V = U + W$ and $U \cap W = \{\theta\}$ or equivalently if each $x \in V$ has a unique representation

$$x = y + z, \, y \in U, \, z \in W.$$

Then, W is called an algebraic complement of U in V and vice versa, and (U, W) is called a *complementary pair* of subspaces in V.

Quotient spaces

Let W be a subspace of linear space V. The coset of an element $x \in V$ with respect to W, denoted by $x + W$, is defined to be the set

$$x + W = \{x + w : w \in W\}.$$

Any two cosets are either disjoint (distinct) or identical and the distinct cosets form a partition of V. Write

$$V/W = \{x + W : x \in V\}.$$

Define the linear operations on V/W by

$$\begin{cases} (x + W) + (y + W) = (x + y) + W \\ \alpha\,(x + W) = \alpha\,x + W \end{cases}$$

where $x, y \in V$ and $\alpha \in \mathbb{K}$. Addition and scalar multiplication in V/W is meaningful. It can be easily verified that V/W, under the linear operations described above, is a linear space over $\mathbb{K}$. The coset W is the additive identity in V/W and $-(x + W) = -x + W$.

1.6.16 Definition. Let W be a subspace of a linear space V. The space V/W is called the *quotient space* (or *factor space*) of V by W.

The function $q : V \rightarrow V/W$ defined by $q(x) = x + W$ is called the *canonical mapping* (or, *quotient mapping*) of V onto V/W. Then, dimension of V/W is called the *co-dimension* of W in V, denoted by codim W. More precisely, we have

$$\text{codim } W = \dim (V/W).$$

Let $V = \mathbb{R}^2$, and we think of vectors in $\mathbb{R}^2$ as the directed line segment whose terminal points are denoted by arrow heads and the common initial point is the origin. Suppose W is a subspace of $\mathbb{R}^2$. A coset $x + W$ is clearly a line parallel to W and V/W is the set of all lines parallel to W. The vector addition and scalar multiplication in the quotient space V/W are indicated in the Fig. 1.1 and Fig. 1.2.

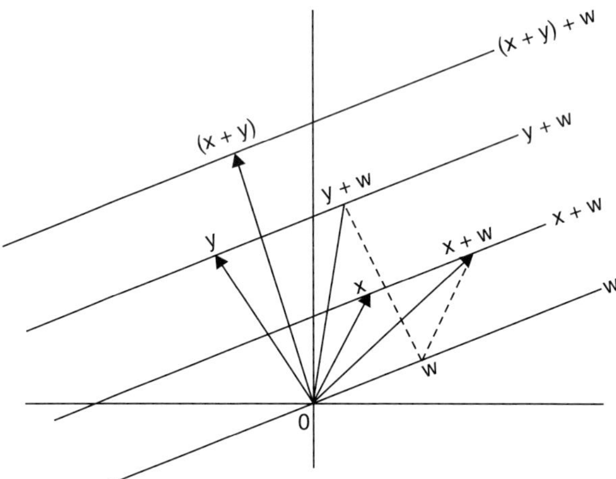

Fig. 1.1

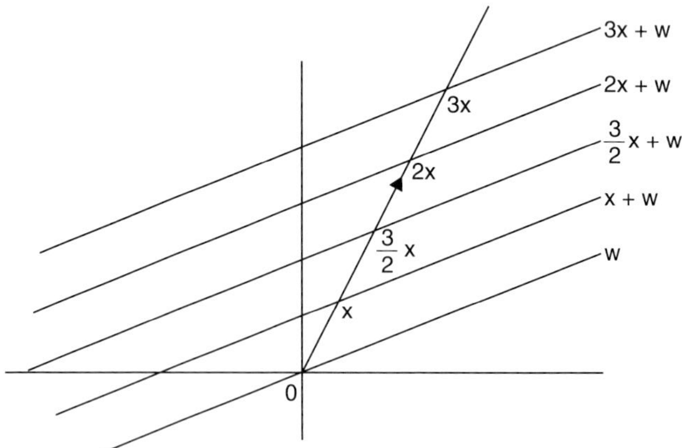

Fig. 1.2

Hyperplanes

A set $M \subset V$ is called a *linear variety* if $M = W + x_0$, where $x_0 \in V$ is fixed and W is a subspace of V. Clearly, M is a translation of W by x_0.

A subspace W of V is called *maximal* if

(*i*) $W \underset{\neq}{\subseteq} V$, and

(*ii*) $\exists$ no subspace W_1 of V such that $W \underset{\neq}{\subseteq} W_1 \underset{\neq}{\subseteq} M$.

In fact, a maximal subspace W of V is of co-dimension 1, *i.e.*, a subspace with $\dim V/W = 1$.

A translation of a maximal subspace is called an *affine hyperplane* or simply a *hyperplane* in V. Equivalently, a hyperplane H in V is a set such that V can be expressed as a direct sum of H and a one-dimensional subspace of V, *i.e.*,

$$V = H + [x],$$

for some $x \in V$.

A hyperplane containing 0 is the same thing as a maximal subspace. All the hyperplanes obtained by translating a particular subspace are said to be parallel to this.

Convex sets

A non-empty subset C of a linear space V is said to be *convex* if

$$x, y \in C, \lambda \geq 0, \mu \geq 0, \lambda + \mu = 1 \quad \Rightarrow \quad \lambda x + \mu y \in C.$$

Geometrically, a line segment joining any two points in C lies wholly in C. The *convex hull* of a non-empty subset A of V is the smallest convex set that contains A or, equivalently, the convex set containing all the linear combinations of vectors from A.

A linear subspace of V and a hyperplane in V are examples of convex sets.

Linear operators

1.6.17 Definition. Let V and $\tilde{V}$ be linear spaces over the same scalar field $\mathbb{K}$ ($\mathbb{R}$ or $\mathbb{C}$). A function $T : V \to \tilde{V}$ is said to be a *linear map* if

(*i*) $T(x + y) = Tx + Ty, \quad \forall\, x, y \in V,$

(*ii*) $T(\alpha x) = \alpha\, T x, \quad \forall\, x \in V$ and $\forall\, \alpha \in \mathbb{K}.$

Notes

1. Linear maps are also called *linear transformations, linear operators* or *homomorphisms.*

2. The linear operations of both the linear spaces V and $\tilde{V}$ are denoted by the same symbol.

3. For linear mappings, it is customary to write the value of T at x by $T\,x$ rather than $T\,(x)$.

1.6.18 Examples

1. Let V and $\tilde{V}$ be any two linear spaces over the same scalar field $\mathbb{K}$. The mapping $O : V \to \tilde{V}$ such that $O\,v = \theta$, $\forall\, v \in V$, is a linear operator, called the *zero operator.*

2. Let V be an arbitrary linear space. The mapping $I : V \to V$ defined by $I\,v = v$, $\forall\, v \in V$, is linear operator, called the *identity operator* on V.

3. Let $A = [a_{ij}]$ be a fixed $m \times n$ matrix with a_{ij} in $\mathbb{K}$ ($\mathbb{R}$ or $\mathbb{C}$). By using the matrix notation for vectors in $\mathbb{K}^m$ and $\mathbb{K}^n$, we can define a mapping $T : \mathbb{K}^n \to \mathbb{K}^m$ by $T\,(X) = A\,X$, where

$$
X = \begin{bmatrix} x_1 \\ x_2 \\ . \\ . \\ . \\ x_n \end{bmatrix}
$$

is a $n \times 1$ matrix. Observe that the product AX is an $m \times 1$ matrix. Thus, T maps $\mathbb{K}^n$ into $\mathbb{K}^m$. Writing

$$
T(X) = \begin{bmatrix} y_1 \\ y_2 \\ . \\ . \\ . \\ y_m \end{bmatrix}
$$

we have

$$T(X) = \begin{bmatrix} y_1 \\ y_2 \\ . \\ . \\ . \\ y_m \end{bmatrix} = \begin{bmatrix} a_{11} & a_{12} \ldots a_{1n} \\ a_{21} & a_{22} \ldots a_{2n} \\ . \\ . \\ . \\ a_{m1} & a_{m2} \ldots a_{mn} \end{bmatrix} \begin{bmatrix} x_1 \\ x_2 \\ . \\ . \\ . \\ x_n \end{bmatrix}$$

Then, T is a linear operator.

We now give some basic properties of linear operators.

1.6.19 Theorem. Let $T : V \to \tilde{V}$ be a linear operator. Then:

(a) $T\theta = \theta$.

(b) $T(-x) = -Tx, \ \forall \ x \in V$.

(c) $T(x - y) = Tx - Ty, \ \forall \ x, y \in V$.

The kernel of T, denoted by ker (T), of a linear operator $T : V \to \tilde{V}$ is defined as

$$\ker (T) = \{x \in V : Tx = \theta\}.$$

Note. ker (T) is also called the *null space* of T and is denoted by $\mathcal{N}(T)$.

1.6.20 Theorem. *Let $T : V \to \tilde{V}$ be a linear operator. Then:*

(a) *$\mathcal{R}(T)$, the range of T, is a subspace of $\tilde{V}$.*

(b) *ker (T), the kernel of T, is a subspace of V.*

Let X, Y and Z be linear spaces over the same scalar field $\mathbb{K}$. Suppose $T : X \to Y$ and $S : Y \to Z$ be linear operators. Then, the product map $ST : X \to Z$ is defined by

$$(S\,T)\,x = S\,(T\,x).$$

Note that $S\,T$ is also a linear operator (Verify !).

1.6.21 Definition. A linear operator $T : V \to \tilde{V}$ is *invertible* if there is a map $S : \tilde{V} \to V$ such that ST and TS are the identity operators I_V on V and $I_{\tilde{V}}$ on $\tilde{V}$, respectively. In this case, S is called the *inverse* (*algebraic*) of T and we write $S = T^{-1}$.

1.6.22 Theorem. *Let $T : V \to \tilde{V}$ be a linear operator. Then:*

(a) *T is bijective $\Rightarrow$ T is invertible $\Rightarrow$ ker $(T) = \{\theta\}$.*

(b) *T^{-1}, if exists, is a linear operator.*

(c) *If dim $V = n < \infty$ and T^{-1} exists, then dim $\tilde{V} = n$.*

We, now introduce more general concepts than the inverse T^{-1} *of T.*

1.6.23 Definition. Let V and $\tilde{V}$ be linear spaces over the same scalar field $\mathbb{K}$. Suppose $T : V \to \tilde{V}$ is a linear operator. If there exists an operator $S : \tilde{V} \to V$ such that

$$ST = I_V,$$

then S is said to be a *left inverse* of T, written T_l^{-1}. Similarly, if there exists an operator $R : \tilde{V} \to V$ such that

$$TR = I_{\tilde{V}},$$

then, R is said to be a *right inverse* of T, written T_r^{-1}.

1.6.24 Theorem. *Let $T : V \to \tilde{V}$ be a linear operator. Then:*

(a) *If T_r^{-1} exists, then T is surjective and the equation $T x = y$ has at least one solution $x = T_r^{-1} y$.*

(b) *If T_l^{-1} exists, then T is injective. Further, if $T x = y$ has a solution then it is unique.*

Let $L(V, \tilde{V})$ be the set of all linear operators of V into $\tilde{V}$. For S, $T \in L(V, \tilde{V})$, define the sum and the scalar multiplication in $L(V, \tilde{V})$ as follows:

$$(S + T)(x) = S x + T x, \quad \forall\, x \in V$$

and

$$(\alpha\, T)\, x = \alpha\, T x, \quad \forall\, x \in V \text{ and } \forall\, \alpha \in \mathbb{K}.$$

1.6.25 Theorem. *$L(V, \tilde{V})$ is a linear space under linear operations given above.*

Write $$L(V) \equiv L(V, V).$$

1.6.26 Theorem. *Let S, T, R be in $L(V)$ and $\alpha \in \mathbb{K}$. Then:*

(a) *$T(S R) = (T S) R$.*

(b) *$T(S + R) = T S + T R$.*

(c) $(T + S)\, R = T\, R + S\, R.$

(d) $\alpha\, (T\, S) = (\alpha\, T)\, S = T\, (\alpha\, S).$

Isomorphisms

Isomorphism of two spaces of the same kind, in fact, describes that the underlying spaces are identical as two copies of the same space in the sense that the structures in them are identically the same whereas the nature of points in them does not matter.

Note that the defining conditions of a linear operator (Definition 1.6.17) of a linear space V into another linear space $\tilde{V}$ express the fact that the linear operations of the spaces V and $\tilde{V}$ are preserved under T. If in addition T is bijective, then the two spaces V and $\tilde{V}$ are the same as far as the linear structures on them are concerned. More precisely

1.6.27 Definition. A bijective linear operator $T : V \to \tilde{V}$, where V and $\tilde{V}$ are linear spaces over the field $\mathbb{K}$, is called an *isomorphism* of V onto $\tilde{V}$. The spaces V and $\tilde{V}$ are said to be *isomorphic* if there is an isomorphism between them. We write $V \cong \tilde{V}$.

One can easily verify that the relation '$\cong$' of isomorphism is an equivalence relation.

We now discuss an isomorphism of the quotient space V/W and the subspace U of V complementary to W. Let $q : V \to V/W$ be the quotient mapping of V onto V/W. It can be easily verified that the restriction mapping $q_{|U} : U \to V/W$ is bijective and preserves linear operations (Verify !). Hence, $q_{|U}$ is an isomorphism of U onto V/W. Thus, we conclude that if U is complementary to W, then U and V/W are isomorphic linear spaces.

The following theorem is most fundamental in the sense that the real space $\mathbb{R}^n$ (or complex space $\mathbb{C}^n$) is the only possible example of an n-dimensional linear space.

1.6.28 Theorem. *Let $V \neq \{\theta\}$ be a finite dimensional linear space of dimension n over the field $\mathbb{K}$. Then V is isomorphic to the linear space $\mathbb{K}^n$.*

Proof. Since V is an n-dimensional linear space, any basis for V contains n vectors. Let $B = \{x_1, x_2, ..., x_n\}$ be a basis of V whose elements are written in the indicated order. If $x \in V$, then x is uniquely expressed in the form.

$$x = \sum_{i=1}^{n} \alpha_i x_i$$

for some uniquely determined scalars α_1, α_2,...., α_n. Define the function $f : V \to \mathbb{K}^n$ given by

$$f(x) = (\alpha_1, \alpha_2,...., \alpha_n).$$

Clearly, f is a well-defined bijective map. Also, it is easy to verify that

$$f(x + y) = f(x) + f(y), \quad \forall\, x, y \in V$$

and

$$f(\lambda\, x) = \lambda\, f(x), \quad \forall\, x \in V \text{ and } \forall\, \lambda \in \mathbb{K}.$$

Hence, f is an isomorphism of V onto $\mathbb{K}^n$. $\blacksquare$

Linear projections

We discuss a particular type of linear operator which is of tremendous importance. Let $V = U \oplus W$ so that each $v \in V$ can be expressed uniquely in the form $v = u + w$, where $u \in U$ and $w \in W$. Since w is uniquely determined by v, we may define a mapping $P : V \to W$ by $P\,v = w$. One can easily verify that

(a) P is linear operator on V, called the linear projection on W along U.

(b) $\mathscr{R}(P) = W$ and $\mathscr{N}(P) = U$.

(c) $P^2 = P$ (Idempotent).

The property of idempotency of P given in (c) is, in fact, a characterising property for a linear operator to be a linear projection. In fact, we have

1.6.29 Theorem. *Let $P : V \to V$ be a linear operator. Then, P is idempotent if and only if there exist subspaces W and U of V such that $V = U \oplus W$ and P is the linear projection on W along U.*

1.6.30 Corollary. *If P is a linear projection on W along U, then $I - P$ is a linear projection on U along W, and conversely.*

Remark. Let W be a given subspace of V. Then, there may be many different subspaces U such that $V = U \oplus W$. Consequently, there may also be many different linear projections on W along U.

A linear projection on W along U can geometrically be described as:

(a) A linear projection P determines a pair of linear subspaces W and U such that $V = U \oplus W$, where $\mathscr{R}(P) = W$ and $\mathscr{N}(P) = U$.

(b) A pair of subspaces W and U with $V = U \oplus W$ determines a linear projection P on W along U.

Thus, the study of linear projection P is equivalent to the study of a pair of subspaces whose direct sum is the given linear space.

Linear functionals

A linear functional on linear space V is a linear operator with domain as V and range in the scalar field $\mathbb{K}$. We denote the linear functionals by lower case letters f, g, h etc. If f and g are any two linear functionals on V, we define the sum $f + g$ and the scalar product αf as follows:

$$(f + g)\, x = f(x) + g(x), \quad \forall\, x \in V$$

and

$$(\alpha f)\,(x) = \alpha f(x), \quad \forall\, \alpha \in \mathbb{K} \text{ and } \forall\, x \in V.$$

In view of Theorem 1.6.25, it follows that $L(V, \mathbb{K})$ is a linear space under the above linear operations. This space is denoted by $V^{\#}$ and is called the *algebraic dual* of V.

Since $V^{\#}$ is a linear space over $\mathbb{K}$, it follows that $L(V^{\#}, \mathbb{K})$ is again a linear space. Note that the elements of $L(V^{\#}, K)$ ($\equiv (V^{\#})^{\#}$) are the linear functionals defined on $V^{\#}$. The algebraic dual $(V^{\#})^{\#}$ of $V^{\#}$ is denoted by $V^{\#\#}$. The linear space $V^{\#\#}$ is called the *second algebraic dual space* of V.

Suppose $x \in V$ is fixed. Define the mapping $\varphi_x : V^{\#} \to \mathbb{K}$ by

$$\varphi_x(f) = f(x), \quad (f \in V^{\#}).$$

Note that φ_x is a linear functional since

$$\varphi_x(f + g) = (f + g)\,(x)$$
$$= f(x) + g(x)$$
$$= \varphi_x(f) + \varphi_x(g), \quad \forall\, f, g \in V^{\#}$$

and
$$\varphi_x(\alpha f) = (\alpha f)\,(x)$$
$$= \alpha f(x)$$
$$= \alpha\, \varphi_x(f), \quad \forall\, f \in V^{\#} \text{ and } \forall\, \alpha \in \mathbb{K}.$$

Hence, $\varphi_x \in V^{\#\#}$. Now, define a mapping $\pi : V \to V^{\#\#}$ by

$$\pi(x) = \varphi_x.$$

Observe that the mapping π is linear since

$$(\pi(x + y))\,(f) = \varphi_{x+y}(f)$$
$$= f(x + y)$$

$$= f(x) + f(y)$$

$$= (\varphi_x + \varphi_y)(f), \quad \forall f \in V^{\#}$$

$$\Rightarrow \qquad \pi(x + y) = \pi(x) + \pi(y),$$

and

$$(\pi(\alpha x))(f) = \varphi_{\alpha x}(f)$$

$$= f(\alpha x)$$

$$= \alpha f(x)$$

$$= \alpha \varphi_x(f)$$

$$= \alpha(\pi(x))(f), \quad \forall f \in V^{\#} \text{ and } \forall \alpha \in \mathbb{K}$$

$$\Rightarrow \qquad \pi(\alpha x) = \alpha \pi(x).$$

1.6.31 Definition. The mapping $\pi : V \to V^{\#\#\#}$ defined above is called the *canonical mapping* (or *canonical embedding*) of V into $V^{\#\#\#}$.

It can be shown that π is injective. If π is surjective also, then $\pi(V) = V^{\#\#\#}$. In that case, V is called an *algebraically reflexive space*.

We give below the concept relating to certain functionals on the cartesian product of two linear spaces.

1.6.32 Definition. A *sesquilinear functional* (or *sesquilinear form*) h on $X \times Y$, where X and Y are linear spaces over the field $\mathbb{K}$ ($\mathbb{R}$ or $\mathbb{C}$), is a mapping

$$h : X \times Y \to \mathbb{K}$$

which satisfies the following conditions:

    (i) $h(x_1 + x_2, y) = h(x_1, y) + h(x_2, y)$

    (ii) $h(x, y_1 + y_2) = h(x, y_1) + h(x, y_2)$

    (iii) $h(\alpha x, y) = \alpha h(x, y)$

    (iv) $h(x, \beta y) = \overline{\beta} h(x, y),$

for $x, x_1, x_2 \in X$; $y, y_1, y_2 \in Y$ and $\alpha, \beta \in \mathbb{K}$.

Note that h is linear in the first argument and conjugate linear in the second one.

Further, h is called *bilinear functional* (or *bilinear form*) if it is linear in both the arguments, *i.e.*,

    (v) $h(x, \beta y) = \beta h(x, y).$

In case X and Y are linear spaces over $\mathbb{R}$, sesquilinear and bilinear functionals are one and the same.

1.6.33 Definition. A *Hermitian sesquilinear form* (or simply, *Hermitian form*) on $X \times X$, where X is a linear space over $\mathbb{K}$, is a mapping $h : X \times X \to \mathbb{K}$ which satisfies the following conditions:

    (i) $h\,(x + y, z) = h\,(x, y) + h\,(y, z)$

    (ii) $h\,(\alpha\, x, y) = \alpha\, h\,(x, y)$

    (iii) $h\,(x, y) = \overline{h\,(y, x)}$,

for $x, y, z \in X$ and $\alpha \in \mathbb{K}$.

The matrix representation of a linear operator

We shall, now, show that every linear operator on finite dimensional linear space can be represented in terms of matrices. First of all we introduce few notions.

Let V be an n-dimensional linear space over the scalar field $\mathbb{K}$ and let $A = \{x_1, x_2,..., x_n\}$ be any ordered basis for V. By an *ordered basis*, we mean a basis whose vectors are arranged in a definite order which we keep fixed. If $x \in V$, then x can be uniquely expressed in the form

$$x = \alpha_1 x_1 + \alpha_2 x_2 + \dots + \alpha_n x_n,$$

where $\alpha_1, \alpha_2,...., \alpha_n$ are in $\mathbb{K}$. We shall refer to

$$[x]_A = \begin{bmatrix} \alpha_1 \\ \alpha_2 \\ \cdot \\ \cdot \\ \cdot \\ \cdot \\ \alpha_n \end{bmatrix}$$

as the coordinate vector of x with respect to the ordered basis A.

Let $T : V \to \tilde{V}$ be a linear operator of an n-dimensional linear space V into an m-dimensional linear space $\tilde{V}$ $(n \neq 0,\ m \neq 0)$. Choose $A = \{x_1, x_2,...., x_n\}$ and $B = \{y_1, y_2,...., y_m\}$ as ordered bases for V and $\tilde{V}$, respectively. Since $Tx_j \in V$ and B is an ordered basis for $\tilde{V}$, we can (uniquely) expresss Tx_j as

$$Tx_j = c_{1j}\, y_1 + c_{2j}\, y_2 + \dots + c_{mj}\, y_m.$$

This means that

$$[Tx_j]_B = \begin{bmatrix} c_{1j} \\ c_{2j} \\ \cdot \\ \cdot \\ \cdot \\ c_{mj} \end{bmatrix}$$

We, now define an $m \times n$ matrix M by choosing $[Tx_j]_B$ as the j-th column of M, i.e.

$$M = \begin{bmatrix} c_{11} & c_{12} \dots & c_{1n} \\ c_{21} & c_{22} \dots & c_{2n} \\ \cdot & & \\ \cdot & & \\ \cdot & & \\ \cdot & & \\ c_{m1} & c_{m2} \dots & c_{mn} \end{bmatrix}$$

This matrix is uniquely determined by T, and is called the representation of T with respect to the ordered bases A and B. Such a matrix satisfies the property given in the following theorem.

1.6.34 Theorem. *If $y = Tx$ for some $x \in V$, then $[\,y\,]_B = M\,[x]_A$, where $[x]_A$ and $[y]_B$ are the coordinate vectors of x and y with respect to the ordered bases A and B, respectively. Moreover, M is the only matrix satisfying this property.*

If $T : V \to V$ is a linear operator on an n-dimensional linear space V, then to obtain a representation of T we fix ordered bases A and B for V, and obtain a matrix M representing T with respect to A and B. However, it is often convenient in this case to choose $A = B$; and we refer to M as the matrix representation of T with respect to the basis A.

1.6.35 Theorem. *Let $L\,(V,\ \tilde{V}\,)$ be the linear space of all linear operators of an n-dimensional linear space V into an m-dimensional linear space $\tilde{V}\ (n \neq 0$ and $m \neq 0)$. Then, $L\,(V,\ \tilde{V}\,)$ is isomorphic to the linear space $\mathbb{R}^{m \times n}$ of all $m \times n$ matrices.*

Problems

2. Let V_1 and V_2 be any two linear spaces over the same scalar field $\mathbb{K}$, and suppose $V = V_1 \times V_2$ be the cartesian product of V_1 and V_2. Define the algebraic operations by

$$(x_1, x_2) + (y_1, y_2) = (x_1 + y_1, x_2 + y_2)$$
$$\alpha\,(x_1, x_2) = (\alpha\,x_1, \alpha\,x_2).$$

Prove that V is a linear space over $\mathbb{K}$.

3. In Example 1.6.3 (10), prove that $C\,[a, b]$ is a real linear space.

4. Verify all the steps in Example 1.6.3 (6) to prove that l^p $(1 \le p < \infty)$ is a linear space.

5. Let V be a non-zero finite dimensional linear space of dimension n. Prove that every set of $n + 1$ vectors in V is linearly dependent.

6. Let V be a non-zero finite dimensional linear space of dimension n. Let $B = \{x_1, x_2, ..., x_n\}$ be a subset of V. Prove that:

(a) B is a basis for $V \Leftrightarrow B$ is a linearly independent set.

(b) B is a basis for $V \Leftrightarrow$ it spans V.

7. Give an example to show that the multiplication of linear operators, in general, is non-commutative.

8. If two finite dimensional linear spaces V and $\tilde{V}$ over the same scalar field $\mathbb{K}$ are isomorphic, prove that V and $\tilde{V}$ have the same dimension. Also establish the converse.

9. Let $T : V \to \tilde{V}$ be a linear operator from an n-dimensional linear space V into a linear space $\tilde{V}$. Let $\{x_1, x_2, ..., x_n\}$ be a basis for V. If x is any vector in V, prove that $T\,x$ is completely determined by $\{Tx_1, Tx_2, ..., Tx_n\}$.

10. If $T : V \to \tilde{V}$ be a linear operator from an n-dimensional linear space V into linear space $\tilde{V}$, prove that

$$\dim \mathcal{N}(T) + \dim \mathcal{R}\,(T) = \dim V.$$

1.7 Metric Spaces*

The term metric is derived from the term metor (measure). The concept of a metric space is essentially due to a French mathematician M. Fréchet, though

---

* For details of the concepts and results presented in this section, one may refer to [25].

the definition presently in use is that given by the German mathematician F. Hausdorff in 1914. Fréchet introduced the notion in his doctoral thesis presented to the University of Paris in 1906.

1.7.1 Definition. Let X be an arbitrary non-empty set. A mapping $d : X \times X \to$ $\mathbb{R}$ is said to be a *metric* on X if it satisfies the following conditions:

   $(i)\ d\,(x, y) \geq 0, \quad \forall\ x, y \in X$

   $(ii)\ d\,(x, y) = 0 \quad \Leftrightarrow \quad x = y, \quad x, y \in X$

   $(iii)\ d\,(x, y) = d\,(y, x), \quad \forall\ x, y \in X$

   $(iv)\ d\,(x, y) \leq d\,(x, z) + d\,(y, z), \quad \forall\ x, y, z \in X.$

If X is a non-empty set and d is a metric on X, then the pair (X, d) is called a *metric space*.

Notes

1. The condition (iii) expresses the fact that d is symmetric in x and y. The inequality (iv) is generally called the *triangle inequality*. Condition (i) to (iv) are also known as the *axioms of the metric space* $(X, d\,)$.

2. If there is no confusion about the metric d on X, we may write simply X for the metric space $(X, d\,)$.

3. The elements of a metric space $(X, d\,)$ are usually called the points of X. A metric d is also called a *distance function*. The real number $d\,(x, y)$ is called the distance between x and y.

   We shall now give those examples of metric spaces which would be used in the subsequent chapters.

1.7.2 Examples

1. Let X be any non-empty set. Define a mapping $d : X \times X \to \mathbb{R}$ by

$$d\,(x, y) = \begin{cases} 0, & x = y \\ 1, & x \neq y \end{cases}$$

   It is a routine matter to verify that d is a metric on X. Such a metric d is called the *discrete metric* for X; and the space (X, d) is called a *discrete metric space*.

2. Let $X = \mathbb{R}$. Define

$$d(x, y) = |\,x - y\,|, \quad \forall\ x, y \in \mathbb{R}.$$

   Then, d is a metric on $\mathbb{R}$. The mapping d is called the *usual metric* on $\mathbb{R}$.

3. Let $X = \mathbb{C}$. The function d defined by

$$d(x, y) = |x - y|,$$

for all $x, y \in \mathbb{C}$, is a metric on $\mathbb{C}$.

Note. The real number $d(x, y)$ is the usual distance between the points x and y in the complex plane. As such, it is also called the usual metric on C.

4. Let $X = \mathbb{R}^2$ and suppose $x = (\xi_1, \xi_2)$, $y = (\eta_1, \eta_2)$ be any two points in $\mathbb{R}^2$. Define the mappings $d_p, d_\infty : X \times X \to \mathbb{R}$ as follows:

$$d_p(x, y) = \left\{ |\xi_1 - \eta_1|^p + |\xi_2 - \eta_2|^p \right\}^{\frac{1}{p}}, \ 1 \leq p < \infty$$

$$d_\infty(x, y) = \max \{ |\xi_1 - \eta_1|, |\xi_2 - \eta_2| \}.$$

Then, d_p for each $1 \leq p < \infty$ and d_∞ are metrics on the same underlying set $X = \mathbb{R}^2$.

Note. Similar metrics are also available of the complex space $\mathbb{C}^2$.

5. Let $X = \{(x, y) \in \mathbb{R}^2 : x^2 + y^2 = 1\}$, *i.e.*, the unit circle in $\mathbb{R}^2$, and let

$$d_1\left((x, y), (x', y')\right) = \sqrt{(x - x')^2 + (y - y')^2}$$

Then, d_1 is a metric on X.

Another possible metric d_2 on X is defined by $d_2(p, p') = $ arc length between the points p and p' (Fig. 1.3).

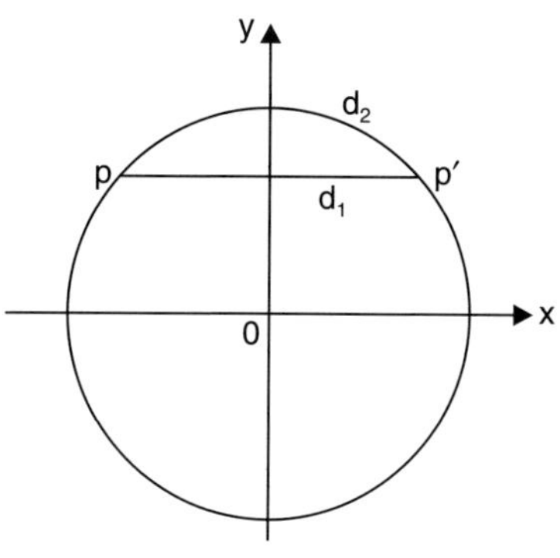

Fig. 1.3

6. Let $X = \mathbb{R}^n$ be the set of all n-tuples of real numbers. Define $d : X \times X \to \mathbb{R}$ by

$$d(x, y) = \sqrt{(\xi_1 - \eta_1)^2 + (\xi_2 - \eta_2)^2 + \dots + (\xi_n - \eta_n)^2}$$

where $x = (\xi_1, \xi_2,...., \xi_n)$ and $y = (\eta_1, \eta_2,...., \eta_n)$ are in $\mathbb{R}^n$. Then, d is a metric on $\mathbb{R}^n$ (Verify the axioms !). The metric space $(\mathbb{R}^n, d)$ is called the *n-dimensional Euclidean space* and is denoted by $\mathbb{R}^n$.

Note. In the case $n = 2$, this gives the coordinate plane of Example 4, and in the case $n = 3$ this gives the usual 3-dimensional space.

7. Let $X = \mathbb{C}^n$ be the set of all n-tuples of complex numbers. Define the function d on $X \times X$ as follows:

$$d(x, y) = \sqrt{|\xi_1 - \eta_1|^2 + |\xi_2 - \eta_2|^2 + ... + |\xi_n - \eta_n|^2} \, ,$$

where $x = (\xi_1, \xi_2,...., \xi_n)$ and $y = (\eta_1, \eta_2,...., \eta_n)$ are in $\mathbb{C}^n$. Then, d is a metric on $\mathbb{C}^n$ (Verify !). This metric on $\mathbb{C}^n$ is called the complex Euclidean metric on $\mathbb{C}^n$; and the metric space $(\mathbb{C}^n, d)$ is called the *n-dimensional unitary space* and is denoted by $\mathbb{C}^n$.

8. The metrics defined in Example 4 above can be extended to the linear space $\mathbb{K}^n$ ($\mathbb{R}^n$ or $\mathbb{C}^n$) by defining

$$d_p (x, y) = \left\{ \sum_{i=1}^{n} |\xi_i - \eta_i|^p \right\}^{\frac{1}{p}} , 1 \leq p < \infty$$

and $\qquad d_\infty (x, y) = \max_{1 \leq i \leq n} \{|\xi_i - \eta_i|\},$

where $x = (\xi_1, \xi_2,...., \xi_n)$ and $y = (\eta_1, \eta_2,..., \eta_n)$ are in $\mathbb{K}^n$.

9. Let $X = \omega$, the set of all sequences (bounded or unbounded) in $\mathbb{K}$ ($\mathbb{R}$ or $\mathbb{C}$). Define the mapping $d : X \times X \to \mathbb{R}$ by

$$d (x, y) = \sum_{i=1}^{\infty} \frac{1}{2^i} \frac{|\xi_i - \eta_i|}{1 + |\xi_i - \eta_i|}$$

where $x = \{\xi_i\}$ and $y = \{\eta_i\}$ are in ω. We shall verify the triangle inequality, and the other axioms being straightforward are left to the reader to verify. Consider the inequality

$$\frac{|a + b|}{1 + |a + b|} \leq \frac{|a|}{1 + |a|} + \frac{|b|}{1 + |b|},$$

where $a, b \in \mathbb{K}$ (Inequality 1.5.2). Write $z = \{\zeta_i\}$. Taking $a = \xi_i - \zeta_i$ and $b = \zeta_i - \eta_i$, we have

$$\frac{|\xi_i - \eta_i|}{1 + |\xi_i - \eta_i|} \leq \frac{|\xi_i - \zeta_i|}{1 + |\xi_i - \zeta_i|} + \frac{|\zeta_i - \eta_i|}{1 + |\zeta_i - \eta_i|}$$

$$\Rightarrow \quad \sum_{i=1}^{\infty} \frac{1}{2^i} \frac{|\xi_i - \eta_i|}{1 + |\xi_i - \eta_i|} \leq \sum_{i=1}^{\infty} \frac{1}{2^i} \frac{|\xi_i - \zeta_i|}{1 + |\xi_i - \zeta_i|} + \sum_{i=1}^{\infty} \frac{1}{2^i} \frac{|\zeta_i - \eta_i|}{1 + |\zeta_i - \eta_i|}$$

$$\Rightarrow \qquad d(x, y) \leq d(x, z) + d(z, y).$$

We thus conclude that ω is a metric space.

10. Let $X = l^p$, $1 \leq p < \infty$, be the set of all sequences $x = \{\xi_i\}$ of scalars (real or complex) such that

$$\sum_{i=1}^{\infty} |\xi_i|^p < \infty.$$

Define the mapping $d : X \times X \to \mathbb{R}$ by

$$d(x, y) = \left(\sum_{i=1}^{\infty} |\xi_i - \eta_i|^p \right)^{\frac{1}{p}},$$

where $x = \{\xi_i\}$ and $y = \{\eta_i\}$ are in l^p.

The first three axioms for d to be a metric on X are simple and the triangle inequality follows by employing the Minkowski inequality. Thus, l^p is a metric space.

11. Let X be the set of all real-valued continuous functions defined on the closed interval $[a, b]$. Define the mappings d_∞ and d_1 on $X \times X$ into $\mathbb{R}$ as follows:

$$d_\infty(x, y) = \max_{t \in [a,b]} |x(t) - y(t)|$$

and
$$d_1(x, y) = \int_a^b |x(t) - y(t)| \, dt.$$

Note that $d_\infty(x, y)$ is precisely the greatest vertical gap between the functions as illustrated in Fig. 1.4 and $d_1(x, y)$ is precisely the area of the region which lies between the functions as shown in Fig. 1.5. It is a routine matter to verify that d_∞ and d_1 are metrices on X. We denote the metric space (X, d_∞) by $C[a, b]$.

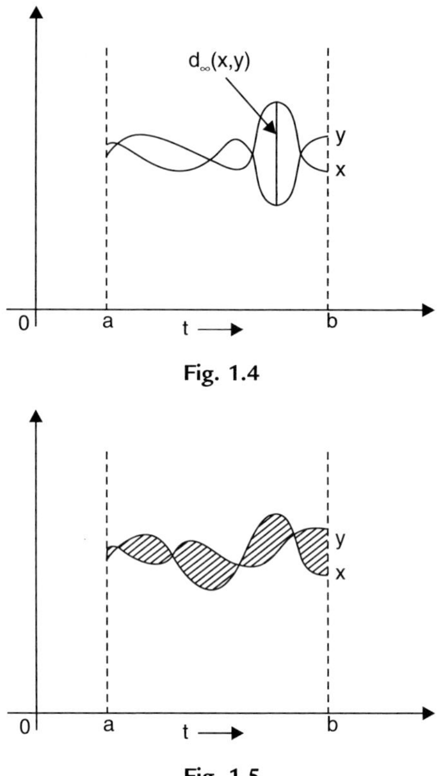

Fig. 1.4

Fig. 1.5

1.7.3 Definition. Let (X, d) be a metric space and Y an arbitrary subset of X. Define the mapping $d_Y : Y \times Y \to \mathbb{R}$ by

$$d_Y (x, y) = d (x, y), \quad \forall \ x, y \in Y.$$

Here d_Y is clearly a metric on Y and so (Y, d_Y) is a metric space. The metric d_Y is called the *relative metric induced* (or simply the *metric induced*) on Y by d. The space (Y, d_Y) is called the *metric subspace* of the metric space (X, d).

The above method of forming subspaces of a given metric space enables us to construct several examples of metric spaces. For example [0, 1] is a subspace of $\mathbb{R}$. Further, the set $\mathbb{Q}$ and the interval $]a, b[$ are also examples of subspaces of $\mathbb{R}$. The real line itself is a subspace of the complex plane.

We give below several concepts which will play an important role in connection with metric spaces.

1.7.4 Definition. Let (X, d) be a metric space.

1. The set $\{x : x \in X, d (x, x_0) < r\}$ is called an *open sphere* (or *open ball*) of radius r about the point x_0. We denote it by $S (x_0 \ ; \ r)$. We speak of the

closed sphere (or *closed ball*) if the inequality $d(x, x_0) < r$ is replaced by $d(x, x_0) \leq r$ and we denote it by $S[x_0 ; r]$.

2. A subset G of X is said to be an *open set* if for each $x \in G$, $\exists$ an $r > 0$ such that $S (x ; r) \subset G$.

3. A subset N of X is said to be a *neighbourhood* of $x \in X$ if $\exists$ some $r > 0$ such that $S (x ; r) \subset N$.

4. A point x in a subset G of X is said to be an *interior point* of G if G is a neighbourhood of x, *i.e.*, if $\exists$ an $r > 0$ such that $S(x ; r) \subset G$. The set of all interior points of G is called the *interior* of G. We denote it by G°.

5. A point x is said to be a *limit point* of a subset A of X if for each $r > 0$,

$$S(x ; r) \cap (A - \{x\}) \neq \phi,$$

that is, if A contains points other than x arbitrarily close to x. Denote the set of all limit points of A by A', called the *derived set* of A.

6. A subset F of X is said to be a *closed set* if F contains all its limit points, *i.e.*, if $F' \subset F$.

7. The closure of a subset A of X, denoted by $\overline{A}$, is the union of A with the set of its limit points, *i.e.*, $\overline{A} = A \cup A'$.

8. A non-empty subset A of X is said to be *bounded* if $\exists$ an element $a \in X$ and $K > 0$ such that $d(a, x) \leq K$, $\forall x \in A$; in other words, if $A \subset S (a ; K)$ of some $a \in X$ and $K > 0$.

1.7.5 Examples

1. Let (X, d) be a discrete metric space (Example 1.7.2 (1)).

   (a) Let $a \in X$. If $0 < r \leq 1$, then $S(a ; r) = \{a\}$.

      If $r > 1$, then $S (a ; r) = X$.

   (b) Every subset of X is an open and closed set.

   (c) Every non-empty subset $A \subset X$ is bounded since for any $a \in X$, we have

$$d (a, x) \leq 1, \quad \forall x \in A.$$

2. An open (or closed) interval in $\mathbb{R}$ with the usual metric is an open (respectively, closed) set in $\mathbb{R}$.

3. Figures below show the unit closed balls for the the metric spaces $(\mathbb{R}^2, d_1)$, $(\mathbb{R}^2, d_2)$ and $(\mathbb{R}^2, d_\infty)$ as discussed in Examples 1.7.2. (4).

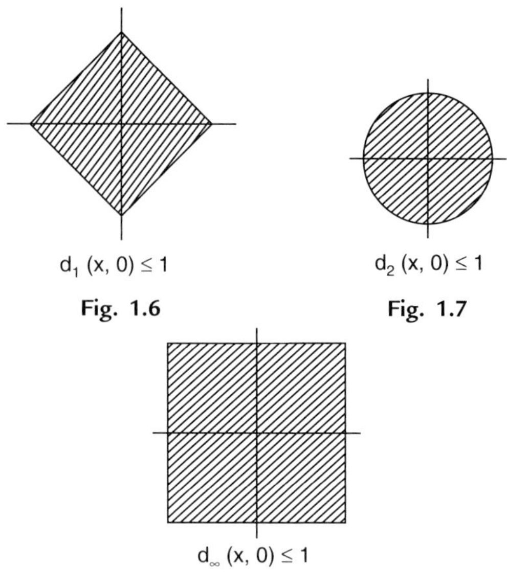

Fig. 1.6   Fig. 1.7

Fig. 1.8

4. In the usual metric space $\mathbb{C}$, the open ball $S (z_0 ; r)$ is an open disc with centre z_0 and radius r.

We, now, present (without proof) a collection of some important elementary results about metric spaces.

1.7.6 Results. *Let (X, d) be a metric space, G a subset of X, $a \in X$ and $r > 0$. Then:*

1. *The empty set ϕ and the full space X both are open as well as closed sets.*

2. *$S(a ; r)$ is an open set and $S [a ; r]$ is a closed set in X. Moreover, $\overline{S(a ; r)} \subset S [a ; r]$.*

3. *An arbitrary union (and a finite intersection) of open sets in X is an open set in X.*

4. *An arbitrary intersection (and a finite union) of closed sets in X is a closed sets in X.*

5. *G is open in X if and only if G^c is closed in X.*

6. *G° is an open set in X and it is the largest open set contained in G. Hence, G is open if and only if $G = G^\circ$.*

7. *$\overline{G}$ is a closed set and it is the smallest closed set containing G. Hence, G is closed if and only if $\overline{G} = G$.*

8. *G is open in X if and only if it is a union of open spheres.*

9. *If G is open in X, then for any $A \subset X$, we have*

$$A \cap G = \phi \quad \Rightarrow \quad \overline{A} \cap G = \phi.$$

10. *If A and B are subsets of X, then*

$$\overline{A \cup B} = \overline{A} \cup \overline{B}$$

and
$$\overline{A \cap B} \subseteq \overline{A} \cap \overline{B}.$$

Distance between sets and diameter of a set

1.7.7 Definition. Let (X, d) be a metric space and let A, B be non-empty subsets of X.

(a) The *distance* between a point $x \in X$ and the set A, denoted by $d(x, A)$, is defined by

$$d(x, A) = \inf \{d(x, y) : y \in A\}.$$

(b) The distance between the sets A and B, denoted by $d(A, B)$, is defined as

$$d(A, B) = \inf \{d(x, y) : x \in A, y \in B\}.$$

(c) The *diameter* of A, denoted by $d(A)$, is defined as

$$d(A) = \sup \{d(x, y) : x, y \in A\}.$$

Note. One may not confuse as we use the same symbol d for the metric (distance between two points), distance of a point from the set A, distance between the sets A and B and the diameter of the set A.

1.7.8 Results. *Let (X, d) be a metric space and A, B be subsets of X. Then:*

1. $x \in A, y \in B \quad \Rightarrow \quad d(A, B) \leq d(x, y)$.

2. $x \in \overline{A} \quad \Leftrightarrow \quad d(\{x\}, A) = 0$.

3. $d(\overline{A}, \overline{B}) = d(A, B)$.

4. $d(A) = 0 \quad \Leftrightarrow \quad A$ *contains atmost one point.*

5. $A \subseteq B \quad \Rightarrow \quad d(A) \leq d(B)$.

6. $d(\overline{A}) = d(A)$.

7. $A \cap B \neq \phi \quad \Rightarrow \quad d(A \cup B) \leq d(A) + d(B)$.

8. $x \in A, y \in B \quad \Rightarrow \quad d(x, y) \leq d(A \cup B)$.

9. $d(A \cup B) \leq d(A) + d(A, B) + d(B)$.

Isometric spaces

1.7.9 Definition. Let (X, d) and (Y, ρ) be two metric spaces. A function $f :$ $X \to Y$ is said to be an *isometry* if

$$\rho\,(f(x), f(y)) = d\,(x, y), \quad \forall\, x, y \in X$$

that is, if f preserves distances. Furthermore, X and Y are said to be *isometric spaces* if $\exists$ a function $f : X \to Y$ such that f is bijective and is an isometry.

Remark. Isometric spaces are essentially identical as metric spaces. Any result concerning only the metric structure of (X, d) will hold in all spaces isometric to it.

Convergence and completeness

In Section 1.3, we have given the concept of convergence of a sequence of real or complex numbers. We note that this notion depends precisely on the concept of distance (*i.e.*, metric), and this, in fact, enables us to define convergence in a metric space.

1.7.10 Definition. Let (X, d) be a metric space. A sequence $\{x_n\}$ in X is said to be:

(a) *Convergent* if there is a point $x \in X$ such that for each $\varepsilon > 0$ there is a positive integer N satisfying

$$d\,(x_n, x) < \varepsilon, \quad \forall\, n \geq N.$$

We write in this case $\qquad\qquad \lim_{n \to \infty} x_n = x \text{ or } x_n \to x.$

(b) *A Cauchy sequence* if for each $\varepsilon > 0$, $\exists$ a positive integer N such that

$$d\,(x_n, x_m) < \varepsilon, \quad \forall\, n, m \geq N.$$

1.7.11 Results. *Let (X, d) be a metric space. Then:*

1. *A convergent sequence $\{x_n\}$ in X is bounded and its limit is unique.*
2. *A Cauchy sequence in X is bounded.*
3. $x_n \to x$ *and* $y_n \to y$ *in* $X \implies d\,(x_n, y_n) \to d\,(x, y).$
4. *A convergent sequence in X is Cauchy but the converse need not be true.*

   Consider for instance $X = \mathbb{R} - \{0\}, x_n = \dfrac{1}{n}$ *($n \in \mathbb{N}$). Then, $\{x_n\}$ is a Cauchy sequence in X with the usual metric on X induced by $\mathbb{R}$ and $\{x_n\}$ converges to $0 \notin X$.*

5. *If $\{x_n\}$ is a convergent sequence in X and has the limit x, then every subsequence $\{x_{n_k}\}$ of $\{x_n\}$ is convergent and has the same limit x.*

6. *If a Cauchy sequence $\{x_n\}$ in X has a convergent subsequence having limit x, then $\{x_n\}$ is convergent with the same limit x.*

1.7.12 Definition. A metric space X is said to be a *complete metric space* if every Cauchy sequence in X converges to a point in X.

1.7.13 Examples

1. The space $\mathbb{R}$ with the usual metric is complete. But the space $\mathbb{Q}$ with the usual metric is not complete.

2. The space $\mathbb{C}$ with the usual metric (*i.e.*, the complex plane) is complete.

3. The space $]0, 1]$ with the usual metric is not complete. However, the space $[0, 1]$ with the usual metric is complete.

4. Every discrete metric space (X, d) is complete.

5. The spaces $\mathbb{R}^n$ and $\mathbb{C}^n$ are complete.

6. The sequence space l^p, $1 \le p < \infty$ is, complete.

7. The sequence space l^∞ is complete.

8. The space $C[a, b]$ with the metric given by

$$d(x, y) = \max_{t \in [a,b]} |x(t) - y(t)|$$

   is complete.

9. The space $C[a, b]$ with the metric given by

$$d(x, y) = \int_a^b |x(t) - y(t)|\, dt$$

   is not complete.

A useful characterisation of the closure of a set in terms of sequences is given in the following.

1.7.14 Theorem. *Let A be a non-empty subset of a metric space (X, d) and $x \in X$. Then, $x \in \overline{A}$ if and only if there is a sequence $\{x_n\}$ in A such that*

$$\lim_{n \to \infty} x_n = x.$$

1.7.15 Corollary. *A non-empty subset A of a metric space X is a closed set if and only if limit of each convergent sequence of points of A is in A.*

1.7.16 Theorem. *Let Y be a subspace of a metric space X. Then, Y is complete $\Rightarrow$ Y is closed.*

Proof. Let $x \in \bar{Y}$. Then, by Theorem 1.7.14, $\exists$ a sequence $\{x_n\}$ in Y such that $\lim_{n \to \infty} x_n = x$. But, every convergent sequence in a metric space is a Cauchy sequence. Since Y is complete, it follows that $\{x_n\}$ converges to x in Y. Hence Y is closed. $\blacksquare$

1.7.17 Theorem. *Let Y be a subspace of a complete metric space X. Then, Y is closed $\Rightarrow$ Y is complete.*

Proof. Let $\{x_n\}$ be a Cauchy sequence in Y. Then, $\{x_n\}$ is also a Cauchy sequence in X. But X being complete, $\{x_n\}$ converges to a point $x \in X$. By Theorem 1.7.14, $x \in \bar{Y}$. But Y being closed, $Y = \bar{Y}$ and hence $x \in Y$. Thus, Y is complete. $\blacksquare$

1.7.18 Corollary. *Let Y be a subspace of a complete metric space X. Then, Y is closed $\Leftrightarrow$ y is complete.*

1.7.19 Theorem (Cantor Intersection). *Let (X, d) be a complete metric space and let $\{F_n\}$ be a decreasing sequence of non-empty closed subsets of X (i.e., $F_1 \supset F_2 \supset .....$) such that $d(F_n) \to 0$ as $n \to \infty$. Then $F = \bigcap_{n=1}^{\infty} F_n$ contains exactly one point.*

Remarks

1. The conclusion of Theorem 1.7.19 may fail to hold if either of the conditions:
   (a) each F_n is closed
   (b) $d(F_n) \to 0$ as $n \to \infty$,
   is dropped.
2. The converse of Theorem 1.7.19 is "If in a metric space (X, d) every sequence $\{F_n\}$ of non-empty closed sets with $d(F_n) \to 0$ as $n \to \infty$ has exactly one point in its intersection,then (X, d) is complete".

Dense sets and separable spaces

1.7.20 Definition. A subset S of a metric space X is *everywhere dense* (or, simply *dense*) in X if $\bar{S} = X$.

1.7.21 Theorem. *Let (X, d) be a metric space and S be a subset of X. Then, the following statements are equivalent:*

(a) *S is dense in X.*

(b) *The only closed superset of S is X.*

(c) *The only open set disjoint from S is ϕ.*

(d) *S intersects every non-empty open set.*

(e) *S intersects every open sphere.*

1.7.22 Definition. A metric space X is said to be *separable* if it has a countable subset which is dense in X.

1.7.23 Examples

1. The metric space $\mathbb{R}$ with its usual metric is separable since the subset $\mathbb{Q} \subset \mathbb{R}$, of all rationals, is countable and dense in $\mathbb{R}$.

2. The space $\mathbb{C}$, of all complex numbers with the usual metric, is separable since the subset $S \subset \mathbb{C}$, where

$$S = \{a + i\,b : a, b \in \mathbb{Q}\},$$

   is countable and $\overline{S} = \mathbb{C}$.

3. The Euclidean space $\mathbb{R}^n$ and the unitary space $\mathbb{C}^n$ are separable (Verify !).

4. The space l^p, $1 \le p < \infty$, is separable.

   Consider the set

   $$S = \{y = (\eta_1, \eta_2, ..., \eta_n, 0, 0, ...) : \eta_i \in \mathbb{Q}, 1 \le i < n, n \in \mathbb{N}\}.$$

Clearly, S is a countable subset of l^p. In order to prove that $\overline{S} = l^p$, let $x = \{\xi_i\}$ be an arbitrary element of l^p. Given $\varepsilon > 0$, there is a positive integer N such that

$$\sum_{i = N+1}^{\infty} |\xi_i|^p < \frac{\varepsilon^p}{2}.$$

Since $\mathbb{Q}$ is dense in $\mathbb{R}$, for each real number, there is a rational number which is as close to it as we wish. Consequently, we can find a $y \in S$ satisfying

$$\sum_{i=1}^{N} |\xi_i - \eta_i|^p < \frac{\varepsilon^p}{2}.$$

Thus

$$[d\,(x, y)]^p = \sum_{i=1}^{N} |\xi_i - \eta_i|^p + \sum_{i=N+1}^{\infty} |\xi_i|^p < \varepsilon^p$$

$$\Rightarrow \quad d\,(x, y) < \varepsilon.$$

Hence $\qquad\qquad\qquad \overline{S} = l^p.\ \blacksquare$

5. The space l^∞ is not separable.

Let $\gamma \in [0, 1]$ be arbitrary. Writing γ in the form of a binary representation, we have

$$\gamma = \sum_{i=1}^{\infty} \frac{\alpha_i}{2^i},$$

where $\alpha_i = 0$ or 1. Thus, we get a sequence $\{\alpha_i\}$ of zeros and ones associated to a $\gamma \in [0, 1]$. We, now, use the fact that the set of points in the interval $[0, 1]$ is uncountable; each $\gamma \in [0, 1]$ has a binary (unique) representation and different γ's have different binary representations. These facts lead to prove that there are uncountably many sequences of zeros and ones. All such sequences are the elements of the space l^∞. The metric on l^∞ shows that any two different sequences must be distance 1 apart. If we let each of

these sequences be the centres of small spheres, say, of radius $\dfrac{1}{3}$, these

spheres do not intersect and we have uncountably many of them. Write $\mathscr{F}$ for the family of all such spheres.

Let A be any dense set in l^∞. Let $x \in l^\infty$ be the centre of some sphere in $\mathscr{F}$.

Since $\overline{A} = l^\infty$, either $x \in A$ or x is a limit point of A. In either case, this sphere must contain a point of A. Thus, it follows that each of the non-intersecting spheres must contain a point of A. Hence, A cannot be countable. But A is any arbitrary dense set in l^∞. This leads to that l^∞ cannot have a countable dense set in it. Hence, l^∞ is not separated. $\blacksquare$

6. The space $C[a, b]$ is separable.

Consider the set S of all polynomials defined on $[a, b]$ with rational coefficients. Clearly, S is a countable set. It is enough to show that $\overline{S} = C[a, b]$.

Note that, according to Weierstrass Approximation Theorem, every function x in $C[a, b]$ can be represented by a limit of uniformly convergent sequence $\{p_n\}$ of polynomials in $P[a, b]$. In fact, uniform convergent is the sequence in the metric space $C[a, b]$, i.e., $p_n \to x$ in $C[a, b]$, as $n \to \infty$. But, any polynomial can be uniformly approximated by a polynomial with rational coefficients since $\mathbb{Q}$ is dense in $\mathbb{R}$. Therefore, $\exists$ a sequence $\{q_n\}$ of polynomials with rational coefficients, i.e., in S such that

$$| p_n(t) - q_n(t) | < \frac{1}{n}, \quad \forall\, t \in [a, b]$$

$$\Rightarrow \qquad d_\infty(p_n, q_n) < \frac{1}{n}, \quad (n = 1, 2, ...).$$

Now

$$d_\infty(x, q_n) \le d_\infty(x, p_n) + d_\infty(p_n, q_n) \to 0 \text{ as } n \to \infty.$$

Hence, $q_n \to x$ in $C[a, b]$. This verifies that $\bar{S} = C[a, b]$. $\blacksquare$

1.7.24 Theorem (*Banach-Mazur*). *Every separable metric space is isometric to some subset of the space $C[a, b]$.*

Intuitively, a dense set S in X means that the points of S are distributed 'thickly' throughout the space X; in other words, S contains points as near as we like to every point of X. We may, now, consider a set S whose closure contains an open ball, i.e., $(\bar{S})^\circ = \phi$. This set is not dense in the entire space X but its closure contains an open ball, i.e., $\bar{S} \ne X$, $(\bar{S})^\circ = \phi$. Such a set is not dense in X. However, it might be thought of being somewhere dense in X. This, motivates the following concept.

1.7.25 Definition. A subset S of a metric space X is said to be *nowhere dense* in X if the interior of its closure is empty, i.e., $(\bar{S})^\circ = \phi$.

1.7.26 Theorem. *Let (X, d) be a metric space and S a subset of X. Then, the following statements are equivalent:*

(a) *S in nowhere dense in X.*

(b) *$\bar{S}$ does not contain any non-empty open set.*

(c) *Each non-empty open set has a non-empty open subset disjoint from $\bar{S}$.*

(d) *Each non-empty open set has a non-empty open subset disjoint from S.*

(e) *Each non-empty open set contains an open sphere disjoint from S.*

1.7.27 Theorem. *A closed set in a metric space (X, d) is nowhere dense in X if and only if its complement is dense (everywhere) in X.*

1.7.28 Examples

1. In $\mathbb{R}$ with the usual metric, each of the following subsets is nowhere dense in $\mathbb{R}$.

    (a) Any set consisting of a single point.

    (b) Any finite set.

    (c) $\mathbb{N}$, the set of natural numbers.

    (d) $\mathbb{Z}$, the set of integers.

    (e) C, the Cantor set.

2. In the Euclidean metric space $\mathbb{R}^2$, the set of points on a line is nowhere dense set.

Baire's category theorem

In 1899, the French mathematician, René Baire, developed a method for classifying the "size", in certain appropriate and suitable sense, of subsets in a metric space; namely, minute, small and large sets.

1.7.29 Definition. A subset S of a metric space X is said to be

    (a) *minute* if it is nowhere dense in X, *i.e.*, if $(\bar{S})^{\circ} = \phi$;

        (These sets are also referred to *rare sets*).

    (b) *small* if S is the union of countably many minute sets in X;

        (These sets are also termed as *meagre sets*, sets of *Baire's first category* or simply of *first category*).

    (c) *large* if it is not small.

        (These sets are often referred to *non-meagre sets* or the sets of *second (Baire) category*).

1.7.30 Results

1. *A finite union of nowhere dense sets is nowhere dense.*

2. *A countable union of nowhere dense sets need not be nowhere dense.*

3. *Every nowhere dense set is meagre but not conversely.*

4. *A countable (finite or infinite) union of meagre sets is a meagre set.*

5. *The complement of a meagre subset of a complete metric space is dense.*

1.7.31 Examples

1. The set $\left\{\dfrac{p}{q}\right\}$, where $\dfrac{p}{q}$ is a rational number, is nowhere dense in $\mathbb{R}$ with the usual metric.

2. The set $\mathbb{Q}$ of all rationals is a meagre set in $\mathbb{R}$ with the usual metric.

3. The set of all vertical lines, intersecting the x-axis in rational points, is a meagre set in the Euclidean metric space $\mathbb{R}^2$ since the set of points on a line in $\mathbb{R}^2$ with the Euclidean metric is nowhere dense.

1.7.32 Theorem. *Every complete metric space is of second category.*

Theorm 1.7.32 is referred to Baire's Category Theorem which can be stated in several equivalent forms. We give below a few of them which are often convenient to use.

1.7.32′ Theorem. *If a complete metric space is the union of a sequence of its subsets, then the closure of at least one of the sets in the sequence must have non-empty interior.*

1.7.32″ Theorem. *In a complete metric space, the intersection of a countable number of dense open sets is itself dense.*

Completion of metric spaces

It is seen that the set of rationals $\mathbb{Q}$ with the usual metric is not a complete metric space but can be enlarged to the set of all reals which is complete. Moreover, the completion $\mathbb{R}$ of $\mathbb{Q}$ is such that $\mathbb{Q}$ is dense in $\mathbb{R}$. Also, recall that in the development of real numbers from the rationals, there is used a construction by way of considering the equivalence classes of Cauchy sequences of rationals. This type of procedure may be generalised to any metric space. It is quite interesting in itself.

1.7.33 Theorem. *Every metric space is isometric to a dense subset of a complete metric space.*

Proof. In a metric space (X, d), two Cauchy sequences $\{x_n\}$ and $\{y_n\}$ are equivalent, written $\{x_n\} \sim \{y_n\}$, if and only if $\lim_{n \to \infty} d(x_n, y_n) = 0$. One can show that this is an equivalence relation. This relation decomposes the set of all Cauchy sequences in X into equivalence classes. Consider a new space $\hat{X}$ consisting of all these equivalence classes at its points.

We define a metric on $\hat{X}$ in the following way:

For any two points (equivalence classes of Cauchy sequences in X) $\hat{x}$ and $\hat{y}$ in $\hat{X}$, we choose (any) sequences $\{x_n\} \in \hat{x}$ and $\{y_n\} \in \hat{y}$, and put

$$\hat{d}\,(\hat{x}, \hat{y}) = \lim_{n \to \infty} d\,(x_n, y_n).$$

Now, one can show that this limit always exists and is independent of the sequences chosen from $\hat{x}$ and $\hat{y}$. In fact, we have

$$d\,(x_n, y_n) \le d\,(x_n, x_m) + d\,(x_m, y_m) + (y_m, y_n)$$

$$\Rightarrow \qquad d\,(x_n, y_n) - d\,(x_m, y_m) \le d\,(x_n, x_m) + d\,(y_m, y_n),$$

and a similar inequality with m and n interchanged. Putting these together, we get

$$\mid d\,(x_n, y_n) - d\,(x_m, y_m) \mid \le d\,(x_n, x_m) + d\,(y_m, y_n).$$

Since $\{x_n\}$ and $\{y_n\}$ are Cauchy sequences, we can make the right hand side as small as we please. This verifies that the limit exists since $\mathbb{R}$ is complete. Moreover, if $\{x_n\} \sim \{x_n'\}$ and $\{y_n\} \sim \{y_n'\}$, then

$$\mid d\,(x_n\,y_n) - d\,(x_n', y_n') \mid \le d\,(x_n, x_n') + d\,(y_n, y_n') \to 0 \text{ as } n \to \infty$$

$$\Rightarrow \qquad \lim_{n \to \infty} d\,(x_n, y_n) = \lim_{n \to \infty} d\,(x_n', y_n').$$

This proves that the limit is independent of the choice of the sequences in $\hat{x}$ and $\hat{y}$.

In order to prove that $\hat{d}$ is a metric on $\hat{X}$, it is sufficient to note that

$$\hat{d}\,(\hat{x}, \hat{y}) = \lim_{n \to \infty} d\,(x_n, y_n)$$

$$\le \lim_{n \to \infty} [d\,(x_n, z_n) + d\,(z_n, y_n)]$$

$$= \hat{d}\,(\hat{x}, \hat{z}) + \hat{d}\,(\hat{z}, \hat{y}).$$

Thus, $(\hat{X}, \hat{d})$ is a metric space. We will now complete the proof of the theorem in the various steps.

Step (i) X is isometric to a subspace of $\hat{X}$.

Define a mapping $\sigma : X \to \hat{X}$ which associates to an $x \in X$, the equivalence class of all Cauchy sequences each of which converges to x. Such a class is not empty, for at least the constant sequence $\{x, x, x,....\}$ is in the class. Clearly, σ is an isometry since

$$\hat{d}(\sigma(x), \sigma(y)) = d(x, y).$$

Since an isometry is injective and $\sigma : X \to \sigma(X)$ is surjective, it follows that X and $\sigma(X) \subset \hat{X}$ are isometric.

Step (ii) $\sigma(X)$ is dense in $\hat{X}$.

Let $\hat{x} \in \hat{X}$ and $\{x_n\} \in \hat{x}$. Then, given $\varepsilon > 0$, $\exists$ a positive integer N such that

$$d(x_n, x_m) < \frac{\varepsilon}{2}, \quad \forall\, n, m \geq N.$$

In particular, for $m = N$, we have

$$d(x_n, x_N) < \frac{\varepsilon}{2}, \quad \forall\, n \geq N.$$

Write $x_N = u$. Clearly, $u \in X$ and the constant sequence

$$\{u, u, u,....\} \in \sigma(u)\,(= \hat{u})$$

By the definition of $\hat{d}$, we have

$$\hat{d}(\hat{x}, \hat{u}) = \lim_{n \to \infty} d(x_n, u) \leq \frac{\varepsilon}{2} < \varepsilon$$

$$\Rightarrow \qquad \hat{x} \in \overline{\sigma(X)}.$$

Hence, $\hat{X} = \overline{\sigma(X)}$, which verifies that $\sigma(X)$ is dense in $\hat{X}$.

Step (iii) $\hat{X}$ is complete.

Let $\{\hat{x}_n\}$ be a Cauchy sequence in $\hat{X}$. Sinse $\overline{\sigma(X)} = \hat{X}$, for each $\hat{x}_n$, $\exists$ a $\hat{y}_n \in \sigma(X)$ such that

$$\hat{d}\,(\hat{x}_n,\ \hat{y}_n) < \frac{1}{n}\,.$$

Then, by triangle inequality, we have

$$\hat{d}\,(\hat{x}_m,\ \hat{y}_n) \le \hat{d}\,(\hat{y}_m,\ \hat{x}_m) + \hat{d}\,(\hat{x}_m,\ \hat{x}_n) + \hat{d}\,(\hat{x}_n,\ \hat{y}_n)$$

$$< \frac{1}{m} + \hat{d}\,(\hat{x}_m,\hat{x}_n) + \frac{1}{n} < \varepsilon,$$

for sufficiently large values of m and n since $\{\hat{x}_n\}$ is a Cauchy sequence. This verifies that $\{\hat{y}_n\}$ is a Cauchy sequence in $\sigma(X)$. Write $y_n = \sigma^{-1}\,(\hat{y}_n)$. This is possible since σ is bijective. But, σ being isometry, $\{\hat{y}_n\}$ is a Cauchy sequence in X. Let $\hat{x}$ be the equivalence class in X such that $\{y_n\} \in \hat{x}$.

Claim $\quad \hat{x}_n \to \hat{x}$ in $\hat{X}$.

Note that

$$\hat{d}\,(\hat{x}_n,\ \hat{x}) \le \hat{d}\,(\hat{x}_n,\ \hat{y}_n) + \hat{d}\,(\hat{y}_n,\ \hat{x})$$

$$< \frac{1}{n} + \hat{d}\,(\hat{y}_n,\hat{x})\,.$$

Since $\hat{y}_n \in \sigma(X)$, it follows that the constant sequence $\{y_n,\ y_n,\ y_n,....\} \in \hat{y}_n$. This together with the fact that $\{y_n\} \in \hat{x}$ gives

$$\hat{d}\,(\hat{x}_n,\hat{x}) < \frac{1}{n} + \lim_{m \to \infty} d\,(y_n,\ y_m) < \varepsilon,$$

for sufficiently large value of n. Thus

$$\lim_{m \to \infty} \hat{x}_n = \hat{x} \text{ in } \hat{X}\,.$$

Hence, $\hat{X}$ is complete.

This completes the proof of the theorem. ∎

1.7.34 Definition. *A completion* of a metric space X is any complete metric space $\hat{X}$ which contains a dense subset to which X is isometric.

1.7.35 Theorem. *All completions of a metric space are isometric.*

1.7.36 Theorem. *The completion of a metric space is separable if and only if the space is separable.*

Continuous mapping

It has been observed that the study of limit behaviour and continuity of real and complex mappings depend precisely on the notion of distance. We, now, generalise the classical concept of continuity of mapping is the general settings of metric spaces.

1.7.37 Definition. Let (X, d) and (Y, ρ) be metric spaces. A mapping $f : X \to Y$ is said to be *continuous at a point* $x_0 \in X$ if for each $\varepsilon > 0$, $\exists\, a\ \delta > 0$ such that

$$x \in X \text{ and } d\,(x, x_0) < \delta \quad \Rightarrow \quad \rho\,(f(x), f(x_0)) < \varepsilon.$$

The mapping f is said to be *continuous on X* if it is continuous at each point of X.

1.7.38 Examples

1. Let (X, d) and (Y, ρ) be metric spaces. The constant function $f : X \to Y$ is continuous on X.

2. Let (X, d) be a metric space. The identity map $I : X \to X$ is continuous on X.

1.7.39 Theorem. *Let (X, d) and (Y, ρ) be metric spaces and let $f : X \to Y$ be a mapping. Then, the following statements are equivalent:*

(a) *f is continuous on X.*

(b) *For each $x \in X$, $f(x_n) \to f(x)$ for every sequence $\{x_n\} \in X$ with $x_n \to x$.*

(c) *$f^{-1}(G)$ is open in X whenever G is open in Y.*

(d) *$f^{-1}(F)$ is closed in X whenever F is closed in Y.*

(e) *$f(\overline{A}) \subset \overline{f(A)}, \quad \forall\, A \subset X.$*

(f) *$\overline{f^{-1}(B)} \subset f^{-1}(\overline{B}), \quad \forall\, B \subset Y.$*

1.7.40 Definition. Let (X, d) and (Y, ρ) be metric spaces. A mapping $f : X \to Y$ is said to a *homeomorphism* if

    (i) f is bijective,

    (ii) f is continuous,

    (iii) f^{-1} is continuous.

If a homeomorphism from X to Y exists, we say that the spaces X and Y are *homeomorphic*.

Note. If a mapping $f: X \to Y$ is continuous together with its inverse f^{-1}, then f is referred to be as bicontinuous.

Remarks

1. If X and Y are isometric, then they are homeomorphic. The converse need not be true.

2. A complete and an incomplete metric space may be homeomorphic.

Compactness

1.7.41 Definitions. Let (X, d) be a metric space.

(a) A collection $\mathscr{C} = \{G_\alpha : \alpha \in \Lambda\}$ of subsets of X is said to be a *cover* of X if
$$\bigcup_{\alpha \in \Lambda} G_\alpha = X.$$

(b) A subclass $\mathscr{C}'$ of a cover $\mathscr{C}$ is said to be a *subcover* of $\mathscr{C}$ if $\mathscr{C}'$ itself covers X.

(c) A subcover $\mathscr{C}'$ of $\mathscr{C}$ is said to a *finite subcover* if $\mathscr{C}'$ has only a finite number of members.

(d) A cover $\mathscr{C}$ of X is said to be an *open cover* of X if every set in $\mathscr{C}$ is an open set.

1.7.42 Definitions. Let (X, d) be a metric space.

(a) The space (X, d) is said to be *compact* if every open cover of X has a finite subcover.

(b) A subspace (Y, d_Y) of (X, d) is said to be *compact* if it is compact as a metric space in its own right.

(c) A subset $Y \subset X$ is said to be *compact* if it is compact as a metric space.

(d) A subset $Y \subset X$ is said to be *relatively compact* (or, precompact) in X if the closure of Y, *i.e.*, $\overline{Y}$, is a compact subset of X.

1.7.43 Example. The usual metric space $\mathbb{R}$ is not compact.

1.7.44 Theorem. *A non-empty closed subset of a compact metric space is compact.*

1.7.45 Theorem. *A compact subset of a metric space is closed and bounded.*

The converse of Theorem 1.7.45 need not be true. However, the converse is true in the special case.

1.7.46 Theorem (*Heine-Borel*). *A subset of the usual metric space $\mathbb{R}$ is closed and bounded if and only if it is compact.*

1.7.47 Theorem. *Let f be a continuous mapping of a metric space (X, d) into a metric space (Y, ρ). If A is a compact subset of X, then $f(A)$ is a compact subset of Y.*

1.7.48 Definitions. Let (X, d) be a metric space.

(a) X is said to have *Bolzano-Weierstrass Property* (*BWP*) if every infinite subset of X has a limit point.

(b) X is said to *sequentially compact* if every sequence in it has a convergent subsequence.

(c) X is said to be *totally bounded* if it has an ε-net for each $\varepsilon > 0$. (For a given $\varepsilon > 0$, by an ε-net of a metric space, we mean a finite subset A of X such that

$$X = \bigcup_{a \in A} S(a ; \varepsilon)).$$

Clearly, if X is totally bounded, then it is also bounded.

1.7.49 Theorem. *In a metric space (X, d), the following statements are equivalent:*

(a) *X is compact.*

(b) *X is sequentially compact.*

(c) *X has BWP.*

(d) *X is complete and totally bounded.*

1.7.50 Theorem. *Any continuous mapping of a compact metric space into a metric space is uniformly continuous.*

1.7.51 Theorem. *A closed subspace of a complete metric space is compact if and only if it is totally bounded.*

1.7.52 Example. The space $C[a, b]$ is not compact since the sequence $\{f_n\}$ in $C[a, b]$, where

$$f_n(t) = n, \quad \forall\, t \in [a, b]$$

has no convergent subsequence.

1.7.53 Definition. A metric space is said to be *locally compact* if each of its points has a neighbourhood with compact closure.

1.7.54 Theorem. *A compact metric space is locally compact. But the converse need not be true.*

1.7.55 Example. The space $\mathbb{R}^n$ with the usual metric is locally compact but not compact.

1.7.56 Example. The space $C\,[a, b]$ is not locally compact.

Problems

11. Prove that a discrete metric space (X, d) is compact if and only if X is finite.

12. Prove that a metric space (X, d) is compact if and only if each family of closed sets, with the property that each finite subfamily has non-empty intersection, has itself non-empty intersection.

13. Prove that if f is one-one and onto continuous mapping of a compact metric space (X, d) into a metric space (Y, ρ), then f^{-1} is continuous on (Y, ρ).

14. For a metric space (X, d) and continuous real-valued mapping f on X, prove that if $A \subset X$ is compact, then f attains a maximum and minimum on A.

15. If f is a real-valued continuous function defined on a metric space (X, d) and $A \subset X$ is compact, prove that f attains a maximum and a minimum on A.

Normed and Banach Spaces   **2**
CHAPTER

In Chapter 1, we have seen that a metric space can be defined on a linear space X. It means that a linear space X can, simultaneously, be a metric space. In this chapter, we establish some kind of relationship, between "algebraic" and "topological" structures of X, by defining a metric d on a linear space X in a special way. Specifically, by employing the algebraic operations of a linear space X, we first introduce the concept of a norm on X, and, further, this norm is used to obtain a metric d of the desired type. This gives rise to a space which we call a *normed space*. With these new concepts, many familiar notions of Chapter 1 about real or complex numbers, such as limits of sequences, Cauchy sequences, completeness and continuous functions can be generalized in normed spaces. It is found that a normed space may not be complete as a metric space. This leads to very interesting and useful spaces called *Banach spaces.*

2.1 Definitions and Elementary Properties

In what follows, $\mathbb{K}$ will denote the field $\mathbb{R}$ (real numbers) or $\mathbb{C}$ (complex numbers). We shall always assume that $\mathbb{R}$ and $\mathbb{C}$ have their usual metrics and that all the linear spaces that we consider will be defined over $\mathbb{K}$ ($\mathbb{R}$ or $\mathbb{C}$).

2.1.1 Definition. Let X be a linear space over the field $\mathbb{K}$. A function $\| \cdot \| : X \to \mathbb{R}$ is said to be a *norm* on X if it satisfies the following conditions:

$(i)\ \| x \| \geq 0, \ \forall\, x \in X$

$(ii)\ \| x \| = 0 \Leftrightarrow x = 0$

$(iii)\ \| x + y \| \leq \| x \| + \| y \|, \ \forall\, x, y \in X$ (the triangle inequality)

$(iv)\ \| \alpha x \| = | \alpha | \, \| x \|, \qquad \forall\, x \in X \text{ and } \forall\, \alpha \in K.$ (homogeneity of norm)

61

A linear space X over $\mathbb{K}$ with a norm $\|\cdot\|$ defined on X is called a *normed linear space* or simply a *normed space* over $\mathbb{K}$, written $(X, \|\cdot\|)$ or simply X. The normed space X is real or complex according as the field $\mathbb{K}$ is $\mathbb{R}$ or $\mathbb{C}$.

Note. The four properties of a norm on a linear space X are called the *axioms of the norm.*

Remarks

1. The property (i) is a consequence of properties (iii) and (iv) in Definition 2.1.1 since $(iv) \Rightarrow \| 0 \| = 0$ and so (iii) and (iv) together give

$$0 = \| x - x \| \leq \| x \| + \| - x \| = 2 \| x \|,$$

which yields $0 \leq \| x \|$.

2. If we regard x in a normed space X as a vector, its length is $\| x \|$ and the length $\| x_1 - x_2 \|$ of the vector $x_1 - x_2$ is the distance between the end points of the vectors x_1 and x_2. Thus, in view of (i) and (ii) in Definition 2.1.1, we say that all vectors in a normed space X have positive lengths except the zero vector which has length zero. The property (iii) states that the length of one side of a triangle cannot exceed the sum of the lengths of the other two sides (Fig. 2.1). The property (iv) means that if we multiply a vector x by a scalar α, then the length of x is multiplied by the absolute value of α.

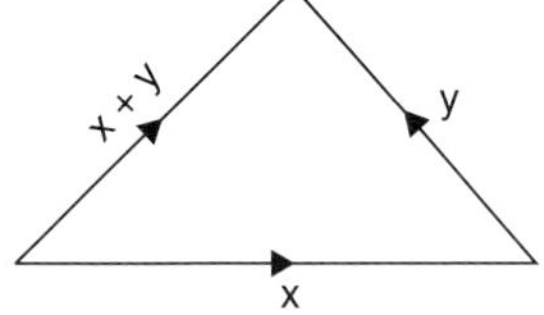

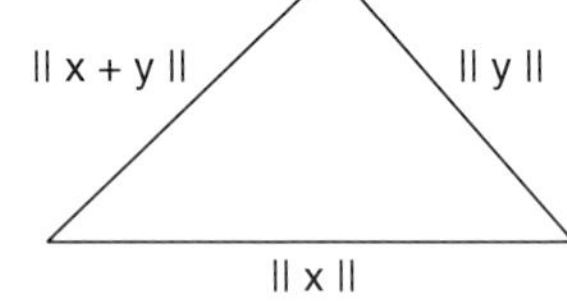

Fig. 2.1

2.1.2 Example. Each of the linear spaces $\mathbb{R}$ and $\mathbb{C}$ with the norm defined by $\| x \| = | x |$ is a normed space.

Note. Whenever $\mathbb{R}$ (or $\mathbb{C}$) is considered as a normed space, the norm will be that defined in Example 2.1.2.

A norm $\|\cdot\|$ on X defines a metric d on X given by

$$(1) \qquad\qquad d (x, y) = \| x - y \|.$$

It is simple to verify, from properties (i)–(iv) of the norm, that d does define a metric on X. The metric thus defined is called the *metric induced by the norm.*

It is easy to check that the metric d induced by a norm on X satisfies the following properties:

$$d(x, y) = d(x - y, 0),$$

and

$$d(\alpha\, x, 0) = |\alpha|\, d(x, 0), \quad x, y \in X \text{ and } \alpha \in \mathbb{K}.$$

The notion of convergence for sequences and the related concepts in normed spaces follow readily from the corresponding definitions in the metric spaces and the fact that we have (1).

2.1.3 Definitions. Let $(X, \|\cdot\|)$ be a normed space.

(i) The set $\{x \in X : \|x - x_0\| < r\}$, denoted by $S\,(x_0\,;\, r)$, is called the *open sphere* (or *open ball*) of radius r with centre x_0.

(ii) The set $\{x \in X : \|x - x_0\| \leq r\}$, denoted by $S\,[x_0\,;\, r]$, is called the *closed sphere* (or *closed ball*) of radius r with centre x_0.

(iii) A sequence $\{x_n\} \subset X$ is said to be a *Cauchy sequence* if for a given $\varepsilon > 0$, $\exists$ a positive integer N such that

$$\|x_m - x_n\| < \varepsilon, \quad \forall\, m, n \geq N.$$

That is, $\{x_n\}$ is a Cauchy sequence in X if and only if

$$\|x_m - x_n\| \to 0 \text{ as } m, n \to \infty.$$

(iv) A sequence $\{x_n\} \subset X$ is said to be *convergent* if $\exists$ an $x \in X$ such that

$$\lim_{n \to \infty} \|x_n - x\| = 0.$$

(v) The space X is said to be *complete* if every Cauchy sequence in X converges to an element in X.

We deduce, from the linear space structure on X, that:

(I) Given $x_0 \in X$ and $r > 0$, then

$$S\,(x_0\,;\, r) = \{x \in X : \|x - x_0\| < r\}$$

$$= \{x_0 + y \in X : \|y\| < r\}, \text{ where } y = x - x_0$$

$$= x_0 + S(0\,;\, r).$$

Thus, a sphere (*open and similarly closed*) *centred at any point in the space X is the translate of a sphere at 0 with the same radius.* Therefore, in a normed space, it is convenient to consider open spheres centred at 0.

(*II*) Given $r > 0$, then

$$S(0 ; r) = \{x \in X : \| x \| < r\}$$

$$= \{x \in X : \| \frac{x}{r} \| < 1\}$$

$$= \{r\, y \in X : \| y \| < 1\}, \text{ where } y = \frac{x}{r},$$

$$= r\, S(0 ; 1).$$

Therefore, in a normed space, without any loss of generality, we can consider the sphere centred at 0 with radius 1. The sphere $S(0 ; 1)$ is called the *unit open sphere* in *X*.

Using Definition 2.1.1, we, now, obtain some elementary properties of a normed space.

2.1.4 Lemma. *In a normed space X,*

$$\big| \| x \| - \| y \| \big| \le \| x - y \|, \ \forall \ x, y \in X.$$

Proof. Observe that

$$\| x \| = \| (x - y) + y \| \le \| x - y \| + \| y \|$$

which gives

$$\| x \| - \| y \| \le \| x - y \|.$$

Further

$$-(\| x \| - \| y \|) \le \| y - x \| = \| x - y \|.$$

Hence, the result follows. ∎

2.1.5 Corollary. *The norm $\| \cdot \|$ is continuous as a mapping from X into $\mathbb{R}$.*

Proof. Let $\{x_n\} \subset X$ be any sequence with $x_n \to x$. Then, by Lemma 2.1.4, it follows that

$$\lim_{n \to \infty} \| x_n \| = \| x \|.$$

This proves that the mapping $\| \cdot \| : X \to \mathbb{R}$ is continuous. ∎

2.1.6 Corollary. *Let X be a normed space over the field $\mathbb{K}$. Let $\{x_n\}$ and $\{y_n\}$ be sequences in X with $\lim_{n \to \infty} x_n = x$ and $\lim_{n \to \infty} y_n = y$, respectively, and let $\{\alpha_n\}$ be a sequence in $\mathbb{K}$ with $\lim_{n \to \infty} \alpha_n = \alpha$. Then:*

(a) $\lim_{n \to \infty} (x_n + y_n) = x + y.$

(b) $\lim_{n \to \infty} \alpha_n x_n = \alpha x.$

Proof. The result follows from

$$\| (x_n + y_n) - (x + y) \| = \| (x_n - x) + (y_n - y) \|$$
$$\leq \| x_n - x \| + \| y_n - y \| \to 0, \text{ as } n \to \infty$$

and

$$\| \alpha_n x_n - \alpha x \| = \| \alpha_n (x_n - x) + \alpha_n - \alpha) x \|$$
$$\leq |\alpha_n| \| x_n - x \| + | \alpha_n - \alpha | \| x \|$$
$$\to 0, \text{ as } n \to \infty. \blacksquare$$

It may be noted that the scalar field $\mathbb{K}$ itself is a normed space, with the absolute value as a norm on it. The operations of vector addition and scalar multiplication in X are functions, respectively, from $X \times X$ and $\mathbb{K} \times X$ into X. Each of these functions is continuous. Thus, we conclude that the operations of addition and scalar multiplication in a normed space X are jointly continuous. In other words, both the structures; namely, algebraic and topological, on X are consistent. This leads, in general, to the concept of a topological linear space. In fact, a *topological linear space* X is a linear space over a field $\mathbb{K}$ ($\mathbb{R}$ or $\mathbb{C}$) together with a topology on X such that the addition and scalar multiplication in X are jointly continuous in the given topology. Normed spaces are the simplest examples of the topological linear spaces.

We recall that a normed space X as a metric space is complete if and only if every Cauchy sequence in X converges to an element in X. A normed space may be incomplete as a metric space. Oftenly, it is necessary to construct elements of a normed space by means of infinite series or integrals, and completeness is then needed to ensure the existence of limits.

2.1.7 Definition. A complete normed linear space X over field $\mathbb{K}$ is called a *Banach space* over $\mathbb{K}$. The Banach space X is called *real* or *complex* according as the underlying field $\mathbb{K}$ is so.

2.1.8 Example. Each of the spaces $\mathbb{R}$ and $\mathbb{C}$ is a Banach space.

Sequences were available to us even in metric spaces. In a normed space, we may proceed an important step further and consider the series as well.

Let $\{x_k\}$ be a sequence in a normed space X. Then, we can associate with $\{x_k\}$, the sequence $\{s_n\}$ of partial sums given by

$$s_n = \sum_{k=1}^{n} x_k \ .$$

And we can discuss the behaviour of the series $\displaystyle\sum_{k=1}^{\infty} x_k$ in regard to its convergence or non-convergence accordingly as the sequence $\{s_n\}$ of its partial sums is so.

2.1.9 Definition. A sequence $\{x_k\}$ in a normed space X is said to be *summable* to the sum s if the sequence $\{s_n\}$ of the partial sums of the series $\displaystyle\sum_{k=1}^{\infty} x_k$ converges to s in X, *i.e.,*

$$\| s_n - s \| \to 0, \ as \ n \to \infty$$

or

$$\left\| \sum_{k=1}^{n} x_k - s \right\| \to 0, \ as \ n \to \infty.$$

In this case, we write

$$s = \sum_{k=1}^{\infty} x_k \ .$$

The sequence $\{x_k\}$ is said to be *absolutely summable* if

$$\sum_{k=1}^{\infty} \| x_k \| < \infty.$$

We know that for a sequence of real numbers (and complex numbers), absolute summability implies summability. But this is not true, in general, for sequences in normed spaces. However, we prove the following which, in fact, gives a characterisation of a Banach space.

2.1.10 Theorem. *A normed space X is a Banach space if and only if every absolutely summable sequence in X is summable in X.*

Proof. Assume that X is a Banach space. Let $\{x_n\}$ be an absolutely summable sequence in X. Then

$$\sum_{n=1}^{\infty} \| x_n \| = M < \infty.$$

Thus, for each $\varepsilon > 0$, $\exists N$ such that

$$\sum_{n=N}^{\infty} \| x_n \| < \varepsilon.$$

Let $s_n = \sum_{k=1}^{n} x_k$ be the partial sums of the series $\sum_{k=1}^{\infty} x_k$. Then

$$\| s_n - s_m \| = \| \sum_{k=m+1}^{n} x_k \|$$

$$\leq \sum_{k=m+1}^{n} \| x_k \|$$

$$\leq \sum_{k=N}^{\infty} \| x_k \|, \quad n > m > N$$

$$< \varepsilon, \qquad n > m > N$$

Thus, $\{s_n\}$ is a Cauchy sequence in X and must converge to some element (say) s in X, since X is complete. Hence, $\{x_n\}$ is summable in X.

Conversely, suppose each absolutely summable sequence in X is summable in X.

Claim. *X is complete.*

Let $\{x_n\}$ be a Cauchy sequence in X. Then, for each k, $\exists$ an integer n_k such that

$$\| x_n - x_m \| < \frac{1}{2^k}, \quad \forall \, n, m \geq n_k.$$

We may choose n_k such that $n_{k+1} > n_k$. Then, $\{x_{n_k}\}$ is a subsequence of $\{x_n\}$. Set

$$y_0 = x_{n_1}$$

$$y_1 = x_{n_2} - x_{n_1}$$

$$\cdots\cdots\cdots\cdots$$

$$\cdots\cdots\cdots\cdots$$

$$\cdots\cdots\cdots\cdots$$

$$y_k = x_{n_{k+1}} - x_{n_k}$$

$$\cdots\cdots\cdots\cdots$$

$$\cdots\cdots\cdots\cdots$$

$$\cdots\cdots\cdots\cdots$$

We note that

$$(i) \sum_{i=0}^{k} y_i = x_{n_{k+1}}$$

$$(ii) \| y_k \| < \frac{1}{2^k}, \ k \geq 1$$

and as such

$$\sum_{k=0}^{\infty} \| y_k \| \leq \| y_0 \| + \sum_{k=1}^{\infty} \frac{1}{2^k} = \| y_0 \| + 1 < \infty.$$

Consequently, the sequence $\{y_k\}$ is absolutely summable and hence summable to some element (say) x in X. Therefore

$$x_{n_k} \to x, \text{ as } k \to \infty.$$

Thus, the Cauchy sequence $\{x_n\}$ in X has a convergent subsequence $\{x_{n_k}\}$ converging to x. Hence, $\lim_{n \to \infty} x_n = x$ (Result 1.7.11 (6)). ∎

Problem 1. Let $X \neq \{0\}$ be a normed space. Prove that X is a Banach space if and only if $\{x \in X : \| x \| = 1\}$ is complete.

2.2 Some Concrete Normed and Banach Spaces

In this section, several examples of normed and Banach spaces have been discussed in detail. We shall also give examples of incomplete normed spaces.

Finally, we shall give an example to show that not every metric on a linear space can be obtained from a norm.

2.2.1 Example (The Euclidean Space $\mathbb{R}^n$). The linear space $\mathbb{R}^n$ (Example 1.6.3 (4)) equipped with the norm given by

$$(2) \qquad \| x \|_2 = \left(\sum_{i=1}^{n} |\xi_i|^2 \right)^{\frac{1}{2}}, \; x = (\xi_1, \xi_2, ..., \xi_n) \in \mathbb{R}^n$$

is a real Banach space.

The triangle inequality follows from the Minkowski Inequality (1.5.6) while the other conditions of the norm are easy to verify. Further, to show that $\mathbb{R}^n$ is complete, let $\{x_m\}$ be a Cauchy sequence in $\mathbb{R}^n$, where

$$x_m = \left(\xi_1^{(m)}, \xi_2^{(m)}, ..., \xi_n^{(m)} \right) \in \mathbb{R}^n.$$

Then, for each $\varepsilon > 0$, $\exists$ a positive integer N such that

$$(3) \qquad \| x_m - x_p \|_2 = \left(\sum_{i=1}^{n} |\xi_i^{(m)} - \xi_i^{(p)}|^2 \right)^{\frac{1}{2}} < \varepsilon, \; \forall \; m, p \geq N.$$

On squaring both sides, we get

$$\sum_{i=1}^{n} |\xi_i^{(m)} - \xi_i^{(p)}|^2 < \varepsilon^2$$

$$\Rightarrow \qquad |\xi_i^{(m)} - \xi_i^{(p)}|^2 < \varepsilon^2$$

$$\Rightarrow \qquad |\xi_i^{(m)} - \xi_i^{(p)}| < \varepsilon, \; \forall \; m, p \geq N \, (i = 1, 2, ..., n).$$

This shows that for each fixed i $(1 \leq i \leq n)$, the sequence $\left\{ \xi_i^{(m)} \right\}_{m=1}^{\infty}$ is a Cauchy sequence in $\mathbb{R}$. Since $\mathbb{R}$ is complete, it converges in $\mathbb{R}$. Let $\xi_i^{(m)} \to \xi_i$ as $m \to \infty$. Using these limits, we define $x = (\xi_1, \xi_2, ..., \xi_n)$. Clearly, $x \in \mathbb{R}^n$. Letting $p \to \infty$ in (3), we obtain

$$\| x_m - x \|_2 < \varepsilon, \quad \forall \; m \geq N$$

$$\Rightarrow \qquad x_m \to x \text{ in } \mathbb{R}^n.$$

Hence, $\mathbb{R}^n$ is complete. ∎

Note. The norm given in (2) is called the *Euclidean norm* on $\mathbb{R}^n$, and the real normed space $\mathbb{R}^n$ is called the *n-dimensional Euclidean space.* The Euclidean plane is the real linear space $\mathbb{R}^2$ with its Euclidean norm: $\| x \|_2 = \sqrt{x_1^2 + x_2^2}$. If $n = 3$, the norm on $\mathbb{R}^n$ is, of course, just the usual distance on $\mathbb{R}^3$. In this case, the closed unit sphere $\{x \in \mathbb{R}^3 : \| x \| \le 1\}$ is simply a solid sphere with centre at the origin and radius 1.

2.2.2 Example (The Unitary Space $\mathbb{C}^n$). The linear space $\mathbb{C}^n$ (Example 1.6.3(5)) equipped with the norm given by

$$\| x \|_2 = \left(\sum_{i=1}^{n} | \xi_i |^2 \right)^{\frac{1}{2}}, \, x = (\xi_1, \xi_2, ..., \xi_n) \in \mathbb{C}^n$$

is a complex Banach space. The space $\mathbb{C}^n$ is called the *n-dimensional unitary space.*

Problem 2. Prove that $\mathbb{C}^n$ is a Banach space.

2.2.3 Example. The linear space $\mathbb{K}^n$ ($\mathbb{R}^n$ or $\mathbb{C}^n$) is a Banach space with each of the norms given by

$$\| x \|_1 = \sum_{i=1}^{n} | \xi_i |$$

and $\qquad \| x \|_\infty = \max \{| \xi_i | : 1 \le i \le n\}$

where $x = (\xi_1, \xi_2, ..., \xi_n\} \in \mathbb{K}^n$.

Note. $\| \cdot \|_1$ is called l^1*–norm* and $\| \cdot \|_\infty$ the *sup norm* (or the *uniform norm*) on $\mathbb{K}^n$.

It has been observed that $\| \cdot \|_1, \| \cdot \|_2$ and $\| \cdot \|_\infty$ are norms on the linear space $\mathbb{K}^n$ ($\mathbb{R}^n$ or $\mathbb{C}^n$). We wish to introduce the general class of norms on $\mathbb{K}^n$ to which these norms relate.

Let $p > 0$ be a real number. Define $\| \cdot \|_p : \mathbb{K}^n \to \mathbb{R}$ by

$$\| x \|_p = \left(\sum_{i=1}^{n} | \xi_i |^p \right)^{\frac{1}{p}}, \ x = (\xi_1, \xi_2, \dots, \xi_n) \in \mathbb{K}^n.$$

It is easy to verify that $\| \cdot \|_p$, for $1 \le p < \infty$, actually defines a norm on $\mathbb{K}^n$, the triangle inequality being proved by Minkowski Inequality (1.5.6). We denote the normed space $(\mathbb{K}^n, \| \cdot \|_p)$ by $l^p(n)$.

2.2.4 Example (The Space $l^p(n)$). The linear space $l^p(n)$, $1 \le p < \infty$, equipped with the norm given by

$$\| x \|_p = \left(\sum_{i=1}^{n} | \xi_i |^p \right)^{\frac{1}{p}}, \ x = (\xi_1, \xi_2, \dots, \xi_n) \in l^p(n)$$

is a Banach space.

Let $\{x_m\}$ be a Cauchy sequence in $l^p(n)$, where

$$x_m = \left(\xi_1^{(m)}, \xi_2^{(m)}, \dots, \xi_n^{(m)} \right) \in \mathbb{K}^n.$$

Then, for each $\varepsilon > 0$, $\exists$ a positive integer N such that

$$(4) \qquad \| x_m - x_k \|_p = \left(\sum_{i=1}^{n} | \xi_i^{(m)} - \xi_i^{(k)} |^p \right)^{\frac{1}{p}} < \varepsilon, \ \forall \ m, k \ge N$$

$$\Rightarrow \qquad | \xi_i^{(m)} - \xi_i^{(k)} | < \varepsilon, \ \forall \ m, k \ge N \ (i = 1, 2, \dots, n).$$

This shows that, for each fixed i $(1 \le i \le n)$, the sequence $\left\{ \xi_i^{(m)} \right\}_{m=1}^{\infty}$ is a Cauchy sequence in $\mathbb{K}$. Since $\mathbb{K}$ is complete, it converges in $\mathbb{K}$. Let $\xi_i^{(m)} \to \xi_i$ as $m \to \infty$. Using these n limits, we define $x = (\xi_1, \xi_2, \dots, \xi_n)$. Clearly, $x \in l^p(n)$. Letting $k \to \infty$ in (4), we obtain

$$\| x_m - x \|_p < \varepsilon, \ \forall m \ge N$$

$$\Rightarrow \qquad x_m \to x \text{ in } l^p(n).$$

Hence, $l^p(n)$ is complete. ∎

Remark. In case $0 < p < 1$, $\| \cdot \|_p$ does not define a norm on $l^p(n)$ unless $n = 1$. For instance, consider $x = (1, 0)$ and $y = (0, 1)$ in $l^p(2)$. Then $\| x \|_p = 1 = \| y \|_p$ and $\| x + y \|_p = \| (1, 1) \|_p = 2^{\frac{1}{p}}$, for each p $(0 < p < \infty)$. But $2^{\frac{1}{p}} > 2$ whenever $0 < p < 1$. Therefore, $\| x + y \|_p > \| x \|_p + \| y \|_p$. However, $l^p(n)$ is a complete metric space with metric ρ defined by

$$\rho\,(x,\,y) = \| x - y \|_p^p, \quad x,\,y \in l^p(n).$$

In order to have a better understanding of the geometry of the spaces $l^p(n)$, for different values of p, let us examine the shapes of the unit spheres (open or closed) for different values of p together with their relationships. For illustration, we give below the shapes of the unit closed spheres S_p in $\mathbb{R}^2$ with norms:

(i) $\quad \| x \|_{\frac{1}{2}} = \left(| \xi_1 |^{\frac{1}{2}} + | \xi_2 |^{\frac{1}{2}} \right)^2$

(ii) $\quad \| x \|_1 = | \xi_1 | + | \xi_2 |$

(iii) $\quad \| x \|_2 = \left(| \xi_1 |^2 + | \xi_2 |^2 \right)^{\frac{1}{2}}$

(iv) $\quad \| x \|_4 = \left(| \xi_1 |^4 + | \xi_2 |^4 \right)^{\frac{1}{4}}$

(v) $\quad \| x \|_\infty = \max \{ | \xi_1 |, | \xi_2 | \}.$

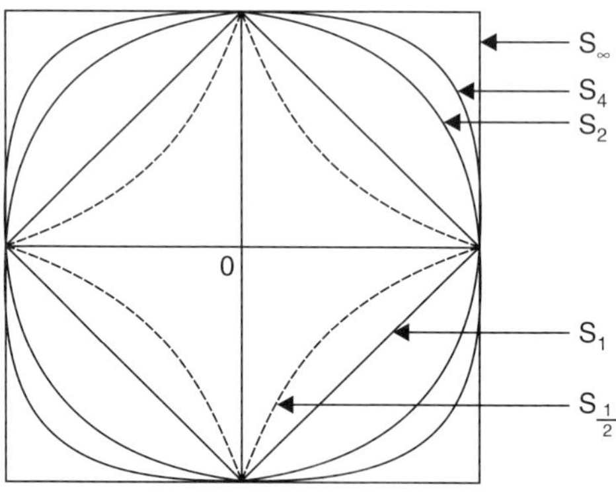

Fig. 2.2

Observations

1. If we allow p to increase from 1 to ∞, then the corresponding sphere swells continuously from the inner square to the outer square, *i.e.*, from S_1 to S_∞ as indicated in Fig. 2.2.

2. Each of the spheres S_p for $1 \le p < \infty$ is convex.

3. The spheres S_p for $p < 1$ are not convex ; for instance, the inner portion of the sphere $\left\{ x \in \mathbb{R}^2 : \| x \|_{\frac{1}{2}} = 1 \right\}$ is starshaped, as shown in Fig. 2.2. In fact, as observed in the remark that follows Example 2.2.4, $\| x \|_{\frac{1}{2}} = 1$ does not define a norm on $\mathbb{R}^2$.

Problems

3. Show that each of the following defines a norm on $\mathbb{R}^2$:

(i) $\| x \|_1 = \dfrac{|x_1|}{a} + \dfrac{|x_2|}{b}$

(ii) $\| x \|_2 = \dfrac{x_1^2}{a^2} + \dfrac{x_2^2}{b^2}$

(iii) $\| x \|_\infty = \max \left\{ \dfrac{|x_1|}{a}, \dfrac{|x_2|}{b} \right\},$

where a and b are two fixed positive real numbers and $x = (x_1, x_2) \in \mathbb{R}^2$. Sketch a closed unit sphere corresponding to each of these norms.

4. Let $\| \cdot \|_1$ and $\| \cdot \|_2$ be norms on a vector space X. Is $\| x \| = \min \left\{ \| x \|_1, \| x \|_2 \right\}$ a norm on X ?

So far only finite dimensional normed and Banach spaces have been discussed. A normed (or Banach) space is said to be *finite* or *infinite dimensional* according as the underlying linear space is so (Definition 1.6.8). We, now, give examples of infinite dimensional spaces.

2.2.5 Example (The Sequence Space l^p). The linear space l^p, $1 \le p < \infty$, (Example 1.6.3 (6)) with the norm given by

$$\| x \|_p = \left(\sum_{i=1}^{\infty} |\xi_i|^p \right)^{\frac{1}{p}}, \ x = \{\xi_i\} \in l^p$$

is a Banach space.

We establish only the triangle inequality and completeness since other properties are straightforward. It follows from the Minkowski Inequality 1.5.10 that for $x = \{\xi_i\}$, $y = \{\eta_i\}$ in l^p,

$$\| x + y \|_p = \left(\sum_{i=1}^{\infty} |\xi_i + \eta_i|^p \right)^{\frac{1}{p}}$$

$$\leq \left(\sum_{i=1}^{\infty} |\xi_i|^p \right)^{\frac{1}{p}} + \left(\sum_{i=1}^{\infty} |\eta_i|^p \right)^{\frac{1}{p}}$$

$$= \| x \|_p + \| y \|_p,$$

which establishes the triangle inequality. We, now, turn to the discussion of the completeness for l^p.

Let $\{x_m\}$ be a Cauchy sequence in l^p, where $x_m = \{\xi_i^{(m)}\}_{i=1}^{\infty}$ is such that

$$\sum_{i=1}^{\infty} |\xi_i^{(m)}|^p < \infty, \quad (m = 1, 2, \ldots).$$

Then, for each $\varepsilon > 0$, $\exists$ a positive integer N such that

$$(5) \qquad \| x_m - x_n \|_p = \left(\sum_{i=1}^{\infty} |\xi_i^{(m)} - \xi_i^{(n)}|^p \right)^{\frac{1}{p}} < \varepsilon, \quad \forall\, m, n \geq N$$

$$\Rightarrow \qquad\qquad |\xi_i^{(m)} - \xi_i^{(n)}| < \varepsilon, \quad \forall\, m, n \geq N \; (i = 1, 2, \ldots).$$

This shows that, for each fixed i $(1 \leq i < \infty)$, the sequence $\left\{\xi_i^{(m)}\right\}_{m=1}^{\infty}$ is a Cauchy sequence in $\mathbb{K}$. Since $\mathbb{K}$ is complete, it converges in $\mathbb{K}$. Let $\xi_i^{(m)} \to \xi_i$ as $m \to \infty$. Using these limits, we define $x = (\xi_1, \xi_2, \ldots)$ and show that $x \in l^p$ and $x_m \to x$.

From (5), we get

$$\sum_{i=1}^{k} |\xi_i^{(m)} - \xi_i^{(n)}|^p < \varepsilon^p, \quad \forall\, m, n \geq N \; (k = 1, 2, \ldots)$$

Letting $n \to \infty$, we obtain

$$\sum_{i=1}^{k} |\xi_i^{(m)} - \xi_i|^p \leq \varepsilon^p, \ \forall \ m \geq N \ (k = 1, 2, \ldots)$$

which, on letting $k \to \infty$, gives

$$(6) \qquad \sum_{i=1}^{\infty} |\xi_i^{(m)} - \xi_i|^p \leq \varepsilon^p, \qquad \forall \ m \geq N.$$

This shows that

$$x_m - x = \{\xi_i^{(m)} - \xi_i\} \in l^p.$$

Since $x_m \in l^p$, it follows by Minkowski Inequality 1.5.10, that

$$x = x_m + (x - x_m) \in l^p.$$

Thus, $x \in l^p$. Furthermore, from (6), we obtain

$$\| x_m - x \|_p < \varepsilon, \quad \forall \ m \geq N$$

which verifies that $x_m \to x$ in l^p. Hence, l^p $(1 \leq p < \infty)$ is a Banach space. ∎

Note. The real and complex spaces l^2 are precisely the infinite dimensional Euclidean and unitary spaces, $\mathbb{R}^\infty$ and $\mathbb{C}^\infty$, respectively.

2.2.6 Example (The Sequence Space l^∞). The linear space l^∞ (Example 1.6.3 (7d)), equipped with the norm given by

$$\| x \|_\infty = \sup_{1 \leq i < \infty} |\xi_i|, x = \{\xi_i\} \in l^\infty$$

is a Banach space.

It is easy to check that l^∞ with the above norm is a normed space. We proceed to show that l^∞ is complete.

Let $\{x_m\}$ be a Cauchy sequence in l^∞, where $x_m = \{\xi_i^{(m)}\}_{i=1}^{\infty}$ is such that

$$\sup_{1 \leq i < \infty} |\xi_i^{(m)}| < \infty, \qquad (m = 1, 2, \ldots).$$

Then, for each $\varepsilon > 0$, $\exists$ a positive integer N such that

$$\| x_m - x_n \|_\infty = \sup_{1 \le i < \infty} | \xi_i^{(m)} - \xi_i^{(n)} | < \varepsilon , \quad \forall \, m, n \ge N.$$

This gives

$$(7) \qquad | \xi_i^{(m)} - \xi_i^{(n)} | < \varepsilon , \quad \forall \, m, n \ge N \quad (i = 1, 2, \ldots).$$

This shows that, for each fixed $i(1 \le i < \infty)$, the sequence $\left\{ \xi_i^{(m)} \right\}_{m=1}^{\infty}$ is a Cauchy

sequence in $\mathbb{K}$. Since $\mathbb{K}$ is complete, it converges in $\mathbb{K}$. Let $\xi_i^{(m)} \to \xi_i$ as $m \to \infty$. Using these limits, we define $x = (\xi_1, \xi_2, \ldots)$ and show that $x \in l^\infty$ and $x_m \to x$.

Letting $n \to \infty$ in (7), we get

$$(8) \qquad | \xi_i^{(m)} - \xi_i | \le \varepsilon , \quad \forall \, m \ge N \quad (i = 1, 2, \ldots).$$

Since $x_m = \left\{ \xi_i^{(m)} \right\}_{i=1}^{\infty} \in l^\infty$, there is a real number K_m such that $| \xi_i^{(m)} | \le K_m , \, \forall i$.

Therefore

$$| \xi_i | = | \xi_i - \xi_i^{(m)} + \xi_i^{(m)} |$$

$$\le | \xi_i^{(m)} - \xi_i | + | \xi_i^{(m)} | \qquad \qquad \text{(triangle inequality)}$$

$$\le \varepsilon + K_m, \quad \forall \, m \ge N \quad (i = 1, 2, \ldots).$$

This inequality being true for each i and the right hand side being independent of i, it follows that $\{\xi_i\}$ is a bounded sequence of numbers. This implies that $x = \{\xi_i\} \in l^\infty$. Furthermore, from (8), we obtain

$$\| x_m - x \|_\infty = \sup_{1 \le i < \infty} | \xi_i^{(m)} - \xi_i | \le \varepsilon , \quad \forall \, m \ge N.$$

This shows that $x_m \to x$ in l^∞. Hence, l^∞ is a Banach space. $\blacksquare$

Note. We may refer to $\| \cdot \|_p$ as the l^p-*norm* on l^p and $\| \cdot \|_\infty$ as the *sup norm* on l^∞.

2.2.7 Example (The Function Space C [a, b]). Consider the linear space $C[a, b]$ of all scalar-valued (real or complex) continuous functions defined on $[a, b]$ (Example 1.6.3 (10)). Define $\|\cdot\|_{\infty} : C[a, b] \to \mathbb{R}$ by

$$\| x \|_{\infty} = \max_{t \in [a,b]} | x(t) |.$$

Then, $(C[a, b], \|\cdot\|_{\infty})$ is a Banach space.

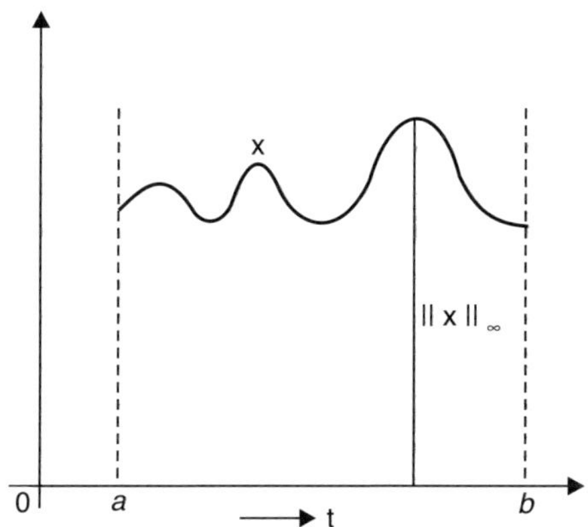

Fig. 2.3

It is easy to check that $\|\cdot\|_{\infty}$ actually defines a norm on $C[a, b]$. To establish the completeness of $C[a, b]$, let $\{x_m\}$ be a Cauchy sequence in $C[a, b]$. Then, for each $\varepsilon > 0$, $\exists$ a positive integer N such that

$$(9) \qquad \| x_m - x_n \|_{\infty} = \max_{t \in [a,b]} | x_m(t) - x_n(t) | < \varepsilon, \quad \forall\, m, n \geq N.$$

Therefore, for any fixed $t = t_0 \in [a, b]$, we get

$$| x_m(t_0) - x_n(t_0) | < \varepsilon, \quad \forall\, m, n \geq N.$$

This shows that $\{x_m(t_0)\}$ is a Cauchy sequence in $\mathbb{K}$. But, $\mathbb{K}$ being complete, this sequence converges. Let $x_m(t_0) \to x(t_0)$ as $m \to \infty$. In this way, we can associate to each $t \in [a, b]$ a unique $x(t) \in \mathbb{K}$. This defines a (pointwise) function x on $[a, b]$. Now, we show that $x \in C[a, b]$ and $x_m \to x$.

From (9), note that

$$| x_m(t) - x_n(t) | < \varepsilon, \qquad \forall\, m, n \geq N \text{ and } \forall\, t \in [a, b].$$

Letting $n \to \infty$, we get

(10) $|x_m(t) - x(t)| \le \varepsilon,\ \forall\ m \ge N$ and $\forall\ t \in [a, b]$.

This verifies that the sequence $\{x_m\}$ of continuous functions converges uniformly to the function x on $[a, b]$ and hence the limit function x is a continuous function on $[a, b]$. As such $x \in C[a, b]$.

Also, from (10), we have

$$\max_{t \in [a,b]}\ |x_m(t) - x(t)| \le \varepsilon,\ \ \forall\ m \ge N$$

$\Rightarrow$ $\|x_m - x\|_\infty \le \varepsilon,\ \ \forall\ m \ge N$

$\Rightarrow$ $x_m \to x$ in $C[a, b]$.

Hence, $(C[a, b], \|\cdot\|_\infty)$ is a Banach space. ∎

Remarks

1. The convergence of a sequence $\{x_n\} \subset C[a, b]$ in the norm $\|\cdot\|_\infty$ is the same as the uniform convergence of $\{x_n\}$ regarded as a sequence of functions defined on $[a, b]$.

2. One may note that the sequence $\{x_n\}$, regarded as a sequence of functions in $C[a, b]$, has another notion of convergence, *e.g.*, pointwise convergence. However, there in no norm on $C[a, b]$ which defines pointwise convergence.

   Let, if possible, $\|\cdot\|$ be a norm on $C[0, 1]$ such that

$$\|x_n - x\| \to 0 \text{ as } n \to \infty$$

if and only if $x_n(t) \to x(t)$, for all $t \in [0, 1]$. Consider the sequence $\{y_n\}$ of functions defined on $[0, 1]$, where

$$y_n(t) = \begin{cases} 2^n t, & 0 \le t \le 2^{-n} \\ 2 - 2^n t, & 2^{-n} < t < 2^{1-n} \\ 0, & \text{otherwise.} \end{cases}$$

Since $y_n \ne 0, \|y_n\| \ne 0, n \in \mathbb{N}$.

Define $x_n = y_n/\|y_n\|, n \in \mathbb{N}$. Then $\|x_n\| = 1, \forall n \in \mathbb{N}$. Therefore, $\{x_n\}$ does not converge to 0 (zero function) with respect to the norm $\|\cdot\|$ on $C[0, 1]$. On the other hand, one can check that

$$x_n(t) \to 0,\ \ \forall\ t \in [0, 1].$$

This is a contradiction.

Notes

1. The function space $C[a, b]$ is real or complex according to $\mathbb{K}$ is $\mathbb{R}$ or $\mathbb{C}$.

2. $\|x\|_\infty$ is precisely the greatest vertical height of the function from t-axis as illustrated in Fig. 2.3.

3. The norm $\|\cdot\|_\infty$ defined on $C[a, b]$ is called the *sup norm* (or *uniform norm*) on $C[a, b]$.

2.2.8 Example (The Function Space $L^p[a, b]$). Consider the linear space L^p $[a, b]$, $0 < p \le \infty$ (Examples 1.6.3 (11) and 1.6.3 (12)). Define the function $\|\cdot\|_p : L^p[a, b] \to \mathbb{R}$, $0 < p \le \infty$, as follows:

$$\|f\|_p = \begin{cases} \left(\displaystyle\int_a^b |f|^p \right)^{\frac{1}{p}}, & 0 < p < \infty \\ \text{ese sup} |f|, & p = \infty. \end{cases}$$

We shall now establish that if $1 \le p \le \infty$, then $\|\cdot\|_p$ defines a norm on $L^p[a, b]$. In case $1 \le p < \infty$, $\|\cdot\|_p$ is called L^p-*norm* (or simply a *p-norm*) on $L^p[a, b]$. It is also referred to *Lebesgue norm*.

Before establishing that $\|\cdot\|_p$ defines a norm on L^p, we give below two important inequalities which are immediately useful in doing so.

Riesz-Hölder Inequality. Let p and q be non-negative extended real numbers such that $\dfrac{1}{p} + \dfrac{1}{q} = 1$. If $f \in L^p[a, b]$ and $g \in L^q[a, b]$, then $f \cdot g \in L^1[a, b]$ and

$$\int_a^b |fg| \le \|f\|_p \|g\|_q.$$

Equality holds if and only if, for some non-zero constants A and B, we have

$$A|f|^p = B|g|^q \text{ a.e.}$$

Justification. When $p = 1$, then $q = \infty$ and the inequality is trivially available. Assume that $1 < p < \infty$ and consequently $1 < q < \infty$. The inequality is trivial if

---

For spaces of the type $L^p[a, b]$, $L^\infty[a, b]$ etc., one need to have the knowledge of measure and integration, in particular, the concepts like measurable sets and functions, *a.e.*, Lebesgue integration. For details, one may refer to [24].

either $f = 0$ a.e. or $g = 0$ a.e. So assume $f \neq 0$ and $g \neq 0$. This gives that $\| f \|_p > 0$ and $\| g \|_q > 0$. Now, using Inequality 1.5.3 with

$$\lambda = \frac{1}{p}, \ \alpha = \left(\frac{|f(t)|}{\| f \|_p} \right)^p, \ \beta = \left(\frac{|g(t)|}{\| g \|_q} \right)^q$$

the result follows.

Riesz-Minkowski Inequality. Let $1 \leq p \leq \infty$. Then, for every pair $f, g \in L^p[a, b]$, we have

$$\| f + g \|_p \leq \| f \|_p + \| g \|_p.$$

Justification. The cases $p = 1, \infty$ are straightforward. We, therefore, assume that $1 < p < \infty$. Since $L^p[a, b]$ is a linear space, $f + g \in L^p[a, b]$. Also, we have

$$\int_a^b |f + g|^p \leq \int_a^b |f + g|^{p-1} |f| + \int_a^b |f + g|^{p-1} |g|.$$

Let $1 < q < \infty$ be such that $\dfrac{1}{p} + \dfrac{1}{q} = 1$. Then, since $(p-1)q = p$, observe that

$$\int_a^b \left(|f + g|^{p-1} \right)^q = \int_a^b |f + g|^p$$

$$\Rightarrow \qquad |f + g|^{p-1} \in L^q[a, b].$$

As such, both $|f + g|^{p-1} |f|$ and $|f + g|^{p-1} |g|$ are in $L^1[a, b]$ and the Riesz-Hölder Inequality leads to

$$\int_a^b |f + g|^{p-1} |f| \leq \| f \|_p \ \| \left(|f + g|^{p-1} \right) \|_q$$

and

$$\int_a^b |f + g|^{p-1} |g| \leq \| g \|_p \ \| \left(|f + g|^{p-1} \right) \|_q.$$

But $$\| (|f+g|^{p-1}) \|_q = \left(\int_a^b |f+g|^{(p-1)q} \right)^{\frac{1}{q}} = (\|f+g\|_p)^{\frac{p}{q}}$$

since $(p-1)\,q = p$. Hence

$$\int_a^b |f+g|^p \le (\|f\|_p + \|g\|_p)\,(\|f+g\|_p)^{\frac{p}{q}}.$$

If $\|f+g\|_p$ is non-zero finite, the result follows on dividing both sides by $\left(\|f+g\|_p\right)^{\frac{p}{q}}$. In case $\|f+g\|_p = 0$, there is nothing to prove while in case $\|f+g\|_p = \infty$, we either have $\|f\|_p = \infty$ or $\|g\|_p = \infty$ in view of the relation $|f+g| \le |f| + g\,|$, and the result is obviously true again. $\blacksquare$

We are now prepared to show that $\|\cdot\|_p$ defines a norm on $L^p[a,b]$. In fact, for $1 \le p \le \infty$, the function $\|\cdot\|_p : L^p[a,b] \to \mathbb{R}$ satisfies the following conditions:

(*i*) $\|f\|_p \ge 0$

(*ii*) $\|f\|_p = 0 \Leftrightarrow f = 0$ a.e.

(*iii*) $\|\alpha f\|_p = |\alpha|\,\|f\|_p,\ \alpha \in \mathbb{R}$

(*iv*) $\|f+g\|_p = \|f\|_p + \|g\|_p.$

The condition (*ii*) follows from the properties of Lebesgue measure and integration while condition (*iv*) is available from the Riesz-Minkowski Inequality.

Unfortunately, the definition of $\|\cdot\|_p$ on $L^p[a,b]$ fails to satisfy the norm requirement that $\|f\|_p = 0 \Rightarrow f = 0$. As such, $\|\cdot\|_p$ is not a norm on $L^p[a,b]$. However, to avoid this difficulty, we do not distinguish between *equivalent function, i.e.,* the functions that are equal almost everywhere. In that situation, we regard the space $L^p[a,b]$ consisting of equivalence classes of functions which can again be verified to be a linear space. Thus $\|\cdot\|_p$ now defines a norm on $L^p[a,b]$, regarded as the space of equivalence classes, and therefore, $L^p[a,b]$ becomes a normed space. If one refers to $L^p[a,b]$ as a normed space, it is in reality the space of equivalence classes $\tilde{f}$ of functions. But this should not pose any problem since, for any $\tilde{f} \in L^p[a,b]$, the norm is given by

$$\|\tilde{f}\|_p = \|f\|_p,$$

where f is any function in the equivalence class $\tilde{f}$, and this definition does not depend on the choice of the function f in the class $\tilde{f}$. In actual practice, the equivalence classes relegated to the background, and the elements of $L^p[a, b]$ are thought of as functions, where any two functions are regarded identical if they are equivalent.

The $\| \cdot \|_p$ on $L^p[a, b]$ induces in a natural way a metric d on it given by

$$d(f, g) = \| f - g \|_p.$$

Finally, we note that the normed spaces $L^p[a, b]$, $1 \leq p \leq \infty$ are complete. This is known as *Riesz-Fisher Theorem*. For the details of the proof of it, one is referred to [24].

Note. The above two inequalities (Riesz-Hölder Inequality and Riesz-Minkowski Inequality) do not hold good for $0 < p < 1$ (see Problems 5 and 6). As such $\| \cdot \|_p$ does not define a norm on $L^p[a, b]$ for $0 < p < 1$.

Problems

5. Let $0 < p < 1$ and q be such that $\dfrac{1}{p} + \dfrac{1}{q} = 1$. If $f \in L^p[a, b]$ and $g \in L^q[a, b]$,

then prove that

$$\int_a^b | f\, g | \geq \left(\int_a^b | f |^p \right)^{\frac{1}{p}} \left(\int_a^b | g |^q \right)^{\frac{1}{q}}$$

provided $\displaystyle\int_a^b | g |^q \neq 0$.

6. Let $0 < p < 1$ and $f, g \in L^p[a, b]$ such that $f \geq 0$ and $g \geq 0$. Prove that
$\| f + g \|_p \geq \| f \|_p + \| g \|_p$.

7. If $0 < p < 1$, then prove that $L^p[a, b]$ is a complete metric space with metric ρ defined by

$$\rho\,(f, g) = \left(\| f - g \|_p \right)^p, \quad f,\ g \in L^p[a, b].$$

8. Prove that

$$L^\infty [a, b] \subset L^p [a, b], \ 1 \le p < \infty.$$

Furthermore, if $f \in L^\infty [a, b]$, then prove that

$$\|f\|_\infty = \lim_{p \to \infty} \|f\|_p.$$

(This gives a relationship between $\|\cdot\|_p$ ($1 \le p < \infty$) and $\|\cdot\|_\infty$ which, in fact, justifies the notation $L^\infty [a, b]$ for the class of essentially bounded functions on $[a, b]$).

We, now, give below some examples of incomplete normed spaces.

2.2.9 Example. The linear space $P\,[a, b]$ (Example 1.6.3 (8)) equipped with the norm given by

$$\|x\|_\infty = \sup_{t \in [a,b]} |x(t)|$$

is an incomplete normed space. Indeed, in view of the Weierstrass Approximation Theorem, the uniform limit of a sequence of polynomials need not be a polynomial.

2.2.10 Example. The real linear space $C\,[-1, 1]$ (Example 1.6.3 (10) with $a = -1, b = 1$) equipped with the norm given by

$$\|x\|_1 = \int_{-1}^{1} |x(t)|\,dt,$$

where integral is taken in the sense of Riemann, is an incomplete normed space. ($\|x\|_1$ is precisely the area of the region as shown in Fig. 2.4).

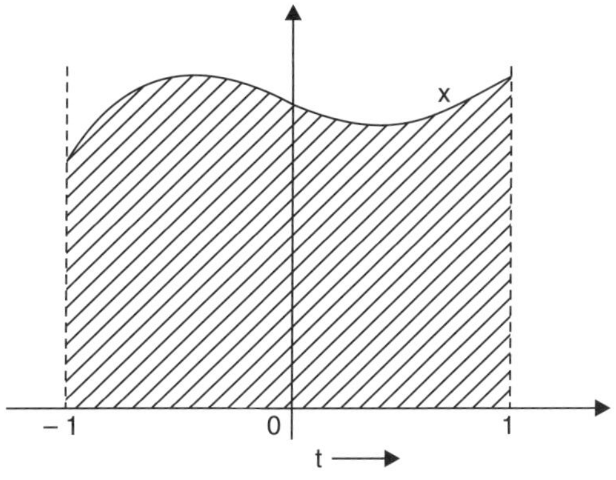

Fig. 2.4

Let $\{x_n\}$ be a sequence in $C[-1, 1]$, where

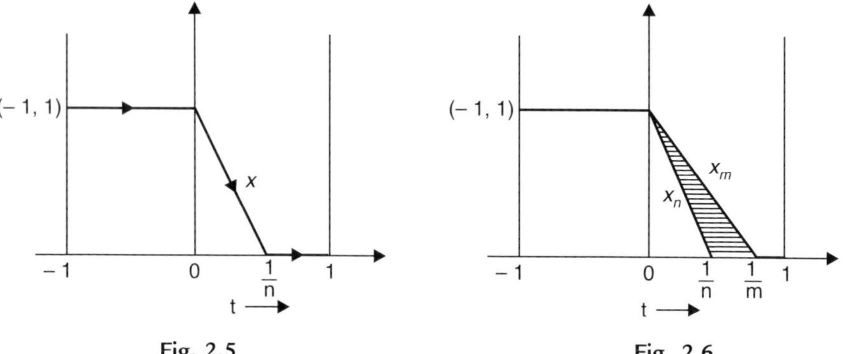

Fig. 2.5   Fig. 2.6

$$
x_n(t) = \begin{cases} 1, & -1 \le t \le 0 \\ 1 - nt, & 0 < t < \dfrac{1}{n} \\ 0, & \dfrac{1}{n} < t \le 1. \end{cases}
$$

It may be observed that $\{x_n\}$ is a Cauchy sequence. Geometrically the functions x_n are shown in Fig. 2.5, and $\|x_n - x_m\|$ represent the area of the triangle shown in Fig. 2.6. Let, if possible, $x_n \to x$ in $C[-1, 1]$. But

$$
\|x_n - x\|_1 = \int_{-1}^{1} |x_n(t) - x(t)|\, dt
$$

$$
= \int_{-1}^{0} |1 - x(t)|\, dt + \int_{0}^{\frac{1}{n}} |x_n(t) - x(t)|\, dt + \int_{\frac{1}{n}}^{1} |x(t)|\, dt.
$$

Since the integrands are non-negative, so is the each integral on the right. Hence, $\|x_n - x\|_1 \to 0$ would imply that each integral on the right approaches zero as $n \to \infty$. Consequently

$$
\begin{cases} \lim_{n \to \infty} \int_{\frac{1}{n}}^{1} |x(t)|\, dt = 0 \\ \int_{-1}^{0} |1 - x(t)|\, dt = 0 \end{cases}
$$

$$\Rightarrow \qquad x(t) = \begin{cases} 1, & -1 \le t \le 0 \\ 0, & 0 < t \le 1. \end{cases}$$

But the function x is not continuous on $[-1, 1]$ and as such $x \notin C[-1, 1]$. This proves that $(C[-1, 1], \|\cdot\|_1)$ is not a Banach space. $\blacksquare$

Problem 9. Consider the linear space $C[a, b]$ with the norm given by

$$\|x\|_p = \left(\int_a^b |x(t)|^p \, dt \right)^{\frac{1}{p}}, \ 1 \le p < \infty.$$

Work out the details to show that this space is an incomplete normed space.

2.2.11 Example. The real linear space $C^1[0, 1]$ of all continuously differentiable functions defined on $[0, 1]$ equipped with the norm given by

$$\|x\|_\infty = \sup_{t \in [0, 1]} |x(t)|$$

is an incomplete normed space.

It is enough to verify that the space is incomplete. Consider the Cauchy sequence $\{x_n\}$ in $C^1[0, 1]$, where

$$x_n(t) = \sqrt{t^2 + \frac{1}{n}}.$$

Note that $\{x_n\}$ converges pointwise to the function x, where

$$x(t) = |t|, t \in [0, 1].$$

But

$$\|x_n - x\|_\infty = \sup_{t \in [0, 1]} |x_n(t) - x(t)|$$

$$= \sup_{t \in [0, 1]} \left| \sqrt{t^2 + \frac{1}{n}} - |t| \right|$$

$$= \frac{1}{\sqrt{n}} \to 0, \text{ as } n \to \infty.$$

This shows that the convergence $x_n \to x$ is uniform on $[0, 1]$. But x is not differentiable at $t = 0$ and hence $x \notin C^1[0, 1]$. This verifies that $C^1[0, 1]$ is not complete. ∎

Problem 10. Prove that the real linear space $C^n[0,1]$ of all n-times continuously differentiable functions defined on $[0, 1]$ equipped with the norm given by

$$\| x \|_\infty = \sup_{t \in [0,1]} | x(t) |$$

is an incomplete normed space.

Note. We shall show in Section 2.5 that an incomplete normed space can be completed. However, the completion of an incomplete normed space may lead to a new kind of elements and it may then be interesting to explore the nature of such elements.

Finally, we present an example to show that not every metric on a linear space can be obtained from a norm. In other words, a linear space can be a metric space without being a normed space.

2.2.12 Example (The Sequence Space ω). Consider the linear space ω of all (bounded or unbounded) sequences in $\mathbb{K}$ with the usual pointwise addition and scalar multiplication of sequences. Let ω be equipped with a metric d given by

$$d(x, y) = \sum_{i=1}^\infty \frac{1}{2^i} \frac{| \alpha_i - \beta_i |}{(1 + | \alpha_i - \beta_i |)}.$$

Let, if possible, d can be induced by a norm $\| \cdot \|$ on the linear space ω. Then, we should have

$$d(\lambda x, \lambda y) = \| \lambda x - \lambda y \| = | \lambda | \, \| x - y \| = | \lambda | \, d(x, y)$$

where $x = \{\alpha_i\}$ and $y = \{\beta_i\}$ are in ω and $\lambda \in \mathbb{K}$. But

$$d(\lambda x, \lambda y) = \sum_{i=1}^\infty \frac{1}{2^i} \frac{| \alpha_i - \beta_i |}{(1 + | \lambda | \, | \alpha_i - \beta_i |)}$$

$$\neq | \lambda | \, d(x, y), \qquad \lambda \neq 0, 1$$

Thus the metric d cannot be obtained from a norm on ω. ∎

Note. For more examples of metric spaces which are not normed spaces, see Problems 23 and 24. However, for conditions under which a metric on a linear space can be obtained by a norm on it, see Problem 25.

Problems

Prove that each of the linear spaces, under the usual algebraic operations, in Problems 11 to 20 is a Banach space with the given norm.

11. The space c of all convergent sequences $x = \{\xi_i\}$ with

$$\| x \|_\infty = \sup_{1 \leq i < \infty} | \xi_i |.$$

12. The space c_o of all sequences $x = \{\xi_i\}$ converging to zero with

$$\| x \|_\infty = \sup_{1 \leq i < \infty} | \xi_i |.$$

13. The space kr of all sequences $x = \{\xi_i\}$ such that the series $\sum_{i=1}^{\infty} \xi_i$ is convergent with $\| x \|_\infty = \sup_{1 \leq n < \infty} | \sum_{i=1}^{n} \xi_i |$.

14. The space bv of all sequences $x = \{\xi_i\}$ of bounded variation with

$$\| x \| = | \xi_i | + V(n), \text{ where } V(n) = \sum_{n=1}^{\infty} | \xi_{n+1} - \xi_n |.$$

15. The space bts of all sequences $x = \{\xi_i\}$ of bounded partial sums with

$$\| x \|_\infty = \sup_{1 \leq n < \infty} | \sum_{i=1}^{n} \xi_i |.$$

16. The space $B\,[a, b]$ of all bounded functions x on $[a, b]$ with

$$\| x \|_\infty = \sup_{t \in [a,b]} | x(t) |.$$

17. The space $C^n\,[a, b]$ of all n-times continuously differentiable functions x on $[a, b]$ with

$$\| x \|_n = \sum_{i=0}^{n} \| x^{(i)} \|_\infty .$$

Here $x^{(i)}$ stands for the i-th derivative of x.

18. The space $C_0(\mathbb{R})$ of all functions x continuous on $\mathbb{R}$ which vanish at ∞ with $\| x \| = \max\limits_{t \in \mathbb{R}} | x(t) |$.

19. The space $BV[a, b]$ of all functions x of bounded variation on $[a, b]$ with $\| x \| = | x(a) | + V(x)$, where $V(x)$ is the total variation of x on $[a, b]$.

20. Let $C_\lambda[0, 1]$ be the set of all functions x defined on the interval $[0, 1]$ and satisfying Hölder (or, Lipschitz) condition with index $\lambda > 0$,

$$| x(t_1) - x(t_2) | \le M | t_1 - t_2 |^\lambda, \qquad t_1, t_2 \in [0, 1]$$

where $M > 0$ is a constant, the norm being defined by the formula

$$\| x \|_\lambda = | x(0) | + \sup_{t_1, t_2 \in [0,1]} \frac{| x(t_1) - x(t_2) |}{| t_1 - t_2 |^\lambda} .$$

(The space $C_\lambda[0, 1]$ is called *Hölder space* with index λ)

Prove that each of the linear spaces, with usual algebraic operations, in Problems 21 and 22 is an incomplete normed space.

21. The space $BV[a, b]$ with $\| x \|_\infty = \sup\limits_{t \in [a,b]} | x(t) |$.

22. The space $C^n[a, b]$ with $\| x \|_{p,n} = \sum\limits_{j=0}^{n} \| x^{(j)} \|_p$.

(The norm $\|\cdot\|_{p,n}$ is called the *Sobolev norm*).

23. Let X be a normed space and a mapping $d^* : X \times X \to \mathbb{R}$ be defined by

$$d^*(x, y) = \min \{1, \| x - y \|\}, x, y \in X.$$

Prove that:

(a) d^* is a metric on X.

(b) There is no norm on X which generates d^*.

24. Let d be the metric induced by a norm on a linear space $X \ne \{0\}$. If d_1 is defined by

$$d_1(x, y) = \begin{cases} 0, & x = y \\ 1 + d(x, y), & x \ne y \end{cases}$$

then prove that d_1 cannot be obtained from a norm on X.

25. Let X be a linear space and d be metric on X such that

$$d\,(x,\,y) = d\,(x - y,\,0)$$

and $d\,(\alpha x,\,0) = |\,\alpha\,|\,d\,(x,\,0)$, $x,\,y \in X$ and $\alpha \in \mathbb{K}$.

Define $\|\,x\,\| = d\,(x,\,0)$, $x \in X$. Prove that $\|\cdot\|$ is a norm on X and that d is the metric induced by the norm $\|\cdot\|$ on X.

2.3 Subspaces

It may be recalled that a non-empty subset Y of a linear space X over $\mathbb{K}$ is a (linear) subspace of X if and only if $x + y \in Y$ and $\alpha x \in y$, $\forall\,x,\,y \in Y$ and $\alpha \in K$ (Definition 1.6.4). We, now, introduce the concept of subspaces of a normed space.

2.3.1 Definition. Let X be a normed space. A non-empty subset Y of X is said to be a *subspace* of X if

(i) Y is a (linear) subspace of X considered as a linear space, and

(ii) Y is equipped with the norm $\|\cdot\|_Y$ induced by the norm $\|\cdot\|$ on X, *i.e.*,

$$\|\,x\,\|_Y = \|\,x\,\|, \quad \forall\,x \in Y.$$

We may denote the subspace $(Y,\,\|\cdot\|_Y)$ simply by Y.

Note. It can be readily seen that a subspace Y of a normed space X is itself a normed space. Further, it is clear that the metric defined on Y by its norm coincides with the restriction to Y of the metric defined on X by its norm and, therefore, Y is a subspace of the metric space X in the sense of Definition 1.7.3.

2.3.2 Definition. A subspace Y of a normed space X is called a *closed subspace* of X if Y is closed in X considered as a metric space.

2.3.3 Definition. Let X be a Banach space. A subspace Y of X considered as a normed space is said to be a *subspace* of Banach space X.

2.3.4 Examples

1. The space c is a closed subspace of l^{∞}.

2. The space c_0 is a closed subspace of c.

3. The space $P\,[a,\,b]$ is a subspace of $C\,[a,\,b]$ which is not closed.

Note. Throughout we shall consider c_0 and c with the norms induced by the norm $\|\cdot\|_{\infty}$ on l^{∞}, unless otherwise indicated.

2.3.5 Theorem. *Let Y be a subspace of a normed space X and $\{x_n\} \subset Y$. If $\{x_n\}$ is Cauchy in X, then it is Cauchy in Y and conversely.*

Proof. Straightforward. ∎

A relationship between the closedness and completeness of a subspace is given in the following.

2.3.6 Theorem. *Let Y be a subspace of a normed space X. Then, Y is complete $\Rightarrow$ Y is closed.*

Proof. Let x be a limit point of Y. Then, every open sphere centred on x contains points (other than x) of Y. In particular, the open sphere $S\left(x; \dfrac{1}{n}\right)$, where n is a positive integer, contains a point x_n of Y, other than x. Thus, $\{x_n\}$ is a sequence in Y such that

$$\| x_n - x \| < \frac{1}{n}, \quad \forall\, n$$

$$\Rightarrow \qquad \lim_{n \to \infty} x_n = x \text{ in } X$$

$$\Rightarrow \qquad \{x_n\} \text{ is a Cauchy sequence in } X \text{ and hence in } Y.$$

But Y being complete, it follows that $x \in Y$. This proves that Y is closed. ∎

2.3.7 Theorem. *Let Y be a subspace of a Banach space X. Then, Y is closed $\Rightarrow$ Y is complete.*

Proof. Let $\{x_n\}$ be a Cauchy sequence in Y. Then, it is so in X. But X being complete, $\exists\, x \in X$ such that $x_n \to x$. Either $x \in Y$, then we are done, or each neighbourhood of x contains points $x_n\ (\neq x)$. As such, x is a limit point of Y. But Y being closed, it follows that $x \in Y$. This completes the proof of the theorem. ∎

2.3.8 Corollary. *Let Y be a subspace of a Banach space X. Then, Y is complete $\Leftrightarrow$ Y is closed.*

2.3.9 Examples

1. Consider the space Φ of sequences

$$x = (\xi_1, \xi_2, \ldots, \xi_n, 0, \ldots)$$

in $\mathbb{K}$, where $\xi_n \neq 0$ for only finitely many values of n. Clearly, $\Phi \subset c_0 \subset l^\infty$, and $\Phi \neq c_0$. Note that c_0 is the closure of Φ in $(l^\infty, \| \cdot \|_\infty)$. Thus, Φ is not closed in l^∞ and hence Φ is an incomplete normed space equipped with the norm induced by the norm $\| \cdot \|_\infty$ on l^∞.

2. For every real number $1 \le p < \infty$, we have

$$\Phi \subset l^p \subset c_0.$$

Note that c_0 is the closure of l^p in c_0 and $l^p \ne c_0$ (Verify !). Thus, l^p is not closed in c_0 and hence l^p is an incomplete normed space when endowed with the norm induced by $\| \cdot \|_\infty$ on c_0.

3. For every real number $1 \le p < \infty$, we have $\Phi \subset l^p$. Note that l^p is the closure of Φ in $(l^p, \| \cdot \|_p)$ and $\Phi \ne l^p$ (Verify !). Thus, Φ is not closed in l^p and hence Φ is an incomplete normed space endowed with the norm induced by $\| \cdot \|_p$ on l^p.

Problems

26. Show that the closure $\overline{Y}$ of a subspace Y of a normed space X is again a subspace of X.

27. Let $1 \le p \le q < \infty$. Prove that:

    (*i*) $\Phi \subset l^p \subset l^q$ and the inclusions are proper.

    (*ii*) l^p regarded as a subspace of $(l^q, \| \cdot \|_q)$ is not closed (and hence not complete) in $(l^q, \| \cdot \|_q)$.

28. If $n \ge m \ge 0$, prove that

$$C^n [a, b] \subset C^m [a, b]$$

and that the space $C^n [a, b]$ with the norm induced by the norm on $C^m [a, b]$ is not closed.

29. Prove that $C [a, b] \subset L^1 [a, b]$ and that the space $C [a, b]$ with L^1-norm is not closed (and hence not complete).

30. Extend the result of Problem 29 to the case of $L^p [a, b]$ with $1 \le p < \infty$.

31. Prove that $C [a, b]$ regarded as a subspace of $L^\infty [a, b]$ is closed.

32. Let $1 \le p < q \le \infty$. Prove that

    (*i*) $L^q [a, b] \subset L^p [a, b]$ and that the inclusions are proper.

    (*ii*) $L^q [a, b]$ regarded as a subspace of $(L^p [a, b], \| \cdot \|_p)$ is not closed (and hence not complete) in $(L^p [a, b], \| \cdot \|_p)$.

(*Note:* Compare this problem with Problem 27).

33. Let X be a Banach space and suppose $X = M + N$, where M and N are closed subspaces of X. Let $x \in X$ have a unique representation: $x = m + n$, $m \in M$ and $n \in N$. Define $\| \cdot \|' : X \to \mathbb{R}$ by

$$\| x \|' = \| m \| + \| n \|, \quad x \in X.$$

Prove that $(X, \| \cdot \|')$ is a Banach space.

34. Prove that the intersection of an arbitrary collection of non-empty closed subspaces of a normed space X is a closed subspace of X.

35. Let A be a subset of a normed space X. The intersection of all closed subspaces of X, each of which contains A is called the *closed linear hull* (or *closed linear span*) of A. Prove that:

(a) The losed linear hull of A is unique.

(b) The closed linear hull of A is the closure of the linear hull of A.

36. A normed space X is said to be *separable* if it is separable as a metric space (Definition 1.7.22) under the metric induced by the norm on X. Prove that:

(a) A finite dimensional normed space is separable.

(b) Each of the spaces c_0 and c is separable.

(c) l^p, $1 \le p < \infty$, is separable.

(d) l^∞ is not separable.

(e) Each of the spaces $C[a, b]$ and $C^n[a, b]$ is separable.

(f) $L^p[a, b]$, $1 \le p < \infty$, is separable.

(g) $L^\infty[a,b]$ is not separable.

(h) Each of the spaces $BV[a, b]$ and $C_\lambda[a, b]$ is separable.

    (All the above spaces are considered with their usual standard norms).

37. Prove that every separable normed space is isometric to a subspace of the space $C[0, 1]$.

2.4 Quotient Spaces

In this section, we shall consider one of the most useful methods of constructing new Banach spaces from the given Banach spaces.

2.4.1 Theorem. *Let X be a normed space over the field $\mathbb{K}$ and let M be a closed subspace of X. Define $\| \cdot \|_q : X/M \to \mathbb{R}$ by*

$$\| x + M \|_q = \inf \{ \| x + m \| : m \in M \}.$$

Then, $(X/M, \| \cdot \|_q)$ is a normed space. Further, if X is a Banach space, then X/M is a Banach space.

Proof. We first show that $\| \cdot \|_q$ defines a norm on X/M. It is obvious from the definition that $\| x + M \|_q \geq 0$, $\forall\, x \in X$. Note that

$$x + M = M$$

$$\Rightarrow \qquad \| x + M \|_q = \| 0 + M \|_q = \| 0 \| = 0.$$

Conversely, let $\| x + M \|_q = 0$ for some $x \in X$. Then $\exists$ a sequence $\{m_k\} \subset M$ such that

$$\lim_{k \to \infty} \| x + m_k \| = 0$$

$$\Rightarrow \qquad - m_k \to x \text{ in } X, \text{ as } k \to \infty$$

$$\Rightarrow \qquad x \in M \qquad\qquad (\because M \text{ is closed})$$

$$\Rightarrow \qquad x + M = M.$$

Thus, $\| x + M \|_q = 0 \;\Rightarrow\; x + M = M$.

Further, for $x, y \in X$, we have

$$\begin{aligned}
\| (x + M) + (y + M) \|_q &= \| (x + y) + M \|_q \\
&= \inf \{\| (x + y) + m \| : m \in M\} \\
&= \inf \{\| (x + m_1) + (y + m_2) \| : m_1, m_2 \in M\} \\
&\leq \inf \{\| x + m_1 \| + \| y + m_2 \| : m_1, m_2 \in M\} \\
&\leq \inf \{ \| x + m_1 \| : m_1 \in M\} + \inf \{\| y + m_2 \| : m_2 \in M\} \\
&= \| x + M \|_q + \| y + M \|_q
\end{aligned}$$

This proves the triangle inequality.

Next, for $x \in X$ and $\alpha \in \mathbb{K}$ with $\alpha \neq 0$, we have

$$\begin{aligned}
\| \alpha\, (x + M) \|_q &= \| \alpha\, x + M \|_q \\
&= \inf \{\| \alpha\, x + m \| : m \in M\} \\
&= \inf \{\| \alpha\, x + \alpha m' \| : m' = \frac{m}{\alpha} \in M\} \\
&= |\alpha| \inf \{\| x + m' \| : m' \in M\} \\
&= |\alpha|\, \| x + M \|_q.
\end{aligned}$$

Thus, we conclude that $(X/M, \|\cdot\|_q)$ is a normed space over $\mathbb{K}$.

Finally, assume that X is a Banach space. To show that X/M is complete, let $\{x_n + M\}$ be a Cauchy sequence in X/M. We shall first construct a convergent subsequence of $\{x_n + M\}$ in X/M. Evidently, it is possible to find a subsequence $\{x_{n_i} + M\}$ of the sequence $\{x_n + M\}$ such that

$$\| (x_{n_2} + M) - (x_{n_1} + M) \|_q < \frac{1}{2},$$

$$\| (x_{n_3} + M) - (x_{n_2} + M) \|_q < \frac{1}{2^2},$$

$$\vdots$$

$$\| (x_{n_{k+1}} + M) - (x_n + M) \|_q < \frac{1}{2^k}.$$

Choose any vector $y_1 \in x_{n_1} + M$. Next select $y_2 \in x_{n_2} + M$ such that

$\| y_2 - y_1 \| < \dfrac{1}{2}$. We then find $y_3 \in x_{n_3} + M$ such that $\| y_3 - y_2 \| < \dfrac{1}{2^2}$. Proceeding

in this way, we get a sequence $\{y_k\}$ in X such that

$$x_{n_k} + M = y_k + M$$

and
$$\| y_{k+1} - y_k \| < \frac{1}{2^k} , \quad (k = 1, 2, \ldots).$$

Let $k > r$. Then

$$\| y_k - y_r \| = \| (y_k - y_{k-1}) + (y_{k-1} - y_{k-2}) + \ldots + (y_{r+1} - y_r) \|$$

$$\leq \| y_k - y_{k-1} \| + \| y_{k-1} - y_{k-2} \| + \ldots + \| y_{r+1} - y_r \|$$

$$< \frac{1}{2^{k-1}} + \frac{1}{2^{k-2}} + \ldots + \frac{1}{2^r} < \frac{1}{2^{r-1}} .$$

Therefore, it follows that $\{y_k\}$ is a Cauchy sequence in X. But X being complete,

$\exists\, y \in X$ such that $\lim\limits_{k \to \infty} \| y_k - y \| = 0$. Since

$$\| (x_{n_k} + M) - (y + M) \|_q = \| (y_k + M) - (y + M) \|_q$$

$$= \| (y_k - y) + M \|_q,$$

$$\leq \| y_k - y \|,$$

it follows that

$$\lim_{k \to \infty} (x_{n_k} + M) = y + M \in X/M.$$

Thus, we have proved that the Cauchy sequence $\{x_n + M\}$ has a convergent subsequence in X/M. Finally, we note that if a subsequence of a Cauchy sequence converges, the sequence itself converges (1.7.11(6)). Hence, the Cauchy sequence $\{x_n + M\}$ converges in X/M and thus X/M is complete.

Note. The norm $\| \cdot \|_q$ on X/M is called the *quotient norm*. Throughout, the quotient space of a normed space will be endowed with the quotient norm. If there is no confusion likely to occur, we may drop the suffix q and denote it simply by $\| \cdot \|$.

Problems

38. Consider the subspace

$$M = \{x \in l^p : x(2n) = 0, n \in \mathbb{N}\}$$

of the space l^p, $1 \leq p \leq \infty$. Show that the quotient space l^p/M is isometrically isomorphic to l^p.

39. Let M be a closed subspace of a normed space X. Prove that the quotient mapping $x \to x + M$ of X onto the quotient space X/M is continuous and that it maps open subsets of X onto open subsets of X/M.

40. Discuss a Banach space X and a closed subspace M of X such that the quotient map $q : X \to X/M$ is not a closed map. Can the quotient map q ever be a closed map.

41. Let $X_1 \equiv (X_1, \| \cdot \|_1)$ and $X_2 \equiv (X_2, \| \cdot \|_2)$ be normed spaces over the same scalar field $\mathbb{K}$. Let $X = X_1 \times X_2$ be the cartesian product of X_1 and X_2. Then X is a linear space over $\mathbb{K}$ (Problem 1.2). Prove that each of the following defines a norm on X:

(a) $\| (x_1, x_2) \|_p = \{ (\| x_1 \|_1)^p + (\| x_2 \|_2)^p \}^{\frac{1}{p}}$, $1 \leq p < \infty$.

(b) $\| (x_1, x_2) \|_\infty = \max \{ \| x_1 \|_1, \| x_2 \|_2 \}$.

42. Let X_1 and X_2 of Problem 41 be Banach spaces. Is X a Banach space? Justify.

43. Generalise the results in Problems 41 and 42 to the case of n spaces X_i, $1 \le i \le n$.

44. Let M be a finite dimensional subspace of a normed space X. Show that:

(a) For $x \in X - M$, $\exists\, y \in M$ such that

$$\| x - y \| = d\,(x, M).$$

(b) For $x + M \in X/M$, $\exists\, y \in x + M$ such that

$$\| y \| = \| x + M \|.$$

45. Let M be a closed subspace of a normed space X. Prove that:

(a) If X is separable, then X/M is separable. Give an example such that X/M is separable but X is not.

(b) If M and X/M are separable, then X is separable.

(c) If any two of the three spaces X, M and X/M are complete, then so is the third.

(This is an example of what is called "two-out of-three result")

2.5 Completion of Normed Spaces

We have seen that there are normed spaces (Examples 2.2.9, 2.2.10 and 2.2.11) which are incomplete. So one is interested to study under what conditions an incomplete normed space becomes complete. Before formulating our result, we give certain related concepts which, otherwise, also are very useful in the subject.

2.5.1 Definition. Let X and Y be normed spaces over the field $\mathbb{K}$.

(i) A mapping $T : X \to Y$ (not necessarily linear) is said to be an *isometry* if it preserves norms, *i.e.*

$$\| Tx \|_Y = \| x \|_X, \quad \forall\, x \in X.$$

Such an isometry is said to *imbed X into Y*.

(ii) The spaces X and Y are said to be *isometric* if $\exists$ a bijective isometry of X to Y. The spaces X and Y are then called *isometric spaces*.

2.5.2 Theorem. *Let X be a normed space. Then, $\exists$ a Banach space $\hat{X}$ and an isometry of X onto W, where W is a dense subspace of $\hat{X}$.*

Proof. By Theorem 1.7.33, $\exists$ a complete metric space $(\hat{X}, \hat{d})$ and an isometry $T : X \to W$, where $W \subset \hat{X}$ is a subspace (subspace of $\hat{X}$ as a metric space) with $\overline{W} = \hat{X}$. In order to prove the theorem, we need

(a) to make $\hat{X}$ into a linear space,

(b) to show W to be a (linear) subspace of $\hat{X}$ regarded as the linear space, and

(c) to define a suitable norm on $\hat{X}$ which induces the metric $\hat{d}$ on $\hat{X}$.

To define on $\hat{X}$, the algebraic operations for $\hat{X}$ being a linear space, we consider $\hat{x}, \hat{y} \in \hat{X}$ and any representatives $\{x_n\} \in \hat{x}$ and $\{y_n\} \in \hat{y}$. We know that $\hat{x}$ and $\hat{y}$ are equivalence classes of Cauchy sequences in X. Set $z_n = x_n + y_n$. Then, $\{z_n\}$ is a Cauchy sequence in X since

$$\| z_n - z_m \| = \| x_n + y_n - (x_m + y_m) \| \le \| x_n - x_m \| + \| y_n - y_m \|.$$

Define the sum $\hat{z} = \hat{x} + \hat{y}$ of $\hat{x}$ and $\hat{y}$ to be the equivalence class for which $\{z_n\}$ is a representative. Thus, $\{z_n\} \in \hat{z}$. This definition is independent of the particular choice of Cauchy sequences belonging to $\hat{x}$ and $\hat{y}$. We know that if $\{x_n\} \sim \{x_n'\}$ and $\{y_n\} \sim \{y_n'\}$, then $\{x_n + y_n\} \sim \{x_n' + y_n'\}$ since

$$\| x_n + y_n - (x_n' + y_n') \| \le \| x_n - x_n' \| + \| y_n - y_n' \|.$$

Similarly, we define the product $\alpha \hat{x} \in \hat{X}$ of a scalar α and $\hat{x}$ to be the equivalence class for which $\{\alpha x_n\}$ is a representative. Moreover, this definition is independent of the particular choice of a representative $\hat{x}$. The zero element of $\hat{X}$ is the equivalence class containing all Cauchy sequences which converge to zero. It is not difficult to see that these two algebraic operations have all the properties required by the definition and as such $\hat{X}$ is a linear space.

Further, it follows that the linear operations on W induced from $\hat{X}$ agree with those induced from X by means of T.

Furthermore, T induces on W a norm $\| \cdot \|_1$, whose value at every $\hat{y} = Tx \in W$ is $\| \hat{y} \|_1 = \| x \|$. The corresponding metric on W (induced by the norm $\| \cdot \|_1$) is the restriction of $\hat{d}$ to W since T is isometry. We can extend the norm $\| \cdot \|_1$ to $\hat{X}$ by setting

$$\| \hat{x} \|_2 = \hat{d} \, (0, \hat{x})$$

for every $\hat{x} \in \hat{X}$. Note that $\| \cdot \|_2$ satisfies all the conditions of it being a norm on $\hat{X}$ which follow from those for $\| \cdot \|_1$ by a limit process. ∎

The space $\hat{X}$ constructed above is sometimes called the *completion* (or *complete enclosure*) of the normed space X. More precisely, we have

2.5.3 Definition. A *completion* of a normed space X is any complete normed space $\hat{X}$ which contains a dense subspace which is isometric to X.

2.5.4 Theorem. *All completions of a normed space are isometric.*

Proof. Let, if possible, $\hat{X}$ and X^* be two completions of a normed space X. In particular, we may assume that $\hat{X}$ and X^* both contain X as a dense subset and are complete. We will now define an isometry T between $\hat{X}$ and X^*. For each $\hat{x} \in \hat{X}$, since X is dense in $\hat{X}$, $\exists$ a sequence $\{x_n\}$ of points of X converging to $\hat{x}$. But we may also consider $\{x_n\}$ as a Cauchy sequence in X^* and , since X^* is complete, it must converge to some point $x^* \in X^*$. Define $T\hat{x} = x^*$ by this construction. We show that this construction is independent of the particular sequence $\{x_n\}$ converging to $\hat{x}$ and gives a one-to-one mapping of $\hat{X}$ onto X^*. Clearly, $Tx = x$, $\forall \, x \in X$. Now, if $x_n \to \hat{x}$ in $\hat{X}$ and $x_n \to x^*$ in X^*, then

$$\| \hat{x} \|_{\hat{X}} = \lim_{n \to \infty} \| x_n \|$$

and
$$\| x^* \|_{X^*} = \lim_{n \to \infty} \| x_n \|.$$

Thus
$$\| T\hat{x} \|_{X^*} = \| \hat{x} \|_{\hat{X}}.$$

Hence, T is an isometry.

2.5.5 Corollary. The space $\hat{X}$ in Theorem 2.5.2 is unique except for isometries.

2.5.6 Examples

1. The completion of the normed space $(P\,[a, b], \|\cdot\|_\infty)$ (Example 2.2.9) is the space $(C\,[a, b], \|\cdot\|_\infty)$.

2. The completion of the normed space $(C\,[a, b], \|\cdot\|_1)$ (Example 2.2.10) is the space $(L^1\,[a, b], \|\cdot\|_1)$.

Problem 46. Discuss the completion of the spaces, $BV\,[a, b]$ and $C^n\,[a, b]$, described in Problems 21 and 22, respectively.

Bounded Linear Operators

3
CHAPTER

If one is interested in the study of a certain class of mathematical objects, it is natural and fruitful to investigate the set of functions (maps, transformations or operators) from one such class to another which preserve some or all of the structures defined on the objects. In this chapter, the study of linear operators from one normed space to another normed space over the same scalar field has been emphasized. These operators preserve the algebraic structure defined on these normed spaces. The fact that a normed space gives rise to a metric, induced by the norm, provides naturally to the extremely important classification of linear operators into continuous and discontinuous ones. In normed spaces, this distinction is facilitated by a very simple criterion for continuity; namely, any linear operator between normed spaces is continuous if and only if it is bounded (Theorem 3.1.3). It is shown that the collection of all these bounded linear operators after introducing suitable operations, is a normed space.

Most of the material in this chapter deals with the bounded linear operators between normed spaces. Completeness of the normed spaces is assumed only where it seems essential. However, completeness is most essential for a few fundamental results.

Most remarkable and useful results proved in this chapter are Open Mapping Theorem, Closed Graph Theorem and Banach-Steinhaus Theorem (also, referred to as the Uniform Boundedness Principle). Various nice applications and related results of these theorems are also presented.

3.1 Definitions, Examples and Basic Properties

A *linear operator* between two normed spaces is defined as a linear operator between their corresponding linear spaces (Definition 1.6.17). Since normed spaces are also metric spaces, it is natural to consider operators on normed spaces which are both linear and continuous. The continuity is understood to be the metric continuity (Definition 1.7.37) given by the norm.

100

Let X and Y be normed spaces over the same field $\mathbb{K}$. An operator $T: X \to Y$ (linear or not) is said to be *continuous at a point* $x_0 \in X$ if $\exists$ a sequence $\{x_n\} \subset X$ such that

$$x_n \to x_0 \text{ in } X \quad \Rightarrow \quad T x_n \to T x_0 \to \text{ in } Y.$$

Equivalently, $T: X \to Y$ is continuous at a point $x_0 \in X$ if for a given $\varepsilon > 0$, $\exists$ a real number $\delta > 0$ such that

$$x \in X, \| x - x_0 \|_X < \delta \quad \Rightarrow \quad \| T x - T x_0 \|_Y < \varepsilon.$$

Further, an operator $T: X \to Y$ is said to be *continuous on X* if it is continuous at each $x \in X$. We prove below an interesting fact that, for a linear operator, continuity at a single point implies the continuity on the whole space X.

3.1.1 Theorem. *Let X and Y be normed spaces over the field $\mathbb{K}$ and $T: X \to Y$ a linear operator. Then, T is continuous on X if and only if T is continuous at a point (any) in X.*

Proof. Let T be continuous at a point $x_0 \in X$ and $x \in X$ be arbitrary. Let $\{x_n\} \subset X$ and $x_n \to x$ in X. Then, $\{x_n - x + x_0\}$ is a sequence in X such that $x_n - x + x_0 \to x_0$. Therefore

$$T (x_n - x + x_0) \to T x_0$$

$$\Rightarrow \qquad\qquad T x_n \to T x.$$

Hence, T is continuous at x.

The converse statement is obvious. ∎

We are now interested to obtain several more useful equivalent forms of continuity.

3.1.2 Definition. Let X and Y be normed spaces over the field $\mathbb{K}$ and $T: X \to Y$ a linear operator. T is said to be *bounded* on X, if $\exists$ a real number $K > 0$ such that

$$\| T x \|_Y \leq K \| x \|_X, \quad \forall\, x \in X.$$

If T is not bounded on X, then it is said to be *unbounded.*

Note. Definition 3.1.2 of a bounded linear operator is not the same as that of an ordinary real or complex function, where a bounded function is one whose range is a bounded set.

The following result shows that a bounded linear operator is the samething as a continuous linear operator.

3.1.3 Theorem. *Let X and Y be normed spaces over the field $\mathbb{K}$ and $T : X \to Y$ a linear operator. Then, T is continuous if and only if T is bounded.*

Proof. Assume that T is bounded. Let $\{x_n\}$ be a sequence in X such that $x_n \to 0$. Then, $\exists$ a real number $K > 0$ such that

$$\| Tx_n \|_Y \leq K \| x_n \|_X \to 0 \text{ , as } n \to \infty$$

$$\Rightarrow \qquad Tx_n \to 0, \text{ as } n \to \infty.$$

Therefore, T is continuous at $x = 0$ and hence T is continuous everywhere (Theorem 3.1.1).

Conversely, let if possible, T be not bounded. Then, for each positive integer n, $\exists \, 0 \neq x_n \in X$ such that

$$\| Tx_n \|_Y > n \| x_n \|_X$$

$$\Rightarrow \qquad \left\| T \left(\frac{x_n}{n \| x_n \|_X} \right) \right\|_Y > 1.$$

Setting $y_n = x_n / (n \| x_n \|_X)$, observe that $y_n \to 0$. But

$$\| Ty_n \|_Y > 1, \qquad \forall \, n \in \mathbb{N}.$$

$$\Rightarrow \qquad Ty_n \nrightarrow 0, \text{ as } n \to \infty.$$

Therefore, T cannot be continuous at the origin which contradicts the hypothesis. This completes the proof. ∎

Recall that a subset S of a normed space X is bounded if and only if $\exists$ some constant $K > 0$ such that $\| x \|_X \leq K, \, \forall \, x \in S$. We next give another characterization of bounded linear operators.

3.1.4 Theorem. *Let X and Y be normed spaces over the field $\mathbb{K}$ and $T : X \to Y$ a linear operator. Then, T is bounded if and only if T maps bounded sets in X into bounded sets in Y.*

Proof. Suppose T is bounded and S is any bounded subset of X. Then, $\exists$ a real number $K_1 > 0$ such that

$$\| Tx \|_Y \leq K_1 \| x \|_X, \qquad \forall \, x \in X$$

and, in particular, $\forall\, x \in S$. The set S being bounded, for some real number $K > 0$, we have

$$\| Tx \|_Y \leq K, \qquad \forall\, x \in S$$

$$\Rightarrow \qquad \{Tx : x \in S\} \text{ is bounded in } Y.$$

Hence, $T(S)$ is a bounded set in Y.

Conversely, for the closed unit sphere $S\,[0;\, 1] = \{x \in X : \| x \|_X \leq 1\}$, the set $T(S\,[0;\, 1])$ is bounded in Y. Therefore, $\exists\, K > 0$ such that

$$\| Tx \|_Y \leq K, \qquad \forall\, x \in S\,[0;\, 1].$$

If $x = 0$, then $Tx = 0$ and the assertion $\| Tx \|_Y \leq K \| x \|_X$ is obviously true. If $x \neq 0$, then $x/\| x \|_X \in S\,[0;\, 1]$, and so

$$\left\| T\left(\frac{x}{\| x \|_X} \right) \right\|_Y \leq K.$$

Hence, it follows that T is bounded. ∎

Remark. The fact established in Theorem 3.1.4 motivates the term "bounded operator".

Our next result summarizes different ways of describing a continuous linear operator (Theorems 3.1.1, 3.1.3 and 3.1.4).

3.1.5 Theorem. *Let X and Y be normed spaces over the field $\mathbb{K}$ and $T : X \to Y$ a linear operator. Then, the following statements are equivalent:*

(a) *T is continuous at the origin.*

(b) *T is continuous.*

(c) *T is bounded*

(d) *T maps bounded sets in X into bounded sets in Y.*

If $T : X \to Y$ is a bounded linear operator, then in view of Theorem 3.1.4, the set $\{Tx : x \in X, \| x \|_X \leq 1\}$ is bounded in Y. The supremum of all non-negative real numbers $\| Tx \|_Y$ such that $\| x \|_X \leq 1$, $(x \in X)$ is denoted by $\| T \|$ and is called the norm of T. More precisely, we have

3.1.6 Definition. Let X and Y be normed spaces over the field $\mathbb{K}$ and $T : X \to Y$ a bounded linear operator. The non-negative real number $\| T \|$ defined by

$$\| T \| = \sup \left\{ \| Tx \|_Y : x \in X, \| x \|_X \leq 1 \right\}$$

is called the norm (or bound) of T.

Note. We shall prove in the next section that $\| T \|$ is indeed a norm on the set of all bounded linear operators from a normed space X into a normed space Y.

3.1.7 Theorem. *Let X and Y be normed spaces over the field $\mathbb{K}$ and $T : X \to Y$ a bounded linear operator. Then, $\| T \|$ can be expressed by any one of the following formulae:*

(a) $\| T \| = \sup \{ \| Tx \|_Y : x \in X, \| x \|_X \leq 1 \}.$

(b) $\| T \| = \sup \{ \| Tx \|_Y : x \in X, \| x \|_X = 1 \}.$

(c) $\| T \| = \sup \left\{ \dfrac{\| Tx \|_Y}{\| x \|_X} : x \in X, x \neq 0 \right\}.$

(d) $\| T \| = \inf \{ K : K > 0 \text{ and } \| Tx \|_Y \leq K \| x \|_X, \forall\, x \in X \}.$

Proof. Assume that (a) gives the definition of $\| T \|$ (Definition 3.1.6). If we denote right sides of the expressions in (b), (c) and (d), respectively, by M_2, M_3 and M_4, then it is enough to prove that

$$\| T \| = M_2 = M_3 = M_4.$$

Since

$$\{ x \in X : \| x \|_X = 1 \} \subset \{ x \in X : \| x \|_X \leq 1 \}$$

it follows that

$$\sup \{ \| Tx \|_Y : x \in X, \| x \|_X = 1 \} \leq \sup \{ \| Tx \|_Y : x \in X, \| x \|_X \leq 1 \}.$$

Thus

$$\| T \| \geq M_2.$$

Again, since T is a linear operator, we can write

$$M_3 = \sup \left\{ \left\| T \left(\frac{x}{\| x \|_X} \right) \right\|_Y : x \in X, x \neq 0 \right\}.$$

If $y = x/\| x \|_X$, then $\| y \|_X = 1$. Therefore, it implies that

$$M_3 = \sup \{ \| Ty \|_Y : y \in X, \| y \|_X = 1 \} = M_2.$$

Thus $$M_3 = M_2.$$

Next, the defining expression for M_3 gives

$$\| Tx \|_Y \le M_3 \| x \|_X, \qquad \forall\, x \in X.$$

This, in view of the expression for M_4, immediately verifies that $M_4 \le M_3$.

Finally, again from the definition of M_4, it follows that

$$\| Tx \|_Y \le M_4 \| x \|_X, \quad \forall\, x \in X$$

$$\Rightarrow \qquad \| Tx \|_Y \le M_4, \ \forall\, x \in X \text{ with } \| x \|_X \le 1.$$

Therefore

$$\| T \| \le M_4.$$

Combining together the assertions obtained above, we have

$$\| T \| \ge M_2 = M_3 \ge M_4 \ge \| T \|. \ \blacksquare$$

As an immediate consequence of Theorem 3.1.7 (c), we have

3.1.8 Corollary. *Let X and Y be normed spaces over the field $\mathbb{K}$ and $T : X \to Y$ a bounded linear operator. Then*

$$\| Tx \|_Y \le \| T \| \, \| x \|_X, \qquad \forall\, x \in X.$$

We give below some examples of bounded and unbounded linear operators.

3.1.9 Examples

1. Let X and Y be normed spaces. Define the function $O : X \to Y$ by $O\,x = 0$, $\forall\, x \in X$, where 0 is the zero vector in Y. It may be readily seen that O is a linear operator and further, O is bounded and $\| O \| = 0$.

2. Let X be a normed space. Define the function $I : X \to X$ by $Ix = x$, $\forall\, x \in X$. Then, I is the identity function. Clearly, I is linear. If $X \ne \{0\}$, then I is bounded and $\| I \| = 1$.

3. Let $T = [a_{ij}]$ be $m \times n$ real matrix. Then $T : \mathbb{R}^n \to \mathbb{R}^m$ is a bounded linear operator. Indeed, if $x = (\xi_1, \xi_2, ..., \xi_n) \in \mathbb{R}^n$, write $\eta_i = \sum_{j=1}^{n} a_{ij} \xi_j$,

   Denote by Tx the transformed finite sequence $y = \{\eta_i\}$. Take the norm on $\mathbb{R}^n$ as

$$\| x \| = \sum_{j=1}^{n} |\xi_j|, \ x = \{\xi_j\} \in \mathbb{R}^n$$

Then

$$\| y \| = \sum_{i=1}^{m} |\eta_i| \le \sum_{i=1}^{m} \sum_{j=1}^{n} |a_{ij}| \, |\xi_j|$$

$$\Rightarrow \qquad \| Tx \| \le K \| x \|.$$

4. Consider the operator $T : l^2 \to l^2$ given by

$$Tx = (x_1, \tfrac{1}{2} x_2, \tfrac{1}{3} x_3, \ldots), \ x = \{x_i\} \in l^2.$$

Then T is a bounded linear operator with $\| T \| = 1$. Indeed, $\| Tx \| \le \| x \|$. Also

$$Te_1 = (1, 0, 0, \ldots) = e_1$$

so that $\qquad \qquad \| Te_1 \| = \| e_1 \| = 1.$

5. Consider the Banach space $(C\,[0,\,1],\, \| \cdot \|_\infty)$ (Example 2.2.7 with $a = 0$, $b = 1$). Assume that a function $k : [0,\,1] \times [0,\,1] \to \mathbb{R}$ is continuous. Define $T : C\,[0,\,1] \to C\,[0,\,1]$ by

$$(Tx)\,(t) = \int_{0}^{1} k\,(t,\,\tau)\, x\,(\tau)\, d\,\tau,\, x \in C\,[0,\,1].$$

Then, it may be easily seen that T is a linear operator. We prove below that T is bounded.

Since k is continuous, it follows that k is bounded on $[0,\,1] \times [0,\,1]$ and thus, $\exists$ a constant $C \ge 0$ such that

$$(1) \qquad \qquad |\, k\,(t,\,\tau)\,| \le C, \qquad \forall\,(t,\,\tau) \in [0,\,1] \times [0,\,1].$$

Also, since $x \in C\,[0,\,1]$, it follows that

$$(2) \qquad \qquad |\, x\,(t)\,| \le \sup_{t \in [0,1]} |\, x\,(t)\,| = \| x \|_\infty.$$

Therefore, by using (1) and (2), we get

$$\| Tx \|_\infty = \sup_{t \in [0,1]} |\,(Tx)\,(t)\,|$$

$$= \sup_{t \in [0,1]} \; |\int_0^1 k\,(t,\tau)\,x\,(\tau)\,d\,\tau|$$

$$\leq \sup_{t \in [0,1]} \int_0^1 |k\,(t,\tau)|\;|x\,(\tau)|\,d\,\tau \leq C\,\|\,x\,\|_\infty.$$

Thus, T is bounded. ∎

6. Consider the normed space $(P\,[a,\,b],\,\|\cdot\|_\infty)$ (Example 2.2.9 with $a = 0$, $b = 1$). Define $T:\; P\,[a,\,b] \to P\,[a,\,b]$ by

$$(Tx)\,(t) = x'\,(t), \qquad t \in [0,\,1].$$

Here prime denotes the differentiation of the function x with respect to t. The function T is called the differentiation operator on X. It is easy to check that T is linear. But T is unbounded. Indeed, let $x_n\,(t) = t^n$, $\forall\,n \in \mathbb{N}$. Then

$$\|\,x_n\,\|_\infty = \sup_{t \in [0,1]}\,|\,t^n\,| = 1$$

and

$$(Tx_n)\,(t) = x'_n\,(t) = nt^{n-1}.$$

Therefore

$$\|\,Tx_n\,\|_\infty = n = n\,\|\,x_n\,\|_\infty, \quad \forall\,n \in \mathbb{N}.$$

Thus, there is no fixed real number $K > 0$ such that

$$\|\,Tx_n\,\|_\infty \leq K\,\|\,x_n\,\|_\infty.$$

Hence, T is unbounded. ∎

We recall that the null space of a linear operator $T : X \to Y$, where X and Y are linear spaces, is defined by

$$\mathcal{N}(T) = \{x \in X : Tx = 0\},$$

where 0 is the zero vector of Y, and it is a subspace of X [Theorem 1.6.20 (b)]. We now prove the following.

3.1.10 Theorem. *Let X and Y be normed spaces-over the field $\mathbb{K}$ and $T : X \to Y$ a bounded (continuous) linear operator. Then the null space $\mathcal{N}(T)$ is closed.*

Proof. Let $x \in \overline{\mathcal{N}(T)}$ be arbitrary. Then, $\exists$ a sequence $\{x_n\} \subset \mathcal{N}(T)$ such that $x_n \to x$ (Theorem 1.7.14). Since T is continuous, $Tx_n \to Tx$. But $x_n \in \mathcal{N}(T)$ gives $Tx_n = 0$, $\forall n$. Therefore, $Tx = 0$ and hence $x \in \mathcal{N}(T)$. ∎

Remark. The range of a bounded linear operator need not be closed.

3.1.11 Examples

1. Consider the operator $T : l^\infty \to l^\infty$ defined by

$$Tx = y, \ x = \{\alpha_j\}, \ y = \left\{\frac{\alpha_j}{j}\right\}.$$

Clearly, T is linear. Moreover, T is bounded since

$$\| T \| = \sup_{\substack{x \in l^\infty \\ \|x\| = 1}} \| Tx \|_\infty = \sup_{\substack{\{\alpha_j\} \in l^\infty \\ \|\{\alpha_j\}\|_\infty = 1}} \left\| \left\{\frac{\alpha_j}{j}\right\} \right\|_\infty = \sup_{1 \le j < \infty} \left| \frac{\alpha_j}{j} \right| = 1.$$

Further, note that range of T is not closed (Verify !). ∎

2. Consider the space $X = (C[0, 1], \|\cdot\|_\infty)$ and the integral operator $K : X \to X$ given by

$$(Kx)(t) = \int_0^t x(\tau)\, d\tau$$

Then K is a bounded linear operator and $\| K \| = 1$ (Verify !).

(An operator of this type with a variable range of integration is called a *Voltra integral operator*).

Remark. Note that $\mathcal{R}(K)$, the range of K, is the set of continuously differentiable functions on $[0, 1]$ which vanish at $t = 0$ and that $\mathcal{R}(T)$ is a linear subspace of $C[0, 1]$ but it is not closed. However, a perturbation of the identity operator by K, results into the closed range.

Problem 1. Let $T = I + K$ be the operator on $C[0, 1]$, where K is as defined in Example 3.1.11 (2). Prove that the range of T is the whole space $C[0, 1]$ and therefore closed.

We recall that an operator $T : X \to Y$, where X and Y are linear spaces, is invertible (in algebraic sense) if $\exists$ an operator $S : Y \to X$ such that ST and TS are the identity operators on X and Y, respectively. In this case, S is called the (algebraic) inverse of T and we write $S = T^{-1}$ (Definition 1.6.21). Further it is well known that T is invertible if and only if T is bijective (Theorem 1.6.22).

In general, T^{-1}, if exists, of a bounded linear operator T need not be bounded; consider for instance, the bounded linear operator T of Example 3.1.11(1). However, we give below a criteria for T^{-1} to be bounded.

3.1.12 Theorem. *Let X and Y be normed spaces over the field $\mathbb{K}$ and $T : X \xrightarrow{\;onto\;} Y$ a linear operator. Then, T^{-1} exists and is a bounded linear operator if and only if $\exists$ a constant $K > 0$ such that*

(3) $\| Tx \|_Y \geq K \| x \|_X$, $\forall\, x \in X$

Proof. Suppose $T^{-1} : Y \to X$ exists and is a bounded linear operator. Then, $\exists$ a constant $K_1 > 0$ such that

$$\| T^{-1}(Tx) \|_X \leq K_1 \| Tx \|_Y, \qquad \forall\, x \in X$$

$$\Rightarrow \quad \| Tx \|_Y \geq K \| x \|_X, \qquad \forall\, x \in X$$

where $K = \dfrac{1}{K_1}$. This proves (3).

Conversely, assume that (3) holds. Then

$$Tx = 0 \quad \Rightarrow \quad x = 0.$$

Therefore, T is injective. Also, T being onto (given), it follows that T^{-1} exists in the algebraic sense. It is easy to prove that T^{-1} is linear.

Finally, we prove that T^{-1} is bounded. Indeed, for each $x \in X$, $\exists$ some $y \in Y$ such that $x = T^{-1} y$ and conversely. Further, note that

$$\| T(T^{-1}y) \|_Y \geq K \| T^{-1} y \|_X, \quad \forall\, y \in Y$$

$$\Rightarrow \quad \| T^{-1} y \|_X \leq \frac{1}{K} \| y \|_Y, \qquad \forall\, y \in Y$$

$$\Rightarrow \quad T^{-1} \text{ is bounded on } Y.$$

This completes the proof. ∎

3.1.13 Definition. Let X and Y be normed spaces over the same field $\mathbb{K}$ and $T : X \to Y$ be a bounded linear operator. If the algebraic inverse, T^{-1}, of T exists and is bounded, the operator T is said to be *invertible* and the operator T^{-1} is called the *topological inverse* (or, simply, inverse) of T.

3.1.14 Theorem. *Let X and Y be Banach spaces over the same field $\mathbb{K}$ and $T : X \to X$ be a bounded linear operator. If there exists a constant $K > 0$ satisfying (3), then $\mathscr{R}(T)$, the range of T, is a closed set.*

Proof. Let $\{y_n\} \subset \mathscr{R}(T)$ such that it converges to y in Y. Then, $\exists \{x_n\} \subset X$ with $Tx_n = y_n$, $n \in \mathbb{N}$. Since

$$\| y_n - y_m \|_Y = \| Tx_n - Tx_m \|_Y = \| T(x_n - x_m) \|_Y$$

$$\geq K \| x_n - x_m \|_X, \qquad \forall\, m, n \in \mathbb{N}.$$

and $\{y_n\}$ is a Cauchy sequence in Y, it follows that $\{x_n\}$ is a Cauchy sequence in X. The space X being complete, $\exists\, x \in X$ such that

$$Tx = \lim_{n \to \infty} Tx_n = \lim_{n \to \infty} y_n = y.$$

Hence $$y = Tx \in \mathscr{R}(T). \blacksquare$$

From Theorems 3.1.12 and 3.1.14, we have the following important characterization of invertible operators.

3.1.15 Theorem. *If X, Y and T are as in Theorem 3.1.14, then T is invertible if and only if $\mathscr{R}(T)$ is donse in Y and $\exists$ a constant $K > 0$ such that (3) is satisfied.*

3.1.16 Corollary. *Let X, Y and T be as in Theorem 3.1.14. Then, the operator T is not invertible if and only if $\mathscr{R}(T)$ is not dense in Y or $\exists$ a sequence $\{x_n\} \subset X$ with $\| x_n \|_X = 1$ for all $n \in \mathbb{N}$, but*

$$\lim_{n \to \infty} Tx_n = 0.$$

Problems

2. Let X be a normed space over the field $\mathbb{K}$ and let $T : l^p(n) \to X$, $1 \leq p < \infty$, be a linear operator. Prove that T is bounded.

   [Hint: Use the natural basis $\{e_1, e_2, ..., e_n\}$ for $l^p(n)$].

3. Let X be a normed space over the field $\mathbb{K}$ and let M be a closed subspace of X. Consider the quotient space X/M (Theorem 2.4.1). Prove that the quotient mapping $T : X \to X/M$ defined by $Tx = x + M$ is a bounded linear operator and $\| T \| \leq 1$.

4. Let X and Y be normed spaces over the field $\mathbb{K}$. Let $T : X \to Y$ be a bounded linear operator and M be its null space. Prove that T induces a natural linear operator $S : X/M \to Y$ with $\| S \| = \| T \|$.

5. For any $x \in C$ [0, 1], let T_x be an operator on L^2 [0, 1] defined by

$$(T_x y)(t) = x(t)\, y(t)$$

Prove that:

(a) T_x is a bounded linear operator on L^2 [0, 1]

(b) If $x(t) = 1 + t$, then T_x is invertible.

(c) If $x(t) = t$, then T is not invertible.

[Hint: Use Corollary 3.1.16]

[$x \in C$ [0, 1] is called the *multiplier*]

6. Let T and $\tilde{T}$ be bounded linear operators from X to Y, where X and Y are Banach spaces. If T has inverse and $\| \tilde{T} - T \| < \| T^{-1} \|$, prove that $\tilde{T}$ has inverse.

3.2 Spaces of Bounded Linear Operators

In Section 3.1, we introduced the concept of a bounded linear operator and presented several nice properties of these operators. Given two normed spaces over the same scalar field, we shall demonstrate in this section that we can construct other normed spaces with the help of bounded linear operators between these spaces.

3.2.1 Theorem. *Let X and Y be normed spaces over the field $\mathbb{K}$. Then, the set $B(X, Y)$ of all bounded linear operators from X into Y is a (linear) subspace of the space $L(X, Y)$.*

Proof. Obviously $B(X, Y) \subset L(X, Y)$ and $B(X, Y) \neq \phi$ (Example 3.1.9). Let T, $S \in B(X, Y)$. Then, $\exists\, K_1, K_2 > 0$ such that

$$\begin{cases} \| Tx \|_Y \le K_1 \| x \|_X \\ \| Sx \|_Y \le K_2 \| x \|_X, \end{cases} \quad \forall\, x \in X$$

$$\Rightarrow \quad \| (T + S)(x) \|_Y \le \| Tx \|_Y + \| Sx \|_Y$$

$$\le K_1 \| x \|_X + K_2 \| x \|_X$$

$$= (K_1 + K_2) \| x \|_X, \quad \forall\, x \in X$$

and

$$\| (\lambda T)(x) \|_Y = | \lambda | \, \| Tx \|_Y \le (| \lambda | K_1 \, \| x \|_X, \quad \forall\, x \in X \text{ and } \lambda \in \mathbb{K}.$$

Thus, $T + S \in B(X, Y)$ and $\lambda T \in B(X, Y)$. Hence, $B(X, Y)$ is a subspace of $L(X, Y)$. $\blacksquare$

It follows from Theorem 3.2.1, that $B(X, Y)$ itself is a linear space. We now proceed to show that $B(X, Y)$ can be made into a normed space. We also prove that $B(X, Y)$ is a Banach space if Y is a Banach space. It is interesting to note that completeness of $B(X, Y)$ does not depend upon that of the normed space X.

3.2.2 Theorem. *Let X and Y be normed spaces over the field $\mathbb{K}$ and let $B(X, Y)$ be the linear space of all bounded linear operators $T : X \to Y$. Define $\|\cdot\| : B(X, Y) \to \mathbb{R}$ by*

$$\| T \| = \sup \{\| Tx \|_Y : x \in X, \| x \|_X \leq 1\}.$$

Then, $(B(X, Y), \|\cdot\|)$ is a normed space. Furthermore, if Y is a Banach space, then $B(X, Y)$ is so.

Proof. Since every $T \in B(X, Y)$ is bounded, the norm $\|\cdot\|$ on $B(X, Y)$ is well defined. Further, observe that:

(i) $\| T \| \geq 0, \ \forall \ T \in B(X, Y)$

(ii) $\| T \| = 0 \quad \Rightarrow \quad \| Tx \|_Y = 0, \ \forall \ x \in X$ with $\| x \|_X \leq 1$

$\qquad\qquad\qquad \Rightarrow \quad Tx = 0, \ \forall \ x \in X$

$\qquad\qquad\qquad \Rightarrow \quad T = 0.$

Conversely, if $T = 0$, then clearly $\| T \| = \| 0 \| = 0$.

(iii) For any $\lambda \in \mathbb{K}$, we have

$$\| \lambda T \| = \sup \{\| (\lambda T)(x) \|_Y : x \in X, \| x \|_X \leq 1\}$$

$$= \sup \{|\lambda| \| Tx \|_Y : x \in X, \| x \|_X \leq 1\}$$

$$= |\lambda| \| T \|.$$

(iv) For $T, S \in B(X, Y)$, we have

$$\| T + S \| = \sup \{\| (T + S)(x) \|_Y : x \in X, \| x \|_X \leq 1\}$$

$$\leq \sup \{\| Tx \|_Y : x \in X, \| x \|_X \leq 1\}$$

$$+ \sup \{\| Sx \|_Y : x \in X, \| x \|_X \leq 1\}$$

$$= \| T \| + \| S \|.$$

Hence, the assertions (i)-(iv) above verify that $\| \cdot \|$ defines a norm on $B(X, Y)$.

Finally, assume that Y is a Banach space. To prove the completeness of $B(X, Y)$ under the norm $\| \cdot \|$, let $\{T_n\}$ be a Cauchy sequence in $B(X, Y)$. Note that

$$\| T_n x - T_m x \|_Y = \| (T_n - T_m)(x) \|_Y$$

$$\leq \| T_n - T_m \| \, \| x \|_X, \qquad \forall \, x \in X.$$

Thus, it follows that $\{T_n x\}$ is a Cauchy sequence in Y for each $x \in X$. But Y being complete, $\{T_n x\}$ converges in Y. Let $T_n x \to Tx$. We shall now prove that $T \in B(X, Y)$.

(a) *T is linear.* For $x, y \in X$ and $\lambda \in \mathbb{K}$, we have

$$T(x + y) = \lim_{n \to \infty} T_n (x + y)$$

$$= \lim_{n \to \infty} (T_n x + T_n y) \qquad (\because \ T_n \text{ is linear})$$

$$= Tx + Ty$$

and
$$T(\lambda x) = \lim_{n \to \infty} T_n (\lambda x) = \lim_{n \to \infty} \lambda \, T_n x = \lambda \, Tx.$$

This verifies that T is linear.

(b) *T is bounded.* Since for each n, the linear operator $T_n : X \to Y$ is bounded, it follows that

$$\| T_n x \|_Y \leq \| T_n \| \, \| x \|_X, \ \forall \, x \in X.$$

Also, the sequence $\{T_n\}$ being Cauchy in $(B(X, Y), \| \cdot \|)$, it is bounded and hence there is a constant $K > 0$ such that

$$\| T_n \| \leq K, \qquad \forall \, n \in \mathbb{N}.$$

Consequently

$$\| T_n x \|_Y \leq K \| x \|_X, \qquad \forall \, x \in X \text{ and } n \in \mathbb{N}.$$

Thus, we have

$$\| Tx \|_Y \leq \| Tx - T_n x \|_Y + \| T_n x \|_Y$$

$$\leq \| Tx - T_n x \|_Y + K \| x \|_Y, \qquad \forall \, x \in X \text{ and } n \in \mathbb{N}.$$

Since n is arbitrary, letting $n \to \infty$ and using $T_n x \to Tx$, we obtain

$$\| Tx \|_Y \leq K \| x \|_X, \quad \forall\, x \in X.$$

Hence, T is bounded.

The assertions (a) and (b) above verify that $T \in B(X, Y)$.

Finally, we show that $T_n \to T$, as $n \to \infty$, in $(B(X, Y), \| \cdot \|)$. Since $\{T_n\}$ is a Cauchy sequence in $B(X, Y)$, for each $\varepsilon > 0$, $\exists$ an integer $N > 0$ such that

$$\| T_n - T_m \| < \varepsilon, \forall\, n, m \geq N$$

$$\Rightarrow \quad \| T_n x - T_m x \|_Y < \varepsilon \| x \|_X, \quad \forall\, x \in X.$$

Taking $m \to \infty$, we get

$$\| T_n x - Tx \|_Y < \varepsilon \| x \|_X, \qquad \forall\, n \geq N \text{ and } x \in X.$$

This gives

$$\| T_n - T \| = \sup \{ \| T_n x - Tx \|_Y : x \in X, \| x \|_X = 1 \}$$

$$\leq \varepsilon, \quad \forall\, n \geq N.$$

Hence, $T_n \to T$, as $n \to \infty$, in $(B(X, Y), \| \cdot \|)$. This completes the proof. ∎

We shall now give a result on the composition of two bounded linear operators in $B(X, Y)$ in a particular situation when $Y = X$.

3.2.3 Theorem. *Let X be a normed space over the field $\mathbb{K}$. If $T, S \in B(X, X)$, then $S T \in B(X, X)$ and*

$$\| S T \| \leq \| S \| \, \| T \|.$$

Proof. Note that $T : X \to X$ is a linear operator since

$$(S T)(\alpha x + \beta y) = S(T(\alpha x + \beta y))$$

$$= S(\alpha\, Tx + \beta\, Ty)$$

$$= \alpha\,(S T)\, x + \beta\,(S T)\, y, \qquad \forall\, x, y \in X;\ \alpha, \beta \in \mathbb{K}$$

Further, since

$$(S T)(x_n) = S(Tx_n) \to S(0) = 0, \text{ as } n \to \infty$$

whenever $x_n \to 0$, it follows that ST is continuous and hence bounded. Thus, $ST \in B(X, X)$.

Finally, we have

$$\| (S\,T)\,(x)\,\|_X = \| S\,(Tx)\,\|_X \leq \| S\,\| \,\| T\,\| \,\| x\,\|_X, \quad \forall\, x \in X.$$

Hence

$$\| S\,T\,\| = \sup\,\{\| (S\,T)\,(x)\,\|_X : x \in X,\, \| x\,\|_X \leq 1\}$$

$$\leq \sup\,\{\| S\,\| \,\| T\,\| \,\| x\,\|_X : x \in X,\, \| x\,\|_X \leq 1\}$$

$$\leq \| S\,\| \,\| T\,\|. \;\blacksquare$$

Notation. We denote the normed space of all bounded linear operators on X by $B\,(X)$ instead of $B\,(X,\,X)$. $\blacksquare$

In view of Theorems 3.2.2 and 3.2.3, we immediately have the following result:

3.2.4 Theorem. *Let X be a normed space over the field $\mathbb{K}$ and let $B\,(X)$ be the set of all bounded linear operators on X. Then*

(a) *$B\,(X)$ is a normed space.*

(b) *$B\,(X)$ is complete if X is complete.*

(c) *If $S,\, T \in B\,(X)$, then $S\,T \in B\,(X)$ and*

$$\| S\,T\,\| \leq \| S\,\| \,\| T\,\|.$$

We conclude this section by first introducing the concept of Banach algebra and then proving that the Banach space $B\,(X)$ is, in fact, a Banach algebra.

3.2.5 Definition. Let A be a linear space over the field $\mathbb{K}$. We define the operation of multiplication in A in such a way that for every ordered pair of elements $x,\, y \in A$, $\exists$ a unique product $x,\, y \in A$. A is said to be an *algebra* over $\mathbb{K}$ if the following conditions are satisfied:

$(i)\ x\,(yz) = (x\,y)\,z$

$(ii)\ x\,(y + z) = xy + xz$

$(iii)\ (x + y)\,z = xz + yz$

$(iv)\ \alpha\,(xy) = (\alpha\,x)\,y = x\,(\alpha\,y)$

for all $x,\, y,\, z \in A$ and $\alpha \in \mathbb{K}$.

An algebra A is said to be *real* or *complex* according as $\mathbb{K}$ is $\mathbb{R}$ or $\mathbb{C}$. An algebra A is said to be *commutative* (or *abelian*) if the multiplication in A is commutative ; *i.e.*, $xy = yx$, $\forall\, x,\, y \in A$. An algebra A is called an *algebra with identity* if A contains an element e, called the identity element, such that

$$xe = x = ex, \quad \forall\, x \in A.$$

3.2.6 Definition. An algebra A is said to be a *normed algebra* if

(i) A is a normed space as a linear space equipped with a norm $\|\cdot\|$; and

(ii) $\|xy\| \le \|x\|\,\|y\|$, $\forall\, x\, y \in A$.

A normed algebra is said to be a *Banach algebra,* if it is complete as a metric space.

3.2.7 Examples

1. $\mathbb{R}$ and $\mathbb{C}$ are commutative Banach algebras with identity $e = 1$.

2. The space $C\,[a, b]$ is a commutative Banach algebra with identity $e = 1$, the product xy being defined as

$$(x\,y)\,(t) = x\,(t)\,y\,(t),\ \forall\, t \in [a, b].$$

3. The space $B\,(X)$ of all bounded linear operators on a complex Banach space $X \ne \{0\}$, is a complex Banach algebra. For $S,\, T \in B\,(X)$, we define

$$(S\,T)\,(x) = S\,(Tx),\ \forall\, x \in X.$$

Then, by Theorem 3.2.3, $S\,T \in B\,(X)$. Further, it can be readily verified that

(i) $R\,(S\,T) = (R\,S)\,T$

(ii) $R\,(S + T) = R\,S + R\,T$

(iii) $(R + S)\,T = R\,T + S\,T$

(iv) $\alpha\,(S\,T) = (\alpha\,S)\,T = S\,(\alpha\,T)$

for all $R,\, S,\, T \in B\,(X)$ and $\alpha \in \mathbb{C}$.

3.3 Equivalent Norms

We begin this section by determining certain special classes of bounded linear operators which motivate the concept of the equivalent norms on the same linear space and also will be quite useful later on in our subsequent studies.

Let X and Y be normed spaces over the field $\mathbb{K}$. Recall that a mapping (not necessarily linear) $T : X \to Y$ is said to be an *isometry* if it preserves norms, *i.e.,*

$$\| Tx \|_Y = \| x \|_X, \quad \forall\, x \in X.$$

3.3.1 Examples

1. The identity operator I on a normed space X is an isometry.

2. The linear operator $S : l^2 \to l^2$ given by

$$S\,(\xi_1, \xi_2, \xi_3, \ldots, \ldots) = (0, \xi_1, \xi_2, \ldots)$$

is an isometry on l^2. Indeed,

$$\| S\,(x) \|^2 = |\,0\,|^2 + |\,\xi_1\,|^2 + |\,\xi_2\,|^2 + \ldots$$

$$= \| x \|^2,\ x = \{\xi_i\} \in l^2.$$

The operator S is called the *unilateral shift operator.*

Note. It is easy to check that the unilateral shift operator does not map l^2 onto l^2.

3.3.2 Theorem.
Let X and Y be normed spaces over the same field $\mathbb{K}$ and $T : X \to Y$ be a linear operator. If T is an isometry, then T is bounded and $\| T \| = 1$.

Proof. Straightforward.

Remark. The converse of Theorem 3.3.2 need not be true.

3.3.3 Example.
Consider the linear operator

$$T : X \to \mathbb{K}, \quad X = (C\,[0, 1], \| \cdot \|_\infty)$$

given by

$$Tx = x\,(0).$$

Then

(a) T is bounded and $\| x \| = 1$. Indeed, we have

$$|\,Tx\,| = |\,x\,(0)\,| \le \sup\,\{|x\,(t)\,| : t \in [0, 1]\} = \| x \|_\infty.$$

Thus T is bounded and $\| T \| \le 1$. On the other hand, if $y : [0, 1] \to \mathbb{K}$ is given by $y\,(t) = t$, for all $t \in [0, 1]$, then $y \in [0, 1]$. Moreover

$$\| y \| = \sup\,\{|\,y\,(t)\,| : t \in [0, 1]\} = 1$$

and $\qquad |\,Ty\,| = |\,y\,(0)\,| = 1$

so that $\qquad 1 = |\,Ty\,| \le \| T \|\,\| y \| = \| T \|.$

(b) T is not an isometry. Indeed, consider the identity function $p\,(t) = t$, $t \in [0, 1]$. Clearly, $p \in C\,[0, 1]$ and $\|\,P\,\|_\infty = 1$ while $|\,Tp\,| = 0$

Remark. In view of Example 3.3.3, one may note that a linear operator being an isometry asserts more than that it has norm 1.

An isometry $T : X \to Y$ is said to be an isometric isomorphism if T is linear and bijective (Definition 2.5.1). Clearly, in this case, T is a homeomorphism between X and Y; in other words, T and T^{-1} both are continuous.

3.3.4 Definition. Two normed spaces X and Y are said to be *isometrically isomorphic* normed spaces if $\exists$ an isometric isomorphism between them.

To introduce more general concepts than isometric isomorphism, we recall that metric spaces $(X, d\,)$ and (Y, d') are homeomorphic (or topologically equivalent) if $\exists$ a mapping $f : X \to Y$ which is bijective and bicontinuous (Definition 1.7.40).

3.3.5 Definition. Let X and Y be normed spaces over the field $\mathbb{K}$. A mapping $T : X \to Y$ is said to be a *topological isomorphism* (or *linear homeomorphism*) if

(*i*) T is a linear operator, and

(*ii*) T is a homeomorphism.

Remark. The relation of "being topologically isomorphic to" is an equivalence relation on the set of all normed spaces over the field $\mathbb{K}$.

It can be easily seen that an isometric isomorphism is a special topological isomorphism. In fact, if two normed spaces are topologically isomorphic, the topological isomorphism sets up a one-to-one correspondence between the points of the spaces and is, at the same time, a linear space isomorphism so that the spaces have identical topological structure and linear structure without corresponding points necessarily having equal norms.

3.3.6 Theorem. *Let X and Y be normed spaces over the field $\mathbb{K}$ and let $T : X \xrightarrow{\;onto\;} Y$ be a linear operator. Then, T is a topological isomorphism if and only if $\exists$ constants $K_1, K_2 > 0$ such that*

$$K_1 \,\|\,x\,\|_X \leq \|\,Tx\,\|_Y \leq K_2 \,\|\,x\,\|_X, \qquad \forall\, x \in X.$$

Proof. Assume that $T : X \to Y$ is a topological isomorphism, Then, T^{-1} exists and is continuous on Y, and so T^{-1} is a bounded linear operator from Y onto X. Therefore, by Theorem 3.1.12, $\exists$ a constant $K_1 > 0$ such that

$$K_1 \| x \|_X \leq \| Tx \|_Y, \quad \forall\, x \in X.$$

But T being bounded, $\exists$ a constant $K_2 \,(= \| T \|) > 0$ such that

$$\| Tx \|_Y \leq K_2 \| x \|_X, \quad \forall\, x \in X.$$

Thus, we have

$$K_1 \| x \|_X \leq \| Tx \|_Y \leq K_2 \| x \|_X, \quad \forall\, x \in X.$$

Conversely, the inequality

$$\| Tx \|_Y \geq K_1 \| x \|_X, \quad \forall\, x \in X.$$

implies that T^{-1} exists and is a bounded linear operator (Theorem 3.1.12).

Similarly, the other inequality

$$\| Tx \|_Y \leq K_2 \| x \|_X, \quad \forall\, x \in X$$

verifies that T is a bounded linear operator. Hence, $T : X \to Y$ is a topological isomorphism. ∎

We know that a norm on a linear space X induces a topology, called the *norm topology,* on X. When different norms on the same linear space X induce the same norm topology on X, we say that they are equivalent norms on X. Alternatively, this concept may be defined by the following equivalent form.

3.3.7 Definition. Two norms $\| \cdot \|$ and $\| \cdot \|'$ on the same linear space X are said to be *equivalent norms* on X if the identity mapping $I_X : (X, \| \cdot \|) \to (X, \| \cdot \|')$ is a topological isomorphism of $(X, \| \cdot \|)$ onto $(X, \| \cdot \|')$.

3.3.8 Theorem. *Two norms $\| \cdot \|$ and $\| \cdot \|'$ on the same linear space X are equivalent if and only if $\exists$ positive constants K_1 and K_2 such that*

$$K_1 \| x \| \leq \| x \|' \leq K_2 \| x \|, \quad \forall\, x \in X.$$

Proof. In view of Definition 3.3.7 , we have

$\| \cdot \|$ and $\| \cdot \|'$ are equivalent norms on X

$\Leftrightarrow$ The identity mapping I_X is a topological isomorphism of $(X, \| \cdot \|)$ onto $(X, \| \cdot \|')$

$\Leftrightarrow \exists$ constants $K_1 > 0$ and $K_2 > 0$ such that

$\quad K_1 \| x \| \leq \| I_X x \|' \leq K_2 \| x \|, \quad \forall\, x \in X$

$\Leftrightarrow K_1 \| x \| \leq \| x \|' \leq K_2 \| x \|, \quad \forall\, x \in X.$

This completes the proof. ∎

Note. The relation 'norm equivalence ' is an equivalence relation among the norms on X.

3.3.9 Example. The norms $\| \cdot \|_1$, $\| \cdot \|_2$ and $\| \cdot \|_\infty$ are equivalent norms on $\mathbb{K}^n$ ($\mathbb{K} = \mathbb{R}$ or $\mathbb{C}$) (Examples 2.2.2 and 2.2.3).

Let $x = (\xi_1, \xi_2, ..., \xi_n) \in \mathbb{K}^n$. Then

$$\| x \|_\infty \leq \sum_{i=1}^{n} | \xi_i | \leq n \max_{1 \leq i \leq n} | \xi_i |.$$

This gives

(4) $$\| x \|_\infty \leq \| x \|_1 \leq n \| x \|_\infty.$$

Thus, $\| \cdot \|_1$ and $\| \cdot \|_\infty$ are equivalent norms on $\mathbb{K}^n$.

Further

$$\| x \|_2 \leq \sum_{i=1}^{n} | \xi_i | \leq \sqrt{n} \left(\sum_{i=1}^{n} | \xi_i |^2 \right)^{\frac{1}{2}}.$$

(Cauchy-Schwartz Inequality (1.5.5))

This gives

(5) $$\| x \|_2 \leq \| x \|_1 \leq \sqrt{n} \, \| x \|_2.$$

Hence, $\| \cdot \|_2$ and $\| \cdot \|_1$ are equivalent norms on $\mathbb{K}^n$ and, in view of the fact that equivalence of norms is an equivalence relation, it follows from the above facts that $\| \cdot \|_1$, $\| \cdot \|_2$ and $\| \cdot \|_\infty$ are equivalent norms on $\mathbb{K}^n$.

Remark. The expressions (4) and (5) in the above example give the following relation between open spheres in case of $\mathbb{R}^2$:

$$\frac{1}{2} S_{\| \cdot \|_\infty} (0 ; 1) \subset S_{\| \cdot \|_1} (0 ; 1) \subset S_{\| \cdot \|_\infty} (0 ; 1)$$

(Fig. 3.1)

and

$$\frac{1}{\sqrt{2}} S_{\| \cdot \|_2} (0 ; 1) \subset S_{\| \cdot \|_1} (0 ; 1) \subset S_{\| \cdot \|_2} (0 ; 1).$$

(Fig. 3.2)

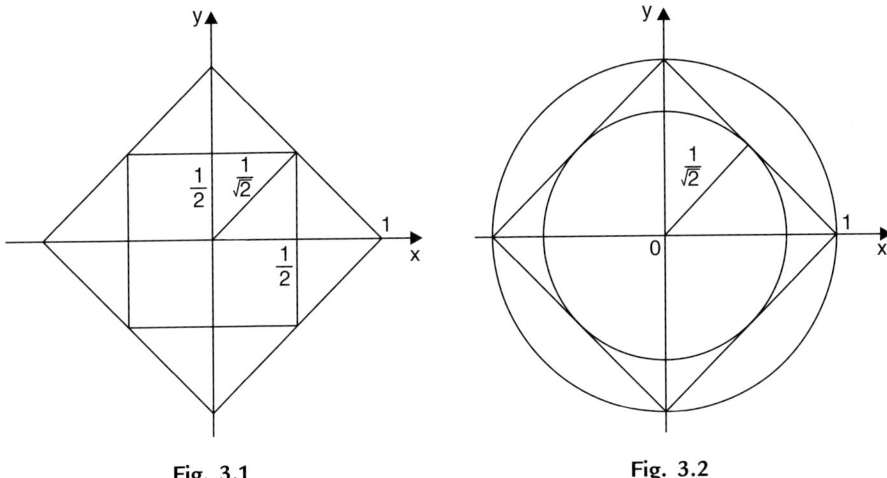

Fig. 3.1   Fig. 3.2

3.3.10 Definition. A normed space property which holds for a linear space under equivalent norms is said to be a *linear topological invariant property* for normed spaces.

Completeness and boundedness of sets are linear topological invariant properties for normed spaces.

3.3.11 Theorem. *Let $\| \cdot \|$ and $\| \cdot \|'$ be equivalent norms on a linear space X. Then:*

(a) *$(X, \| \cdot \|)$ is a Banach space if and only if $(X, \| \cdot \|')$ is so.*

(b) *A set is bounded in $(X, \| \cdot \|)$ if and only if it is bounded in $(X, \| \cdot \|')$.*

Proof. (a) Suppose $(X, \| \cdot \|)$ is a Banach space. Let $\{x_n\}$ be a Cauchy sequence in $(X, \| \cdot \|')$. Then, for each $\varepsilon > 0$, $\exists$ an integer $N > 0$ such that

$$\| x_n - x_m \|' < \varepsilon, \ \forall \, n, m \geq N$$

Since $\| \cdot \|$ and $\| \cdot \|'$ are equivalent norms on X, $\exists$ a constant $K_1 > 0$ such that

$$K_1 \| x_n - x_m \| \leq \| x_n - x_m \|' \qquad \text{(Theorem 3.3.5)}$$

$$< \varepsilon, \qquad \forall \, n, m \geq N$$

$\Rightarrow$ $\{x_n\}$ is a Cauchy sequence in $(X, \| \cdot \|)$.

But $(X, \| \cdot \|)$ being complete, $\{x_n\}$ converges in $(X, \| \cdot \|)$. Let $\| x_n - x \| \to 0$, as $n \to \infty$. Again, in view of the equivalence of the norms $\| \cdot \|$ and $\| \cdot \|'$ on X, for some $K_2 > 0$, we have

$$\| x_n - x \|' \leq K_2 \| x_n - x \|.$$

Consequently,

$$\| x_n - x \|' \to 0, \text{ as } n \to \infty.$$

Hence, $(X, \| \cdot \|')$ is a Banach space.

The converse follows similarly by interchanging the roles of $\| \cdot \|$ and $\| \cdot \|'$.

(b) Let $A \subset X$ be a bounded set in $(X, \| \cdot \|)$. Then, $\exists$ a constant $M > 0$ such that

$$\| x \| \leq M, \quad \forall \, x \in A.$$

Therefore, by Theorem 3.3.5, we have

$$\| x \|' \leq K_2 M, \quad \forall \, x \in A.$$

Hence, A is bounded in $(X, \| \cdot \|')$.

The converse follows similarly by applying the other inequality of Theorem 3.3.8. ∎

Problems

7. Let X and Y be normed spaces over the field $\mathbb{K}$ and let $T : X \to Y$ be a topological isomorphism. Define, for each $x \in X$

$$\| x \| = \| Tx \|_Y$$

and for each $y \in Y$

$$\| y \|' = \| T^{-1} y \|_X.$$

Prove that:

(a) $\| \cdot \|$ defines a norm, equivalent to $\| \cdot \|_X$, on X.

(b) $\| \cdot \|'$ defines a norm, equivalent to $\| \cdot \|_Y$, on Y.

8. Let X and Y be normed spaces over the field $\mathbb{K}$ and let $T : X \to Y$ be a topological isomorphism of X onto Y. If $\| \cdot \|_X'$ and $\| \cdot \|_Y'$ are norms on X and Y, respectively, that are equivalent to $\| \cdot \|_X$ and $\| \cdot \|_Y$, then show that T is a topological isomorphism of $(X, \| \cdot \|_X')$ onto $(X, \| \cdot \|_Y')$.

9. Let $\| \cdot \|$ and $\| \cdot \|'$ be equivalent norms on a linear space X. Show that:

(a) The Cauchy sequences in $(X, \| \cdot \|)$ and $(X, \| \cdot \|')$ are the same.

(b) The convergent sequences in $(X, \| \cdot \|)$ and $(X, \| \cdot \|')$ are the same.

10. Let there be two norms on the same linear space which are not equivalent. Then:

(a) Show that at least one of them is discontinuous on the unit sphere of the other.

   (b) Can each norm be discontinuous when restricted to the unit sphere of the other?

11. Given two norms on the same linear space such that one is complete and the other is incomplete. Can these two norms be equivalent? Justify your answer.

12. Let $(X, \| \cdot \|')$ be a normed space and $Y \subset X$ be a subspace as a linear space. Suppose $\| \cdot \|'$ is a norm on Y which is equivalent to $\| \cdot \|_Y$. Show that there is a norm $\| \cdot \|_1$ on X which is equivalent to $\| \cdot \|$ such that $\| \cdot \|_Y$ is precisely $\| \cdot \|'$.

3.4 Finite Dimensional Normed Spaces and Compactness

Finite dimensional normed spaces form a special and priviledged class of normed spaces. These spaces are very useful, particularly, in the study of spectral theory, approximation theory, etc. We first investigate the properties of these spaces and then obtain a characterization of such spaces in terms of compactness.

Recall that a nonempty subset of the usual metric space $\mathbb{R}_u$ (and hence of the usual normed space) is compact if and only if it is closed and bounded (Theorem 1.7.46). The next theorem extends this result to $(\mathbb{K}^n, \| \cdot \|_2)$. (Note that $\mathbb{K}^n$ is a finite dimensional normed space).

3.4.1 Theorem (Borel-Lebesgue). *A nonempty subset of the normed space* $(\mathbb{K}^n, \| \cdot \|_2)$ *is compact if and only if it is closed and bounded.*

Proof. Let A be a closed and bounded subset of $(\mathbb{K}^n, \| \cdot \|_2)$. One may observe that the metric spaces $(\mathbb{C}^n, \| \cdot \|_2)$ and $(\mathbb{R}^{2n}, \| \cdot \|_2)$ are isometric. Therefore, we may suppose that $\mathbb{K} = \mathbb{R}$. By Example 1.7.13(1) and Corollary 1.7.18, the subspace A is complete. Therefore, by Theorem 1.7.51, it is sufficient to prove that A is totally bounded. Since A is bounded, $\exists$ a positive integer N such that

$$A \subset A_N = \{(\xi_1, \xi_2, ..., \xi_n) \in R^n : | \xi_j | \leq N \text{ for } j = 1, 2, ..., n\}.$$

If we can prove that A_N is totally bounded it will follow that A is also totally bounded since a non-empty subset of a totally bounded set is totally bounded.

Let $\varepsilon > 0$ and let m be the least positive with $m\, \varepsilon \geq \sqrt{n}$, and let

$$E_m = \left\{ \left(\frac{p_1}{m}, \frac{p_2}{m}, ..., \frac{p_n}{m} \right) : p_j \in \mathbb{Z}, | p_j | \leq m\, N \text{ for } 1 \leq j \leq n \right\}.$$

There are exactly $(1 + 2\, m\, N)^n$ distinct points in the set E_m. Let $(\xi_1, \xi_2, ..., \xi_n) \in A_N$. Then, for $j = 1, 2, ..., n$, we have $|\xi_j| \leq N$ and hence we can find an integer p_j with $|p_j| \leq mN$ and $|m\,\xi_j - p_j| < 1$. Since

$$\| (\xi_1, \xi_2, ..., \xi_n) - \left(\frac{p_1}{m}, \frac{p_2}{m}, ..., \frac{p_n}{m} \right) \| = \left(\sum_{j=1}^{n} \left(\xi_j - \frac{p_j}{m} \right)^2 \right)^{\frac{1}{2}} < \frac{\sqrt{n}}{m} \leq \varepsilon ,$$

E_m is a finite ε-net in A_N. This verifies our assertion.

The converse follows from Theorem 1.7.45. ∎

The following theorem relates the topological structures for n-dimensional normed spaces over the same scalar field.

3.4.2 Theorem. *Any two n-dimensional normed spaces over the same scalar field are topologically isomorphic.*

Proof. Let X and Y be any two n-dimensional normed spaces over the same scalar field $\mathbb{K}$. The case $n = 0$ is trivial, and so we assume $n \geq 1$. Since the relation of "topological isomorphism" is an equivalence relation, it will suffice to prove that X is topologically isomorphic to $(\mathbb{K}^n, \| \cdot \|_2)$, where the norm $\| \cdot \|_2$ is given by

$$\| x \|_2 = \left(\sum_{i=1}^{n} |\xi_i|^2 \right)^{\frac{1}{2}} , \, x = (\xi_1, \xi_2, ..., \xi_n) \in \mathbb{K}^n .$$

Let $\{e_1, e_2, ..., e_n\}$ be a basis for X. Define the mapping $T : \mathbb{K}^n \to X$ by

$$T (\xi_1, \xi_2, ..., \xi_n) = \sum_{i=1}^{n} \xi_i e_i .$$

Then, it is easy to check that T is a bijective linear operator. Therefore, T is an isomorphism of $\mathbb{K}^n$ onto X. Now, it remains to show that T and T^{-1} are continuous.

Note that

$$\| Tx \|_X = \| \sum_{i=1}^{n} \xi_i e_i \|_X$$

$$= \sum_{i=1}^{n} |\xi_i| \, \| e_i \|_X$$

$$\leq \left(\sum_{i=1}^{n} |\xi_i|^2 \right)^{\frac{1}{2}} \sum_{i=1}^{n} \| e_i \|_X$$

$$\Rightarrow \qquad \| Tx \|_X \leq K \| x \|_2, \quad \text{where } K = \sum_{i=1}^{n} \| e_i \|_X > 0.$$

This is true for all $x \in X$. Thus, T is bounded and hence continuous on $(\mathbb{K}^n, \| \cdot \|_2)$.

Further, in order to show that T^{-1} is continuous on X, define the mapping $f : \mathbb{K}^n \to \mathbb{K}$ by

$$f(x) = \| Tx \|_X.$$

Then, $f = \| \cdot \|_X \circ T$. Since T and $\| \cdot \|_X$ are continuous mappings, it follows that $f : \mathbb{K}^n \to \mathbb{K}$ is continuous. Now consider the unit sphere $S = \{ x \in \mathbb{K}^n : \| x \|_2 \leq 1 \}$ in $(\mathbb{K}^n, \| \cdot \|_2)$. Clearly, S is a closed and bounded subset of $\mathbb{K}^n$. Therefore, by the Borel Lebesgue Theorem (Theorem 3.4.1), S is a compact subset of $\mathbb{K}^n$. Since f is continuous, it follows that $f(S)$ is a compact subset of $\mathbb{K}$ (Theorem 1.7.47). Let

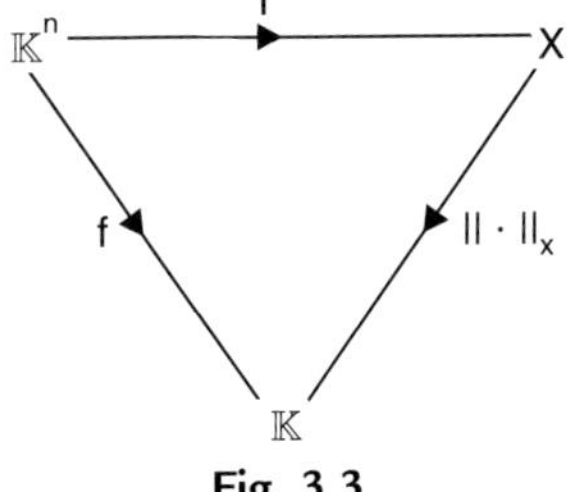

Fig. 3.3

$$m = \inf \{ f(x) : x \in S \}.$$

Then

$$m = \inf \{ \| Tx \|_X : x \in S \}$$

$$\Rightarrow \qquad 0 \leq m \leq \| Tx \|_X, \quad \forall \, x \in S.$$

Also, $\exists \, x_0 \in S$ such that $m = f(x_0)$. If $m = 0$, then

$$\| Tx_0 \|_X = f(x_0) = 0$$

$$\Rightarrow \quad Tx_0 = 0.$$

This contradicts the fact that T is one-to-one and $\| x_0 \| = 1$. Therefore

$$0 < m \leq \| Tx \|_X, \qquad \forall \, x \in S$$

$$\Rightarrow \qquad \| Tx \|_X \geq m \| x \|_2, \ \forall \ x \in \mathbb{K}^n$$

$$(\because \ x \neq 0 \quad \Rightarrow \quad \| (x/\| x \|_2) \|_2 = 1)$$

$\Rightarrow \quad T^{-1}$ is bounded (and hence continuous) in view of Theorem 3.1.13.

We thus conclude that T is a topological isomorphism between X and $(\mathbb{K}^n, \| \cdot \|_2)$. ∎

Recall that $\| \cdot \|_1$, $\| \cdot \|_2$ and $\| \cdot \|_\infty$ are equivalent norms on linear space $\mathbb{K}^n$ (Example 3.3.9). The following is a more general result.

3.4.3 Corollary. *All norms on a finite dimensional linear space are equivalent.*

Proof. Let X be an n-dimensional linear space over the field $\mathbb{K}$ and let $\| \cdot \|$ and $\| \cdot \|'$ be any two norms on X. Then, $(X, \| \cdot \|)$ and $(X, \| \cdot \|')$ are topologically isomorphic. So $\| \cdot \|$ and $\| \cdot \|'$ induce the same norm topology on X. Hence $\| \cdot \|$ and $\| \cdot \|'$ are equivalent norms on X. ∎

Remark. The convergence or divergence of a sequence (or a series) in a finite dimensional normed space does not depend on the particular choice of a norm on that space.

3.4.4 Corollary. *A finite dimensional normed space is a Banach space.*

Proof. Let X be an n-dimensional normed space over the field $\mathbb{K}$. Then, X is topologically isomorphic to $(\mathbb{K}^n, \| \cdot \|_2)$ (Theorem 3.4.2). But $(\mathbb{K}^n, \| \cdot \|_2)$ is complete (Examples 2.2.1 and 2.2.2). Also, completeness is a linear topological invariant property for normed spaces (Theorem 3.3.11). Thus, it follows that X is complete. ∎

3.4.5 Corollary. *A finite dimensional subspace Y of a normed space X is a closed subset of X.*

Proof. Since Y is a finite dimensional subspace of X, Y itself is a Banach space (Corollary 3.4.4). Hence Y is closed (Theorem 2.3.6). ∎

Remark. An infinite dimensional subspace of a normed space need not be closed.

3.4.6 Example. Take $X = C[0, 1]$ (Example 2.2.7 with $a = 0$, $b = 1$) and $Y =$ span $\{x_0, x_1, x_2, ...\}$ with $x_n (t) = t^n$ $(n = 0, 1, 2, ...)$. Clearly, Y is the set of all polynomials and there are functions in $\overline{Y}$ which are not in Y. For instance, given a non-polynomial function $x \in C[0, 1]$, $\exists$ a sequence $\{p_n\} \subset Y$ converging uniformly to x. Thus, $x \in \overline{Y}$ and $x \notin Y$. Hence, Y is not closed. ∎

3.4.7 Corollary. *In a finite dimensional normed space, every non-empty closed bounded set is compact.*

Proof. Let X be an n-dimensional normed space over the field $\mathbb{K}$ and let A be a non-empty closed bounded subset of X. Then, X and $(\mathbb{K}^n, \| \cdot \|_2)$ are topologically isomorphic spaces (Theorem 3.4.2). Let T be one such topological isomorphism. Since closedness is a topological invariant property for metric spaces and boundedness is a linear topological invariant property for normed spaces, it follows that $T(A)$ is a closed bounded subset of $(\mathbb{K}^n, \| \cdot \|_2)$. Therefore, by Theorem 3.4.1, $T(A)$ is a compact subset of $(\mathbb{K}^n, \| \cdot \|_2)$. Hence in view of $A = T^{-1}(T(A))$ and T^{-1} is continuous, it follows that A is compact. ∎

It is natural to think whether the result in Corollary 3.4.7 is true for every normed space (finite or infinite dimensional). In general, it is not true.

3.4.8 Example. Consider the space $X = (C[0, 1], \| \cdot \|_\infty)$. The unit sphere $S[0; 1]$ is closed and bounded, but it is not compact. For instance, consider the sequence of functions defined by

$$x_n(t) = t^n.$$

Then $x_n \in S[0; 1]$, for all $n \in \mathbb{N}$. Since the convergence in $C[0, 1]$ is the uniform convergence (Example 2.2.7), the sequence $\{x_n\}$ does not have a convergent subsequence.

We next show that a normed space with the property that "all closed bounded subsets are compact" must necessarily be finite dimensional. In fact, this will lead us to one of the fundamental differences between finite dimensional and infinite dimensional normed spaces. We first state and prove a lemma.

3.4.9 Lemma (Riesz). *Let X be a normed space over the field $\mathbb{K}$, Y a proper subspace of X. Then:*

(a) *If Y is closed, then for a given $\varepsilon > 0$, $\exists$ a point $x_0 \in X$ such that $\| x_0 \| = 1$ and*

$$d(x_0, Y) = \inf\{ \| y - x_0 \| : y \in Y \} \geq 1 - \varepsilon.$$

(b) *If Y is finite dimensional, then $\exists$ a point $x_0 \in X$ such that $\| x_0 \| = 1$ and $d(x_0, Y) = 1$.*

Proof. (a) Since Y is a proper subspace of X, $\exists x \in X - Y$. Let

$$d = \inf\{ \| x - y \| : y \in Y \} = d(x, Y).$$

Then, $d > 0$ since Y is closed and $x \notin Y$. Thus, given $\eta > 0$, $\exists\, y_0 \in Y$ with $y_0 \neq x$ such that

$$\| x - y_0 \| < d + \eta.$$

Setting

$$x_0 = \frac{x - y_0}{\| x - y_0 \|},$$

it is clear that $x_0 \in X$ and $\| x_0 \| = 1$. Also

$$\| y - x_0 \| = \left\| y - \frac{x - y_0}{\| x - y_0 \|} \right\|$$

$$= \frac{1}{\| x - y_0 \|} \| (y \| x - y_0 \| + y_0) - x \|$$

$$= \frac{1}{\| x - y_0 \|} \| y_1 - x \|, \quad \forall\, y \in Y$$

where $y_1 = y \| x - y_0 \| + y_0 \in Y$. Hence

$$\| y - x_0 \| \geq \frac{d}{d + \eta} = 1 - \varepsilon, \quad \varepsilon = \frac{\eta}{d + \eta}.$$

(b) Since Y is finite dimensional, infimum is attained. Then, $\exists\, y_0 \in Y$ such that $\| x - y_0 \| = d$. Following arguments as in (a), the result in (b) is established.∎

3.4.10 Theorem (Riesz). *A normed space X is finite dimensional if and only if the closed unit sphere in X is compact.*

Proof. Let X be a finite dimensional normed space over the field $\mathbb{K}$. Then, by Corollary 3.4.7 the closed unit sphere in X is compact.

Conversely, suppose that the closed unit sphere

$$S\,[0\,;1] = \{x \in X : \| x \| \leq 1\}$$

in the normed space X is compact. Let, if possible, X be not finite dimensional. Then, $\exists$ a (infinite) sequence $\{x_n\}$ of linearly independent vectors in X. For each positive integer k, let M_k denote the finite dimensional subspace spanned by $\{x_1, x_2, ..., x_k\}$. Then, for each k, M_k is a closed subspace of M_{k+1}. By Riesz Lemma (Lemma 3.4.9), $\exists$ a point $y_k \in M_{k+1}$ such that

$$\| y_k \| = 1 \text{ and } \| x - y_k \| \geq \frac{1}{2}, \quad x \in M_k.$$

This holds for each positive integer k. Therefore, the sequence $\{y_k\} \subset S\,[0;1]$ and the set $S\,[0;1]$ being compact, $\{y_k\}$ has a convergent subsequence (Theorem 1.7.49). On the other hand, note that

$$\| y_k - y_{k+1} \| \geq \frac{1}{2}, \quad \forall\, k.$$

As such, $\{y_k\}$ has no convergent subsequence. This contradicts the hypothesis that X is infinite dimensional. Hence, we conclude that X is finite dimensional. ∎

3.4.11 Theorem. *A linear operator on a finite dimensional normed space is continuous.*

Proof. Let X be an n-dimensional normed space over the scalar field $\mathbb{K}$ and let $\{e_1, e_2, ..., e_n\}$ be a basis for X. Suppose $T : X \to Y$ be a linear operator, where Y is any normed space over $\mathbb{K}$. Then, any $x \in X$ can be uniquely represented by

$$x = \sum_{i=1}^{n} \lambda_i e_i \,,$$

where $\lambda_1, \lambda_2, ..., \lambda_n$ are in $\mathbb{K}$ and n is some positive integer. Now

$$\| T x \|_Y = \| \sum_{i=1}^{n} \lambda_i \, T\, e_i \|_Y$$

$$\leq \sum_{i=1}^{n} |\lambda_i| \, \| T\, e_i \|_Y$$

$$\leq K \max_{1 \leq i \leq n} |\lambda_i|, \quad K = \sum_{i=1}^{n} \| T\, e_i \|_Y$$

$$\Rightarrow \qquad \| Tx \|_Y \leq K \, \| \{\lambda_i\}_1^n \|_\infty.$$

But X and $(\mathbb{K}^n, \|\cdot\|_\infty)$ are topologically isomorphic normed spaces (Theorem 3.4.2) and so $\exists\, A > 0$ such that

$$\|\,\{\lambda_i\}_1^n\,\|_\infty \le A\,\|\,x\,\|_X.$$

Hence

$$\|\,Tx\,\|_Y \le KA\,\|\,x\,\|_X, \quad \forall\, x \in X$$

which verifies that T is bounded (and hence continuous). $\blacksquare$

Remark. If, instead of the domain, the range of a linear operator is finite dimensional, then the operator need not be bounded.

3.4.12 Example. Consider the linear operator $T : P\,[0,\,1] \to \mathbb{C}$ defined by

$$T\,(x) = x'\,(1),$$

where x' is the derivative of x and $P\,[0,\,1]$ is the normed space as given in Example 2.2.9. The operator T is not bounded where as the range $\mathbb{C}$ is finite dimensional.

Remark. The result in Theorem 3.4.11 is not necessarily true for the case of infinite dimensional normed spaces. In fact, we have the following.

3.4.13 Theorem. *Let X be a normed space with dim $X = \infty$. Then, $\exists$ an unbounded linear operator $T : X \to Y$, where $Y \ne \{0\}$ is any normed space.*

Problems

13. Let M be a finite dimensional proper subspace of a normed space X. Show that:

    (a) $\exists\, a\; y \in X$ such that $d\,(y,\, M) = 1$.

    (b) For $x \in X - M$, $\exists\, y \in M$ such that $\|\,x - y\,\| = d\,(x,\, M)$.

    (c) For $x + M \in X/M$, $\exists\, y \in x + M$ such that

$$\|\,y\,\| = \|\,x + M\,\|.$$

14. Let $\{X_i\}$ be a (strictly) decreasing sequence of finite dimensional subspaces of a normed space (*i.e.*, $X_{i+1} \subset X_i$, the inclusions being proper). Prove that there are unit vectors $x_n \in X_n$ such that $d\,(x_n,\, x_{n-1}) = 1,\; \forall\, n \ge 2$.

15. Prove Theorem 3.4.11.

16. Prove that a Banach space cannot have a countably infinite Hamel basis.

Hence or otherwise prove that the space $\Phi = [\{\alpha_i\} \subset \mathbb{K} : \alpha_i \neq 0$ for at most finitely many i's] [Example 2.3.9 (i)] equipped with the norm given by

$$\| \{\alpha_i\} \|_\infty = \sup_{1 \leq i < \infty} |\alpha_i|$$

is an incomplete normed space.

17. Let X and Y be n-dimensional normed spaces. Show that $\exists$ a $T_0 \in B(X, Y)$ such that $T_0^{-1} \in B(Y, X)$ and that

$$d(X, Y) = \inf \{\| T \| \| T^{-1} \| : T \in B(X, Y)\} = \| T_0 \| \| T_0^{-1} \|$$

[$d(X, Y)$ is called *Banach-Mazur distance* between X and Y.]

18. Let X be a normed space. Show that the following statements are equivalent:

(a) X is finite dimensional.

(b) Every closed and bounded subset of X is compact.

(c) The closed unit sphere $S[0; 1]$ in X is compact.

(d) Every bounded subset of X is relatively compact.

(e) $\exists$ in X a relatively compact neighbourhood of zero.

[A set G in a normed space X is said to be a *neighbourhood of zero* if it contains an open sphere centred at zero.]

3.5 Open Mapping Theorem and its Consequences

Open mapping theorem is one of the basic theorems for the development of the general theory of normed spaces. This theorem gives conditions under which a linear operator is open. In this theorem, we begin to appreciate the importance of the completeness condition for normed spaces.

A mapping of a metric space into another metric space is said to be an *open map* if it maps open sets to open sets. Neither all continuous mappings are open nor all open mappings are continuous.

3.5.1 Example. Consider the metric spaces $(\mathbb{R}, u)$ and $(\mathbb{R}, d)$, where u and d are, respectively, the usual and the discrete metrics on $\mathbb{R}$ [Examples 1.7.2 (1) and 1.7.2 (2)], and also consider the identity map $I : \mathbb{R} \to \mathbb{R}$. Then $I : (\mathbb{R}, d) \to (\mathbb{R}, u)$ is continuous but not open. On the other hand, $I : (\mathbb{R}, u) \to (\mathbb{R}, d)$ is open but not continuous (Verify !)

3.5.2 Theorem (Open Mapping). *Let X and Y be Banach spaces over the field $\mathbb{K}$ and $T : X \xrightarrow{onto} Y$ be a bounded linear operator. Then, T is an open mapping.*

We first state and attempt to prove two lemmas which are needed in the proof of Theorem 3.5.2.

3.5.3 Lemma. *Let X, Y and T be as in Theorem 3.5.2. Let $B_n \equiv S_X (0; 2^{-n})$ and $B_n' \equiv S_Y \left(0; \dfrac{\varepsilon}{2^n} \right)$, $\varepsilon > 0$, be open spheres, respectively, in X and Y. Then*

$$\overline{T (B_n)} \supset B'_n, \quad (n = 0, 1, 2, \dots).$$

Proof. Consider the open sphere $B_1 \equiv S_X \left(0; \dfrac{1}{2} \right)$ in X. Now, if $x \in X$, we may write $x = m \left(\dfrac{x}{m} \right)$, where $m = [2 \parallel x \parallel_X] + 1$ and the square bracket denotes the integral part of $2 \parallel x \parallel_X$. Then, $\dfrac{x}{m} \in B_1$ and so $x \in m B_1$. Hence

$$X = \bigcup_{m = 1}^{\infty} m B_1.$$

Since $T (X) = Y$ and T is linear and bounded, we have

$$Y = \bigcup_{m = 1}^{\infty} m T (B_1).$$

The space Y being complete, by Baire Category Theorem (Theorem 1.7.32), $\exists$ at least one positive integer m such that $m T (B_1)$ is somewhere dense. Hence, $\overline{m T (B_1)}$ (and, therefore, $\overline{T (B_1)}$) must contain a non-empty open sphere in Y. Consequently, there is a point $y_0 \in Y$ and a number $\varepsilon > 0$ such that

$$\{ y \in Y : \parallel y - y_0 \parallel < \varepsilon \} \subset \overline{T (B_1)} .$$

Therefore, it follows that

$$(6) \qquad\qquad B_0' \subset \overline{T (B_1)} - y_0.$$

We next prove that

(7) $$\overline{T(B_1)} - y_0 \subset T(B_0),$$

where B_0 is the open unit sphere in X.

Let $y \in \overline{T(B_1)} - y_0$. Then, there is a sequence $\{u_n\}$, where $u_n = T(u_n') \in T(B_1)$, such that

$$\lim_{n \to \infty} u_n = y + y_0.$$

Again, $y_0 \in S_Y(y_0 ; \varepsilon) \subset T(B_1)$ gives that $y \in T(B_1)$. Therefore, $\exists$ a sequence $\{v_n\}$, where $v_n = T(v_n') \in T(B_1)$, such that $\lim_{n \to \infty} v_n = y_0$. Thus, in view of the fact that u_n', $v_n' \in B_1$, it follows that

$$\| u_n' - v_n' \|_X \leq \| u_n' \|_X + \| v_n' \|_X < \frac{1}{2} + \frac{1}{2} = 1$$

$$\Rightarrow \qquad u_n' - v_n' \in B_0.$$

Now, since

$$\lim_{n \to \infty} T(u_n' - v_n') = \lim_{n \to \infty} (Tu_n' - Tv_n') \qquad (\because T \text{ is linear})$$

$$= \lim_{n \to \infty} u_n - \lim_{n \to \infty} v_n = y,$$

it follows that $y \in \overline{T(B_0)}$. Thus, we have shown that

$$y \in \overline{T(B_1)} - y_0 \quad \Rightarrow \quad y \in \overline{T(B_0)}.$$

This proves (7).

Combining (6) and (7), we get

(8) $$B_0' \subset \overline{T(B_0)}.$$

Noting that $B_n \equiv S_X(0 ; 2^{-n}) = 2^{-n} B_0$ and that T is linear, we have

$$\overline{T(B_n)} = 2^{-n} \overline{T(B_0)}.$$

Therefore, using (8), we obtain

$$B_n' = S_Y\left(0\,;\frac{\varepsilon}{2^n}\right) = 2^{-n}\,B_0' \subset 2^{-n}\,T\,(B_0) = T\,(B_n).$$

This completes the proof of the lemma. ∎

3.5.4 Lemma. *Let X, Y, T, B_0 and B_1' be as in Lemma 3.5.3. Then*

$$T\,(B_0) \supset B_1'$$

for some $\varepsilon > 0$.

Proof. Let $y \in B_1'$. Then, from Lemma *3.5.3 with $n = 1$*, we have

$$y \in \overline{T\,(B_1)}$$

$$\Rightarrow\quad \exists\,x_1 \in B_1 \text{ such that } y_1 = Tx_1 \text{ and } \| y - y_1 \|_Y < \frac{\varepsilon}{2^2}$$

$$\Rightarrow\quad y - y_1 \in B_2'$$

$$\Rightarrow\quad y - y_1 \in \overline{T\,(B_2)}\quad \text{(Lemma 3.5.3 with } n = 2)$$

$$\Rightarrow\quad \exists\,x_2 \in B_2 \text{ such that } y_2 = Tx_2 \text{ and } \| (y - y_1) - y_2 \|_Y < \frac{\varepsilon}{2^3}$$

$$\Rightarrow\quad y - y_1 - y_2 \in B_3' \subset \overline{T\,(B_3)}\,.$$

By induction, we obtain a sequence $\{x_n\} \subset X$ such that $x_n \in B_n$, $y_n = Tx_n$ and

$$(9)\qquad\qquad \| y - \sum_{k=1}^{n} y_k \|_Y < \frac{\varepsilon}{2^{n+1}}\,,\ (n = 1, 2,...).$$

Write $s_n = \sum_{k=1}^{n} x_k$. Then, for $n \geq m$, we have

$$\| s_n - s_m \|_X = \|x_{m+1} + x_{m+2} + ... + x_n \|_X$$

$$\leq \sum_{k=m+1}^{n} \| x_k \|_X.$$

Since $\| x_n \|_X < \dfrac{1}{2^n}$ $\forall n \in \mathbb{N}$, it follows that

$$\| s_n - s_m \|_X \to 0, \text{ as } m \to \infty$$

and hence $\{s_n\}$ is a Cauchy sequence in X. But X being complete, $\exists\, x \in X$ such that $s_n \to x$ in X, and

$$\| x \|_X = \sum_{k=1}^{\infty} \| x_k \|_X < \sum_{k=1}^{\infty} \frac{1}{2^k} = 1.$$

This verifies that $x \in B_0$. Since T is continuous, we have

$$Tx = T\left(\lim_{n \to \infty} s_n \right) = \lim_{n \to \infty} T\, s_n = \lim_{n \to \infty} \left(\sum_{k=1}^{n} y_k \right) = y.$$

Thus, $y \in T(B_0)$. Hence, the result follows. $\blacksquare$

Proof of Theorem 3.5.2. Let $G \subset X$ be an open set. If $G = \phi$, then $T(G) = T(\phi) = \phi$ and thus it follows that $T(G)$ is an open set in Y. Now, suppose that $G \neq \phi$. Since G is open, for every $x_0 \in G$, $\exists\, \alpha > 0$ such that

$$S_X(x_0\,;\,\alpha) \subset G$$

$$\Rightarrow \qquad x_0 + \alpha\, B_0 \subset G$$

$$\Rightarrow \qquad K(G - x_0) \supset B_0, \qquad \text{where } K = \frac{1}{\alpha}$$

$$\Rightarrow \quad K(T(G) - Tx_0) \supset T(B_0) \qquad (\because\ \ T \text{ is linear})$$

Using Lemma 3.5.4, we obtain

$$K(T(G) - Tx_0) \supset B_1'$$

$$\Rightarrow \quad T(G) \supset Tx_0 + \frac{1}{K}\, B_1' = S_Y\left(Tx_0\,;\,\frac{\varepsilon}{2K}\right)$$

$$\Rightarrow \quad T(G) \supset \{ y \in Y : \| y - Tx_0 \|_Y < \frac{\varepsilon}{K} \}$$

$$\Rightarrow \quad T(G) \text{ is open.}$$

This completes the proof of the theorem. $\blacksquare$

Remark. The condition of completeness in Theorem 3.5.2 is essential.

3.5.5 Example. Consider the Banach space $X = (C\,[0, 1], \|\cdot\|_\infty)$, Example 2.2.7 with $a = 0$, $b = 1$, (incomplete) normed space $Y = (C\,[0, 1], \|\cdot\|_1)$, Problem 2.9 with $a = 0$, $b = 1$ and $p = 1$, and identity operator $I : X \to Y$. Note that I is continuous (hence bounded) since

$$\|\,Ix\,\|_1 = \|\,x\,\|_1 \le \|\,x\,\|_\infty, \quad \forall\, x \in X.$$

Also, the operator I being bijective, we can consider its inverse mapping

$$I^{-1} : Y \to X,$$

which is again linear. We, now, claim that I^{-1} is not continuous. Let $\{x_n\}$ be a sequence in $C\,[0, 1]$, where

$$x_n(t) = \begin{cases} 1 - nt, & 0 \le t \le \dfrac{1}{n} \\[2mm] 0, & \dfrac{1}{n} < t \le 1. \end{cases}$$

Note that $x_n \to 0$, as $n \to \infty$, in Y. But

$$\|\,I^{-1}(x_n)\,\|_\infty = \|\,x_n\,\|_\infty$$

$$= \sup_{0 \le t \le 1} |\,x_n(t)\,| = 1$$

$$\Rightarrow \qquad\qquad I^{-1}(x_n) \not\to 0 \text{ in } X.$$

This verifies that I^{-1} is not continuous and hence I is not an open mapping.

As a consequence of the Open Mapping Theorem, we get an important criterion of homeomorphism for linear operators between complete normed spaces.

3.5.6 Corollary. *Let X and Y be Banach spaces and $T : X \to Y$ a bounded linear operator. If T is bijective, then T^{-1} is a bounded linear operator from Y onto X.*

Proof. Since $T : X \to Y$ is bijective, its inverse T^{-1} exists and can easily be seen that it is linear. Further, under the hypothesis, by Open Mapping Theorem, it follows that $T(G)$ is an open set in Y, whenever G is open in X. Also, note that

$$(T^{-1})^{-1}(G) = T(G).$$

This verifies that T^{-1} is continuous (Theorem 1.7.39) and hence bounded. ∎

Note. The result in Corollary 3.5.6 is known as (Banach) *inverse mapping theorem.*

We may restate Corollary 3.5.6 in a slightly different form.

3.5.6′ Corollary. *Let X and Y be Banach spaces and $T : X \to Y$ a bounded linear operator. If T is bijective, then T is a topological isomorphism between X and Y.*

3.5.7 Corollary. *Let $X_1 \equiv (X, \| \cdot \|_1)$ and $X_2 \equiv (X, \| \cdot \|_2)$ be two Banach spaces with the same underlying linear space X. Suppose $\exists$ a constant $K > 0$ such that*

$$\| x \|_1 \leq K \| x \|_2, \quad \forall\, x \in X.$$

Then, $\| \cdot \|_1$ and $\| \cdot \|_2$ are equivalent norms.

Proof. Consider the identity operator $I : X_2 \to X_1$, i.e., $Ix = x$, $\forall\, x \in X$. Clearly, I is a bijective linear operator. Also, under the hypothesis, I is bounded. Therefore, by Corollary 3.5.6, it follows that $I^{-1} : X_1 \to X_2$ is a bounded linear operator. Consequently, by the definition of boundedness, $\exists$ a constant $K_1 > 0$ such that

$$\| I^{-1}x \|_2 \leq K_1 \| x \|_1, \ \forall\, x \in X$$

$$\Rightarrow \qquad \| x \|_2 \leq K_1 \| x \|_1, \quad \forall\, x \in X.$$

This together with the hypothesis gives

$$K_1^{-1} \| x \|_2 \leq \| x \|_1 \leq K \| x \|_2, \quad \forall\, x \in X.$$

Hence, $\| \cdot \|_1$ and $\| \cdot \|_2$ are equivalent norms on X (Theorem 3.3.5). $\blacksquare$

Problems

19. Consider the normed space $(\Phi, \| \cdot \|_\infty)$ [Example 2.3.9 (1)] where elements in Φ are sequences of complex numbers $x = \{\xi_i\}$. Let $T : \Phi \to \Phi$ be defined by

$$Tx = \left(\xi_1, \frac{1}{2}\xi_2, \frac{1}{3}\xi_3, \ldots \right).$$

Prove that T is linear and bounded but T^{-1} is unbounded. Does this contradict Open Mapping Theorem?

20. Define from the Open Mapping Theorem that the normed space $(C\,[0, 1], \| \cdot \|_1)$ is not complete.

21. A mapping T on $C\,[0, 1]$ is defined by

$$(Tx)\,(t) = \int_0^t x\,(\tau)\,d\tau.$$

Prove that T is bounded bijective mapping from $(C\,[0, 1], \|\cdot\|_\infty)$ onto the linear subspace $C_0^1\,[0, 1]$, where

$$C_0^1[0, 1] = \{\, x \in C^1\,[0, 1] : x\,(0) = 0\}.$$

Further, show that T^{-1} is not bounded on $(C_0^1\,[0, 1], \|\cdot\|_\infty)$. Explain why this does not contradict the Open Mapping Theorem.

22. Let X and Y be Banach spaces over the field $\mathbb{K}$ and $T : X \to Y$ be a bounded linear operator. If T is bijective prove that there are positive real numbers K_1 and K_2 such that

$$K_1\,\|\,x\,\|_X \le \|\,Tx\,\|_Y \le K_2\,\|\,x\,\|_X, \ \forall\ x \in X.$$

[*Hint:* Use Corollary 3.5.6 and Theorem 3.3.3.]

23. Let $X_1 \equiv (X, \|\cdot\|_1)$ and $X_2 \equiv (X, \|\cdot\|_2)$ be two Banach spaces with the same underlying linear space X. Let $\{x_n\} \subset X, \|\,x_n\,\|_1 \to 0 \ \Rightarrow \ \|\,x_n\,\|_2 \to 0$. Prove that:

(a) The convergence in X_1 implies convergence in X_2; and conversely.

(b) $\|\cdot\|_1$ and $\|\cdot\|_2$ are equivalent norms on X.

24. Let X be a Banach space with respect to each of the norms $\|\cdot\|_1$ and $\|\cdot\|_2$. Suppose a sequence $\{x_n\}$ in X converges to y in $(X, \|\cdot\|_1)$ and to z in $(X, \|\cdot\|_2)$ such that $y = z$. Prove that the norms $\|\cdot\|_1$ and $\|\cdot\|_2$ are equivalent.

3.6 Closed Graph Theorem and its Consequences

The Closed Graph Theorem is derived directly from the Open Mapping Theorem; still it is considered sometimes to be more useful than the Open Mapping Theorem.

It can be observed that given linear spaces X and Y over the field $\mathbb{K}$, the cartesian product $X \times Y$ is again a linear space over $\mathbb{K}$ under the algebraic operations given by

$$(x_1, y_1) + (x_2, y_2) = (x_1 + x_2, y_1 + y_2), \quad (x_1, y_1), (x_2, y_2) \in X \times Y$$

$$\lambda\,(x, y) = (\lambda x, \lambda y), \quad (x, y) \in X \times Y, \lambda \in \mathbb{K}.$$

In case X and Y are normed spaces, then a norm $\|\cdot\|$ is immediately available for $X \times Y$; namely,

$$\|\,(x, y)\,\| = \max\,\{\|\,x\,\|_X, \|\,y\,\|_Y\}, \quad (x, y) \in X \times Y.$$

Also, we can have two other norms on $X \times Y$ given by

$$\| (x, y) \| = \| x \|_X + \| y \|_Y,$$

and

$$\| (x, y) \| = \| x \|_X^p + \| y \|_Y^p, \quad (1 \le p < \infty).$$

Note that each one of the above three, indeed defines a norm on $X \times Y$ and they are equivalent norms. Finally, we note that if X and Y are Banach spaces over the field $\mathbb{K}$ then so is the normed space $X \times Y$. Indeed, if $\{z_n\}$ with $z_n = (x_n, y_n)$ is a Cauchy sequence in $X \times Y$, then $\{x_n\}$ and $\{y_n\}$ are Cauchy sequences, respectively, in X and Y since

$$\| z_n - z_m \| = \| (x_n - x_m, y_n - y_m) \|$$

$$= \max \{ \| x_n - x_m \|_X, \| y_n - y_m \|_Y \}.$$

Also, since each of the spaces X and Y is complete, it follows that there are points $x \in X$ and $y \in Y$ such that $x_n \to x$ in X and $y_n \to y$ in Y, and hence $z_n \to z$, where $z = (x, y)$ in $(X \times Y, \| \cdot \|)$.

3.6.1 Definition. Let X and Y be linear spaces over the field $\mathbb{K}$ and $T : X \to Y$ be a linear operator. The set given by

$$G(T) = \{(x, Tx) : x \in X\}$$

is called the *graph* of T.

Clearly, $G(T) \subset X \times Y$ is a subspace (linear) of $X \times Y$. If X, Y are normed spaces and $G(T)$ is closed in the normed space $X \times Y$, then T is called a *closed linear operator.*

Note. Technically speaking, T as a function and its graph are identical.

We first give a characterization of closed linear operators.

3.6.2 Theorem. *Let X and Y be normed spaces over the field $\mathbb{K}$. Then, a linear operator $T : X \to Y$ is closed if and only if, for every sequence $\{x_n\} \subset X$ with*

$$\lim_{n \to \infty} x_n = x \text{ and } \lim_{n \to \infty} Tx_n = y, \text{ we have } x \in X \text{ and } Tx = y.$$

Proof. We assume that T is closed and hence its graph $G(T)$ is a closed subspace of the normed space $(X \times Y, \| \cdot \|)$, where we take the norm given by

$$\| (x, y) \| = \max \{ \| x \|_X, \| y \|_Y \}, \quad (x, y) \in X \times Y.$$

Let $\{x_n\}$ be a sequence in X with $x_n \to x$ and $Tx_n \to y$. Then

$$\| (x_n, Tx_n) - (x, y) \| = \| (x_n - x, Tx_n - y) \|$$

$$= \max \{\| x_n - x \|_X, \| Tx_n - y \|_Y\} \to 0, \text{ as } n \to \infty$$

$$\Rightarrow \qquad \lim_{n \to \infty} (x_n, Tx_n) = (x, y) \text{ in } (X \times Y, \| \cdot \|)$$

$$\Rightarrow \qquad (x, y) \in \overline{G(T)}.$$

But $G(T)$ being closed, $(x, y) \in G(T)$. Hence, it follows that $x \in X$ and $Tx = y$.

Towards the converse, let (x, y) be a limit point of $G(T)$. Then, $\exists$ a sequence $\{(x_n, Tx_n)\}$ in $G(T)$ such that

$$\lim_{n \to \infty} (x_n, Tx_n) = (x, y)$$

$$\Rightarrow \qquad \lim_{n \to \infty} \| (x_n - x, Tx_n - y) \| = 0$$

$$\Rightarrow \qquad \max \{\| x_n - x \|_X, \| Tx_n - y \|_Y\} \to 0, \text{ as } n \to \infty$$

$$\Rightarrow \qquad \lim_{n \to \infty} x_n = x \text{ and } \lim_{n \to \infty} Tx_n = y.$$

By the hypothesis, it follows that $x \in X$ and $Tx = y$. Consequently, the limit point $(x, y) \in G(T)$. Therefore, $G(T)$ and hence T is closed. This completes the proof of the theorem. ∎

We are concerned with the relationship between closed linear operators and bounded (continuous) linear operators. In general, we observe that closedness does not imply boundedness, and boundedness does not imply closedness.

3.6.3 Example. Consider the Banach space $(C[0, 1], \| \cdot \|_\infty)$ (Example 2.2.7 with $a = 0$ and $b = 1$) and the normed space $(C^1[0, 1], \| \cdot \|_\infty)$ (Example 2.2.11). Define the mapping $T : C^1[0, 1] \to C[0, 1]$ by

$$(Tx)(t) = x'(t), \quad x \in C^1[0, 1], t \in [0, 1].$$

Then:

(a) T is a linear operator.

(b) T is not bounded.

Let $\{x_n\}$ be a sequence in $C^1[0, 1]$, where

$$x_n(t) = t^n, \quad n \in \mathbb{N}$$

so that

$$(Tx_n)(t) = x_n'(t) = n\, t^{n-1}.$$

Then

$$\begin{cases} \| x_n \|_\infty = \sup_{t \in [0,1]} | x_n(t) | = \sup_{t \in [0,1]} | t^n | = 1, \\[2mm] \| Tx_n \|_\infty = \sup_{t \in [0,1]} | x_n'(t) | = \sup_{t \in [0,1]} | nt^{n-1} | = n, \ \forall\, n \end{cases}$$

$$\Rightarrow \qquad \| Tx_n \|_\infty = n = n \| x_n \|_\infty.$$

Therefore

$$\| T \| = \sup_{\substack{x \in C^1[0,1] \\ x \neq 0}} \frac{\| Tx \|_\infty}{\| x \|_\infty} \geq \sup_{1 \leq n < \infty} \frac{\| Tx_n \|_\infty}{\| x_n \|_\infty} = \infty.$$

Hence, T is not bounded.

(c) T is closed *i.e.*, $G(T)$ is a closed subset of $C^1[0, 1] \times C[0, 1]$.

Let $\{x_n\}$ be a sequence in $C^1[0, 1]$ with

$$\lim_{n \to \infty} x_n = x \text{ and } \lim_{n \to \infty} Tx_n = y.$$

Then, for each $t \in [0, 1]$, we have

$$\int_0^t y(\tau)\, d\tau = \lim_{n \to \infty} \int_0^t (Tx_n)(\tau)\, d\tau$$

$$= \lim_{n \to \infty} \int_0^t x_n'(\tau)\, d\tau$$

$$= \lim_{n \to \infty} [x_n(t) - x_n(0)]$$

$$= x(t) - x(0).$$

Since $y \in C[0, 1]$, it follows from the Fundamental Theorem of calculus, that x is differentiable on $[0, 1]$ and

$$y\,(t) = x'\,(t), \quad \forall\; t \in [0,\,1]$$

$$\Rightarrow \qquad\qquad x \in C^1\,[0,\,1] \text{ and } Tx = y.$$

In view of Theorem 3.6.2, T is closed. ∎

3.6.4 Example. Consider the identity operator $I : C^1\,[0,\,1] \to C^1\,[0,\,1]$, where $C^1\,[0,\,1]$ is a normed space of Example 2.2.11. Then:

(a) I is a bounded linear operator [Example 3.1.9 (2)].

(b) I is not closed. Indeed, if we take an $x \in C\,[0,\,1] - C^1\,[0,\,1]$ and a sequence $\{x_n\} \subset C^1\,[0,\,1]$ with $x_n \to x$, which is possible in view of Weierstrass Approximation Theorem, then the assertion follows immediately in view of Theorem 3.6.2. ∎

There are special settings when the bounded linear operators are closed and vice versa. We discuss one of such results in the Closed Graph Theorem.

3.6.5 Theorem (Closed Graph). *Let X and Y be Banach spaces over the field $\mathbb{K}$ and $T : X \to Y$ be a linear operator. Then, T is closed if and only if T is bounded.*

Proof. Assume that T is bounded. Let $\{x_n\} \subset X$ be such that $x_n \to x$ and $Tx_n \to y$. Then, $x \in X$ since X is complete and $Tx_n \to Tx$ since T is continuous. As such, $Tx = y$. Hence by Theorem 3.6.2, T is closed.

Conversely, assume that T is closed. Then, the graph $G\,(T\,)$ of T is a closed linear subspace of the normed space $X \times Y$. But $X \times Y$ is a Banach space. Therefore, $G\,(T\,)$ is a Banach space under the induced operations and the norm (Theorem 2.3.7).

Now, consider the mapping $P : G\,(T\,) \to X$ defined by $P\,\big((x,\,Tx)\big) = x$, $x \in X$. Clearly, P is linear. We next show that P is bijective. Since P is linear, we need only to prove that $\mathcal{N}(P) = \{0\}$, *i.e.*, the null space of P is $\{0\}$ (Theorem 1.6.22). Indeed, if $P\,[(x,\,Tx)] = 0$, then $x = 0$ and so $Tx = 0$. Also, P is bounded since

$$\|\,P\,((x,\,Tx))\,\| = \|\,x\,\|_X \leq \max\,\{\|\,x\,\|_X,\, \|\,Tx\,\|_Y\} = \|\,(x,\,Tx)\,\|.$$

Thus, P is a bijective bounded linear operator from $G\,(T\,)$ onto X. Therefore, by a consequence of Open Mapping Theorem (Corollary 3.5.6) the map $P^{-1} : X \to G\,(T\,)$ defined by

$$P^{-1}\,(x) = (x,\,Tx)$$

is a bounded linear operator. Then

$$\| Tx \|_Y \leq \max \{\| x \|_X, \| Tx \|_Y\}$$

$$= \| (x, Tx) \|$$

$$= \| P^{-1}(x) \|$$

$$\leq \| P^{-1} \| \| x \|, \ \forall \ x \in X.$$

Hence, T is bounded. ∎

Remark. It is clear, from the first part of the proof of Theorem 3.6.5, that boundedness of a linear operator implies closedness in case the domain space is complete. Therefore, the closed linear operators are considered to be the next best to the bounded (continuous) linear operators.

Problem 25. Let $T_1 \in B(X_1, X_3)$ and $T_2 \in B(X_2, X_3)$, where X_1, X_2, X_3 are Banach spaces. If the equation $T_1 x = T_2 x$ has a unique solution $y = Tx$ for every x, prove that $T \in B(X_1, X_2)$.

We, now, give an application of Closed Graph Theorem to a geometric characterization of the projections on a Banach space. In fact, we discuss the decomposition of a Banach space as the sum of two subspaces and the related problem of the existence of linear operators which are projections onto subspaces. These ideas are, in fact, essential to the discussion of spectral theory. Recall that a linear operator $P : X \to X$, where X is a linear space, is said to be a linear projection operator on X, if P is an idempotent, *i.e.*, $P^2 = P$. Further, if M and N are, respectively, the range and null spaces of P, then X can be expressed as an algebraic direct sum of M and N, *i.e.*

$$X = M \oplus N. \qquad \text{(Theorem 1.6.29)}$$

In case X is a normed space and P is bounded on X, then one may observe that M and N are both closed subspaces of X. Indeed, the null space of any bounded linear operator on a normed space is closed (Theorem 3.1.10) and $M = \mathcal{N}(I - P), N = \mathcal{N}(P)$. Towards the converse of this assertion, we prove:

3.6.6 Theorem. *Let X be a Banach space over the field $\mathbb{K}$ and let M and N be closed linear subspaces of X such that $X = M \oplus N$. Then, there is a unique bounded linear projection operator P on X whose range and null spaces are M and N, respectively.*

Proof. In view of Theorem 1.6.29, $\exists$ a unique linear projection operator P on X whose range and null spaces are M and N, respectively.

We first prove that P is closed. Consider a sequence $\{x_n\} \subset X$ such that

$$\lim_{n \to \infty} x_n = x \text{ and } \lim_{n \to \infty} Px_n = y.$$

But, $\{Px_n\}$ being a sequence in a closed set M, it follows that $y \in M$. Therefore $Py = y$.

Further, since $\qquad x_n - Px_n = (I - P)(x_n) \in N$

and $\qquad\qquad\qquad\qquad \lim_{n \to \infty} (x_n - Px_n) = x - y,$

the closedness of N verifies that $x - y \in N$. Consequently, $P(x - y) = 0$ and so $Px = Py = y$. Thus, P is closed (Theorem 3.6.2). Hence, by Closed Graph Theorem, P is bounded. This completes the proof. ∎

Problems

26. Let X and Y be normed spaces and $T : X \to Y$ be a linear operator.

Prove that:

(a) If T has a closed graph, then T has a closed null space.

(b) By examining the mapping $T : (l^1, \| \cdot \|_\infty) \to (c_0, \| \cdot \|_\infty)$ defined by

$$T(\{\lambda_1, \lambda_2, \lambda_3,....\}) = \{(\lambda_1 + \lambda_2 + \lambda_3 + ...),$$

$$\frac{1}{2}(-\lambda_1 + \lambda_2 + \lambda_3 + ...), \frac{1}{3}(\lambda_1 - \lambda_2 + \lambda_3 + ...), ...\}$$

or, otherwise, the converse of (a) is not true in general.

27. Let X and Y be normed spaces and $T : X \to Y$ be a closed linear operator. Prove that:

(a) The inverse T^{-1}, if exists, is a closed linear operator.

(b) If Y is compact, then T is bounded.

(c) The image, under T, of a compact set in X is closed in Y.

(d) The inverse image, under T, of a compact subset of Y is closed in X.

(e) If two sequences $\{x_n\}$ and $\{x'_n\}$ in X converge to the same limit in X, the sequences $\{T x_n\}$ and $\{T x'_n\}$, if both of them converge, have the same limit in Y.

28. Let X and Y be normed spaces. If $T_1 : X \to Y$ be a closed linear operator and $T_2 \in B\,(X,\,Y)$, prove that $T_1 + T_2$ is a closed linear operator.

29. Let M and N be closed linear subspaces of a Banach space X such that $M \cap N = \{0\}$. Prove that $M \oplus N$ is closed in X if and only if there is a constant $K > 0$ such that $\|\,x - y\,\| \geq K$, $\forall\, x \in M$ and $y \in N$ with $\|\,x\,\| = \|\,y\,\| = 1$.

30. A closed linear subspace M of a normed space X is said to be *invariant* under a linear operator $T : X \to X$ if and only if $T\,(M) \subset M$. The closed linear subspaces M and N are said to *reduce* a linear operator $T : X \to X$ if and only if $X = M \oplus N$ and M and N are invariant under T.

Prove that if P is a bounded linear projection operator on a normed space X with M and N as its range and null space, then M and N reduce T if and only if $PT = TP$.

3.7 Uniform Boundedness Principle

The Uniform Boundedness Principle was obtained by S. Banach and H. Steinhaus in 1927. So it is also known as Banach-Steinhaus Theorem. This theorem, like Open Mapping Theorem and Closed Graph Theorem, requires completeness. Further, this theorem is derived from Baire's Category Theorem. The Principle of Uniform Boundedness asserts that if a sequence of bounded linear operators $T_n \in B\,(X,\,Y)$, where X is a Banach space and Y a normed space, is pointwise bounded, then the sequence $\{T_n\}$ is uniformly bounded. In fact, it enables us to determine whether the norms of a given family of bounded linear operators have a finite least upper bound.

We recall that $B\,(X,\,Y)$ is the normed linear space of all bounded linear operators from a normed space X into a normed space Y (Theorem 3.2.2).

3.7.1 Definition. A set $\mathscr{F}$ of bounded linear operators from a normed space X into a normed space Y is said to be:

(a) *Pointwise bounded* if for each $x \in X$, the set $\{\,Tx : T \in \mathscr{F}\,\}$ is a bounded set in Y.

(b) *Uniformly bounded* if $\mathscr{F}$ is a bounded set in the normed space $B\,(X,\,Y)$.

Observe that if $\mathscr{F}$ is a uniformly bounded set, then $\mathscr{F}$ is a pointwise bounded set. Indeed, by Definition 3.7.1 (b), the boundedness of the set $\mathscr{F}$ in the normed space $B\,(X,\,Y)$ implies that there is a constant $K > 0$ such that

$$\|\,T\,\| \leq K, \quad \forall\, T \in \mathscr{F}.$$

Let $x \in X$ be given. Then

$$\| Tx \|_Y \leq \| T \| \, \| x \|_X \leq K \| x \|_X, \quad \forall \, T \in \mathscr{F}.$$

Hence, it follows that $\mathscr{F}$ is pointwise bounded in Y. However, the converse of this assertion may not hold good.

3.7.2 Example. Consider the normed space $(\Phi, \| \cdot \|_\infty)$, which of course is not complete, (see Example 2.3.9 (1)) and the Banach space $(\mathbb{K}, |\cdot|)$. Consider the family $\mathscr{F} = \{T_n : n \in \mathbb{N}\} \subset L(\Phi, \mathbb{K})$, where $T_n : \Phi \to \mathbb{K}$ is defined by

$$T_n x = n\lambda_n, \; x = \{\lambda_i\} \in \Phi.$$

Clearly, $T_n \in B(\Phi, \mathbb{K})$ for each n. Now, let x_0 be any given point in Φ. Then, $\exists \, n_0 \in \mathbb{N}$ such that $\lambda_n = 0, \; \forall \, n > n_0$. Therefore

$$\| T_n x_0 \| = | T_n x_0 | = \begin{cases} n | \lambda_n |, & n \leq n_0 \\ 0, & n > n_0 \end{cases}$$

$$\leq n_0 \| x_0 \|_\infty, \; \forall \, n \in \mathbb{N}.$$

Thus, the family $\mathscr{F}$ is pointwise bounded.

On the other hand, for a given $n \in \mathbb{N}$, we have

$$\| T_n \| = \sup \{ \| T_n x \| : x \in \Phi, \| x \|_\infty \leq 1 \} = n.$$

This is true for each n. Hence, $\mathscr{F}$ is not uniformly bounded. ∎

In Example 3.7.2, we have noted that for a set of bounded linear operators on a normed space X, pointwise boundedness need not imply uniform boundedness. However, the Uniform Boundedness Principle asserts that pointwise boundedness does imply uniform boundedness if the space X is a Banach space.

3.7.3 Theorem (Uniform Boundedness Principle). *Let X be a Banach space and let Y be a normed space over the field $\mathbb{K}$. If a set $\mathscr{F}$ of bounded linear operators from X into Y is pointwise bounded, then it is uniformly bounded.*

Proof. For each $n \in \mathbb{N}$, write

$$F_n = \{x \in X : \| Tx \|_Y \leq n, \forall \, T \in \mathscr{F} \}.$$

We first show that F_n is a closed subset of X. Let $x \in \overline{F_n}$ and $\{x_k\}$ be a sequence in F_n such that $x_k \to x$. Then

$$\| Tx_k \|_Y \le n, \quad \forall \, T \in \mathscr{F}.$$

Also, T being continuous and $x_k \to x$, we have

$$\lim_{k \to \infty} \| Tx_k \|_Y = \| Tx \|_Y.$$

Therefore

$$\| Tx \|_Y \le n, \quad \forall \, T \in \mathscr{F}$$

$$\Rightarrow \qquad x \in F_n.$$

Hence, F_n is closed.

Now, since $\mathscr{F}$ is pointwise bounded, for each $x \in X$, the set $\{ Tx : T \in \mathscr{F} \}$ is a bounded set in the normed space Y (Definition 3.7.1 (a)). Thus, each $x \in X$ is in some F_n and therefore

$$X = \bigcup_{n=1}^{\infty} F_n.$$

But X being complete, Baire's Category Theorem (Theorem 1.7.32) implies that $\exists$ an $n_0 \in \mathbb{N}$ such that F_{n_0} is not nowhere dense in X. Consequently, $\exists$ a non-empty open sphere $S(x_0 ; r_0)$ such that

$$S(x_0 ; r_0) \subset F_{n_0}$$

$$\Rightarrow \qquad \| Tx \|_Y \le n_0, \quad \forall \, x \in S(x_0 ; r_0) \text{ and } \forall \, T \in \mathscr{F}.$$

For the sake of convenience, we write the preceding expression in the form

$$\| T(S(x_0 ; r_0)) \|_Y \le n_0, \ \forall \, T \in \mathscr{F}.$$

Note that

$$S(x_0 ; r_0) = x_0 + r_0 \, S(0 ; 1).$$

Also, since $x_0 \in S(x_0 ; r_0)$, we have

$$\| Tx_0 \|_Y \le n_0, \ \forall \, T \in \mathscr{F}.$$

Therefore

$$\| T(S(0 ; 1)) \|_Y = \left\| T \left(\frac{S(x_0 ; r_0) - x_0}{r_0} \right) \right\|_Y$$

$$\le \frac{1}{r_0} \left(\| T(S(x_0 ; r_0)) \|_Y + \| Tx_0 \|_Y \right)$$

$$\leq \frac{2n_0}{r_0}, \ \forall \, T \in \mathscr{F}.$$

$$\Rightarrow \qquad \| \, Tx \, \|_Y \leq \frac{2n_0}{r_0}, \quad \forall \, x \in S \, (0 \, ; 1) \text{ and } \forall \, T \in \mathscr{F}.$$

$$\Rightarrow \qquad \| \, T \, \| = \sup_{x \in S \, (0 \, ; 1)} \| \, Tx \, \|_Y \leq \frac{2 \, n_0}{r_0}, \ \forall \, T \in \mathscr{F}.$$

Hence, $\mathscr{F}$ is a bounded subset of $B \, (X, \, Y)$. This proves that $\mathscr{F}$ is uniformly bounded. $\blacksquare$

An alternative statement of Theorem 3.7.3 is

3.7.3′ Theorem. *Let* X *be a Banach space and let* Y *be a normed space over the field* $\mathbb{K}$. *If* $\mathscr{F} \subset B \, (X, \, Y)$ *be such that, for each* $x \in X$, *the set* $\{Tx : T \in \mathscr{F}\}$ *is bounded in* Y, *i.e.*

$$\| \, Tx \, \|_Y \leq K_x, \ \forall \, T \in \mathscr{F}$$

where K_x *is a positive real number (the number* K_x *vary, in general, with x and does not depend on* T *). Then, the set* $\{\| \, T \, \| : T \in \mathscr{F}\}$ *is bounded, i.e.,* $\exists$ *a constant* $K > 0$ *such that*

$$\| \, T \, \| \leq K, \ \forall \, T \in \mathscr{F}.$$

The Uniform Boundedness Principle is one of the most useful results in functional analysis. It has several applications. We conclude this section by giving some of its interesting applications. However, many more important applications are presented in Chapter 4 and subsequent chapters.

3.7.4 Theorem. *Let X be a Banach space and let Y be a normed space over the field* $\mathbb{K}$. *If* $\{T_n\} \subset B \, (X, \, Y)$ *be a sequence such that*

$$(10) \qquad \lim_{n \to \infty} T_n x = Tx, \ (x \in X)$$

exists, then $T \in B \, (X, \, Y)$.

Proof. Note that (10) defines a mapping $T : X \to Y$ which is, clearly, a linear operator. We need only to show that T is bounded.

Since, for each $x \in X$, $\{T_n x\}$ is a convergent sequence in Y, it follows that $\{T_n x\}$ is bounded in Y. By the Principle of Uniform Boundedness (Theorem 3.7.3), the set $\{\| \, T_n \, \|\}$ is bounded, *i.e.*

$$\| T_n \| \leq K, \qquad \forall\, n \in \mathbb{N}$$

for some $K > 0$. Therefore

$$\| T_n x \|_Y \leq \| T_n \| \, \| x \|_X \leq K \| x \|_X, \ \forall\, x \in X \text{ and } \forall\, n \in \mathbb{N}.$$

This, in view of (10), gives

$$\| Tx \|_Y \leq K \| x \|_X, \ \forall\, x \in X.$$

Hence, T is bounded. ∎

Theorem 3.7.4 can be restated as:

3.7.4′ Theorem. *Let X be a Banach space and let Y be a normed space over the field $\mathbb{K}$. If $\{T_n\} \subset B\,(X, Y)$ be a sequence such that $T_n \to T$ pointwise in Y, then $T \in B\,(X, Y)$.*

Remark. Theorem 3.7.4 (or Theorem 3.7.4′) may not hold in case X is an incomplete normed space.

3.7.5 Example. Take $X = (l^1, \| \cdot \|_\infty)$ and $Y = (\mathbb{K}, |\cdot|)$. Consider the sequence $\{T_n\}$ of mappings, where $T_n : l^1 \to \mathbb{K}$ is defined by

$$T_n x = \sum_{i=1}^{n} \lambda_i, \ (x = \{\lambda_i\} \in l^1).$$

Clearly, T_n is a linear operator. Also, T_n is bounded for each n since

$$\| T_n x \| = |\, T_n x\,| \leq n \| x \|_\infty, \quad \forall\, x \in X.$$

Thus, $\{T_n\} \subset B\,(X, Y)$. Further, note that

$$\left\| T_n x - \sum_{i=1}^{\infty} \lambda_i \right\| = |\, \lambda_{n+1} + \lambda_{n+2} + \,....\,| \leq \sum_{i=n+1}^{\infty} |\lambda_i|.$$

Writing $Tx = \sum_{i=1}^{\infty} \lambda_i$, we find that $T : X \to Y$ is a linear operator such that

$T_n x \to Tx$, for each $x \in X$ and as such $T_n \to T$ pointwise in Y. However, T is not bounded (Verify !). Also, note that $(l^1, \| \cdot \|_\infty)$ is not complete (Example 2.3.9 (2)).

Problems

31. Let X be a Banach space and let Y be a normed space. If $\{T_n\}$ be a sequence of bounded linear operators from X into Y and is such that $\{T_n x\}$ is Cauchy in Y for each $x \in X$, prove that $\{\|T_n\|\}$ is bounded.

[*Hint:* Note that a Cauchy sequence is bounded. Apply Uniform Boundedness Principle.]

32. Let X and Y be normed spaces and $T : X \to Y$ be a linear operator with the property that the set $\{\|Tx_n\| : n \in \mathbb{N}\}$ is bounded whenever $x_n \to 0$ in X. Prove that $T \in B(X, Y)$.

33. Let $\mathscr{F} \subset B(X, Y)$, where X is a Banach space and Y is a normed space. Suppose that the set $\{Tx_0 : T \in \mathscr{F}\}$ is unbounded for some point $x_0 \in X$. Prove that the subset of X consisting of those points $x \in X$ for which the set $\{Tx : T \in \mathscr{F}\}$ is bounded in Y, is of first category in X.

[*Hint:* Examine the proof of Uniform Boundedness Principle.]

34. Let the operator $S : l^2 \to l^2$ be defined by

$$S(\{\xi_1, \xi_2, ...\}) = \{\xi_3, \xi_4, ...\}.$$

Let $T_n = S^n$. Then, find a bound for $\|T_n x\|$. Also, find

(a) $\lim_{n \to \infty} \|T_n x\|,$

(b) $\|T_n\|,$

(c) $\lim_{n \to \infty} \|T_n\|.$

35. Let $y = \{\eta_i\}, \eta_i \in \mathbb{C}$, be such that $\displaystyle\sum_{i=1}^{\infty} \xi_i \overline{\eta}_i$ converges for every $x = \{\xi_i\} \in c_0$,

where $c_0 \subset l^{\infty}$ is the subspace of all complex sequences converging to zero. Prove that

$$\sum_{i=1}^{\infty} |\eta_i| < \infty.$$

[*Hint:* Use Uniform Boundedness Principle.]

36. Let $T_n \in B(X, Y), n \in \mathbb{N}$, where X and Y are Banach spaces. Prove that the following statements are equivalent:

(a) $\{\|T_n\|\}$ is bounded.

(b) $\{\|T_n x\|\}$ is bounded for all $x \in X$.

4
CHAPTER

Bounded Linear Functionals

While studying an abstract mathematical system, we generally study the structure preserving mappings from one system to another system of the same type. With this basic principle of strategy in mind we, in Chapter 3, have investigated fundamental properties of bounded linear operators from normed spaces into normed spaces. We, now, wish to explore a special case arising when the codomain of a bounded linear operator is the simplest non-trivial normed space. Since $\mathbb{R}$ and $\mathbb{C}$ are the simplest of all normed spaces, we in the present chapter study the bounded linear operators from arbitrary normed spaces into the special normed spaces $\mathbb{R}$ or $\mathbb{C}$. Such bounded linear operators are called *bounded linear functionals*. As such all general theorems proved in the preceding chapter for bounded linear operators are also valid for bounded linear functionals (Theorem 4.1.3). For instance, the collection of all bounded linear functionals over a normed space X is itself a normed space and is called the *dual space* of X. Furthermore, this space is a Banach space irrespective of whether the original space X is a Banach space or not. Moreover, in this chapter, we consider some special problems, the solutions of which are much simpler for functionals than for arbitrary operators.

We shall demonstrate a few non-trivial examples of bounded linear functionals on some concrete normed spaces. The existence of non-zero bounded linear functionals on an arbitrary normed space was first established by Hahn in 1927. Banach, in 1929, established a more general theorem, which is now known as Hahn-Banach Theorem (Theorem 4.3.3). This theorem has wide ranging applications in the study of normed and Banach spaces. In fact, like three fundamental theorems; namely, Open Mapping Theorem, Closed Graph Theorem and Uniform Boundedness Principle proved in Chapter 3, the Hahn-Banach Theorem is also one of the most remarkable and fundamental theorem in functional analysis.

151

4.1 Definitions, Examples and Basic Properties

Let X be a normed space over the field $\mathbb{K}$ ($\mathbb{R}$ or $\mathbb{C}$). Recall that $\mathbb{K}$ is a normed space over $\mathbb{K}$ with the usual metric. A linear operator from X into $\mathbb{K}$ is called a linear functional on X. More precisely

4.1.1 Definition. Let X be a normed space over the field $\mathbb{K}$. A mapping $f : X \to \mathbb{K}$ is said to be a *linear functional* on X if

$$f(\alpha x + \beta y) = \alpha f(x) + \beta f(y), \quad \forall\ x, y \in X \text{ and } \alpha, \beta \in \mathbb{K}.$$

A linear functional is said to be *real* or *complex* according as the field $\mathbb{K}$ is $\mathbb{R}$ or $\mathbb{C}$, respectively.

Throughout this chapter, $\mathbb{K}$ denotes the field of scalars, it may be either the real or complex field. We denote functionals by lower case letters like f, g, h, ... and the value of f at a point x in the domain of f by $f(x)$.

Note. In fact, functional analysis was initially the analysis of functionals.

Now, referring to the definition of a bounded linear operator $T : X \to Y$ (Definition 3.1.2), in the particular case when $Y = \mathbb{K}$, we have

4.1.2 Definition. Let X be a normed space over the field $\mathbb{K}$. A linear functional $f : X \to \mathbb{K}$ is said to be *bounded* if $\exists$ a real number $K > 0$ such that

$$|f(x)| \le K \| x \|, \quad \forall\ x \in X.$$

Since, a bounded linear functional is a special case of a bounded linear operator, most of the concepts and general theorems studied in Chapter 3 for bounded linear operators are also valid for bounded linear functionals. For the sake of completeness, we now state (without proof) some of the important facts about bounded linear functionals.

4.1.3 Theorem. *Let f be a linear functional on a normed space X. Then:*

(a) *f is continuous if and only if f is continuous at a point (any) in X (Theorem 3.1.1).*

(b) *f is continuous if and only if f is bounded (Theorem 3.1.3).*

(c) *If f is bounded, then $\| f \|$, norm of f, is a non-negative real number, which can be expressed by any one of the following equivalent formulae:*

$$\text{(i) } \| f \| = \sup \{ |f(x)| : x \in X, \| x \| \le 1 \}.$$

$$\text{(ii) } \| f \| = \sup \{ |f(x)| : x \in X, \| x \| = 1 \}.$$

(iii) $\|f\| = \sup \{\dfrac{|f(x)|}{\|x\|} : x \in X$ and $x \neq 0\}$.

(iv) $\|f\| = \inf \{K : K > 0$ and $|f(x)| \leq K \|x\|, \forall \, x \in X\}$.

(Theorem 3.1.7)

(d) *If f is bounded, then*

$$|f(x)| \leq \|f\| \, \|x\|, \forall \, x \in X. \qquad\qquad (Corollary\ 3.1.8)$$

Equivalently, if $x \neq 0$ then

$$\|f\| \geq \frac{|f(x)|}{\|x\|}.$$

4.1.4 Examples

1. Let $\mathbb{R}^n$ be the real normed space with the usual norm (Example 2.2.1) and let $a = (a_1, a_2, ..., a_n)$ be a fixed non-zero vector in $\mathbb{R}^n$. Define the functional $f : \mathbb{R}^n \to \mathbb{R}$ by

$$f(x) = x \cdot a$$

where $x = (\xi_1, \xi_2, ..., \xi_n) \in \mathbb{R}^n$ and $x \cdot a$ denotes the familiar scalar product of x and a. Then, f is a bounded linear functional on $\mathbb{R}^n$ with $\|f\| = \|a\|$.

Clearly, f is a linear functional and it is bounded since, by virtue of Cauchy-Schwartz Inequality (1.5.5), we have

$$|f(x)| = |x \cdot a| \leq \|x\| \, \|a\|, \forall \, x \in \mathbb{R}^n.$$

Furthermore, we have

$$\|f\| = \sup \{|f(x)| : x \in X, \|x\| \leq 1\} \leq \|a\|.$$

But, by Theorem 4.1.3 (d), we get

$$\|f\| \geq \frac{|f(a)|}{\|a\|} = \frac{\|a\|^2}{\|a\|} = \|a\|.$$

Hence $\qquad\qquad \|f\| = \|a\|.$ ∎

2. Consider the Banach space $(l^1, \|\cdot\|_1)$ (Example 2.2.5 with $p = 1$). Define the linear functional $f : l^1 \to \mathbb{R}$ by

$$f(x) = \sum_{i=1}^{\infty} \xi_i, \, (x = \{\xi_i\} \in l^1).$$

Then

$$|f(x)| \le \sum_{i=1}^{\infty} |\xi_i| = \|x\|_1, \, \forall \, x \in l^1.$$

This implies that f is bounded on l^1 and $\|f\| \le 1$. Also, for $x = e_1 \in l^1$, Theorem 4.1.3 (d) yields

$$\|f\| \ge \frac{|f(e_1)|}{\|e_1\|_1} = 1.$$

Hence, f is a bounded linear functional with $\|f\| = 1$. ∎

3. Let $\alpha = \{\alpha_i\} \in l^{\infty}$ be fixed. Define the linear functional

$$f_\alpha : l^p \to \mathbb{K}, \, 1 \le p < \infty, \text{ by}$$

$$f_\alpha(x) = \left\{ \sum_{i=1}^{\infty} |\alpha_i \, \xi_i|^p \right\}^{\frac{1}{p}}, \, x = \{\xi_i\} \in l^p.$$

Then

$$|f_\alpha(x)| \le \left\{ \sup_{1 \le i < \infty} |\alpha_i|^p \sum_{i=1}^{\infty} |\xi_i|^p \right\}^{\frac{1}{p}}$$

$$= M \|x\|_p, \, M = \sup_{1 \le i < \infty} |\alpha_i| < \infty.$$

Hence f_α is a bounded linear functional on l^p and $\|f_\alpha\| = \|\alpha\|$.

4. Let $X = (C[a, b], \|\cdot\|_\infty)$ be the Banach space (Example 2.2.7). Take a fixed point $t_0 \in \,]a, b[$. Define the functional $f_{t_0} : X \to \mathbb{R}$ by

$$f_{t_0}(x) = x(t_0), \, x \in C[a, b].$$

Then, f_{t_0} is a bounded linear functional with $\|f_{t_0}\| = 1$.

The linearity of f is obvious. It is bounded, since, we have

$$| f_{t_0}(x) | = | x(t_0) | \leq \sup_{t \in [a,b]} | x(t) | = \| x \|_\infty, \ \forall \ x \in X$$

$\Rightarrow \quad f_{t_0}$ is bounded and $\| f_{t_0} \| \leq 1.$

On the other hand, by Theorem 4.1.3 (d), note that

$$\| f_{t_0} \| \geq \frac{| f_{t_0}(x) |}{\| x \|_\infty}, \ \forall \ x \in C[a, b].$$

In particular, for $x = x_0$, where $x_0(t) = 1$ for all $t \in [a, b]$, we have

$$\| f_{t_0} \| \geq \frac{| f_{t_0}(x_0) |}{\| x_0 \|_\infty} = \frac{| x_0(t_0) |}{\| x_0 \|_\infty} = 1.$$

Hence $\qquad \| f_{t_0} \| = 1.$ ∎

Note. The functional presented in the preceding example is generally found in quantum mechanics where it is usually written in the form

$$\delta_{t_0}(x) = \int_a^b x(t)\, \delta(t - t_0)\, dt$$

where $\delta(t) \equiv 0$ everywhere except at the point $t = 0$ and is such that its integral equals unity. This functional is called the *Dirac δ-function,* in fact, *Dirac distribution.*

We next consider on $C[a, b]$ a linear functional of some other type.

4.1.5 Example. Let X be the same as in Example 4.1.4 (3). Define the functional $f: X \to \mathbb{R}$ by

$$f(x) = \int_a^b x(t)\, dt, \ x \in X.$$

Then f is a bounded linear functional with $\| f \| = b - a.$

It can be readily seen that f is linear. Also, we have

$$| f(x) | = \left| \int_a^b x(t)\, dt \right|$$

$$\leq \sup_{t \in [a,b]} | x(t) | (b - a)$$

$$= (b - a) \, \| x \|_\infty, \quad \forall \, x \in X$$

$$\Rightarrow \quad f \text{ is bounded and } \| f \| \leq b - a.$$

On the other hand, if $x = x_0$, where $x_0 (t) = 1$ for all $t \in [a, b]$, then

$$\| x_0 \|_\infty = 1 \text{ and } | f (x_0) | = \int_a^b dt = b - a$$

and so

$$\| f \| \geq \frac{| f (x_0) |}{\| x_0 \|_\infty} = b - a.$$

Hence $\qquad\qquad \| f \| = b - a. \ \blacksquare$

Problem 1. Let x_0 be a fixed continuous function on $[a, b]$. Prove that the functional defined on the Banach space $(C \, [a, b], \| \cdot \|_\infty)$ (Example 2.2.7) by

$$f (x) = \int_a^b x (t) \, | x_0 (t) | \, dt, \ x \in C \, [a, b]$$

is a bounded linear functional with

$$\| f \| = \int_a^b | x_0 (t) | \, dt \ .$$

4.1.6 Examples (Unbounded Linear Functionals)

1. The linear functional f on the normed space $(l^1, \| \cdot \|_\infty)$ defined by

$$f (x) = \sum_{i = 1}^\infty \xi_i, \ (x = \{ \xi_i \} \in l^1)$$

is unbounded.

Choose the sequence $\{ x_n \} \subset l^1$, where $x_n = \sum_{i = 1}^n e_i$. For each $n \in \mathbb{N}$, note that

$$\begin{cases} \| x_n \|_\infty = \| \sum_{i = 1}^n e_i \|_\infty = 1, \\ f (x_n) = n. \end{cases}$$

$$\Rightarrow \qquad |f(x_n)| = n \parallel x_n \parallel_\infty, \ n \in \mathbb{N}.$$

Therefore, it follows that $\parallel f \parallel = \infty$ and hence f is an unbounded linear functional on $(l^1, \parallel \cdot \parallel_\infty)$.

2. Let $X = (C[a, b], \parallel \cdot \parallel_1)$ be the normed space (Problem 2.8 with $p = 1$) and let $\delta_{t_0} : C[a, b] \to \mathbb{R}$ be the linear functional as in Example 4.1.4 (4). Then δ_{t_0} is unbounded on X.

A linear functional may be bounded or unbounded. But there are bounded functionals which need not be linear.

4.1.7 Example. Let $(X, \parallel \cdot \parallel)$ be a normed space over $\mathbb{R}$. Then, the mapping defining the norm $\parallel \cdot \parallel : X \to \mathbb{R}$ itself is a bounded functional on X since $\parallel \cdot \parallel$ is continuous at every point of X (Corollary 2.1.5). But $\parallel \cdot \parallel$ is not linear since, in general, we have

$$\parallel x + y \parallel \neq \parallel x \parallel + \parallel y \parallel, \quad \forall\, x, y \in X.$$

Now, for linear functionals, we shall prove that the property of the kernel being closed characterises continuity.

4.1.8 Theorem. *Let X be a normed space. A linear functional f on X is bounded (i.e. continuous) if and only if $\ker(f)$ is closed.*

Proof. Let f be bounded (and therefore continuous). Then, since

$$\ker(f) = \{\, x \in X : f(x) = 0 \} = f^{-1}(\{0\})$$

and $\{0\}$ is a closed subset of X, it follows that $\ker(f)$ is closed.

For the converse, assume that $\ker(f)$ be a closed set. If $f \equiv 0$, then f is clearly continuous and hence bounded. Suppose f is not the zero functional. Clearly, $X - \ker(f) \neq \phi$ and $X - \ker(f)$ is open. Then, $\exists\, a \in X - \ker(f)$ such that $f(a) \neq 0$. Letting $b = \dfrac{a}{f(a)}$, note that $f(b) = 1$ and $b \in X - \ker(f)$. Since $X - \ker(f)$ is open, $\exists\, r > 0$ such that

$$S(b\, ;\, r) \subset X - \ker(f).$$

We shall first prove that

$$(1) \qquad |f(x)| < 1, \qquad \forall\, x \in S(0\, ;\, r).$$

Suppose the contrary. Then, $\exists$ some $x_1 \in S(0\;;\,r)$ such that

$$|f(x_1)| \geq 1.$$

Then

$$y = \frac{-x_1}{f(x_1)} \in S(0\;;\,r)$$

and $\qquad\qquad f(b+y) = f(b) + f(y) = 1 - 1 = 0.$

Therefore, $b + y \in \ker(f)$ and $b + y \in S(b\;;\,r)$. Thus

$$\ker(f) \cap S(b\;;\,r) \neq \phi.$$

This contradicts the fact that $S(b\;;\,r) \subset X - \ker(f)$ proving (1).

Now, take any $x \neq 0$. Then

$$\left\| \frac{rx}{2\|x\|} \right\| = \frac{r}{2} < r.$$

Therefore, by the assertion (1) established above, we have

$$\left| f\left(\frac{rx}{2\|x\|} \right) \right| < 1$$

$$\Rightarrow \qquad\qquad |f(x)| < \left(\frac{2}{r} \right) \|x\|.$$

Also, if $x = 0$, then

$$|f(x)| = 0 = \|x\|$$

$$\Rightarrow \qquad\qquad |f(x)| \leq \left(\frac{2}{r} \right) \|x\|.$$

Thus, we have shown that

$$|f(x)| \leq \left(\frac{2}{r} \right) \|x\|, \quad \forall\, x \in X.$$

Hence, f is bounded. ∎

Remark. The result in Theorem 4.1.8 need not hold, in general, for linear operators between normed spaces.

4.1.9 Example. Consider the normed spaces $X = (C[0, 1], \| \cdot \|_1)$, $Y = (C[0, 1], \| \cdot \|_\infty)$ and the identity mapping $I : X \to Y$. Note that I is a linear operator which is not bounded (Example 3.5.5). On the other hand, we have

$$\ker (I) = I^{-1} (\{0\}) = \{0\}$$

and $\{0\}$ is a closed set. ∎

We further investigate the nature of the kernels of linear functionals on a normed space.

4.1.10 Theorem. *Let X be a normed space and let f be a linear functional on X. Then, the ker (f) is either dense or closed in X.*

We first state and prove a lemma.

4.1.11 Lemma. *Let f be a non-zero linear functional on a linear space X and let $x_0 \in X - \ker (f)$. Then, any $x \in X$ can be expressed uniquely in the form $x = p + \alpha x_0$, where $p \in \ker (f)$ and α is some scalar.*

Proof. Since $x_0 \in X - \ker (f), f(x_0) \neq 0$. Setting

$$\alpha = \frac{f(x)}{f(x_0)},$$

define

$$p = x - \frac{f(x)}{f(x_0)} x_0.$$

Then, $x = p + \alpha x_0$ and clearly $f(p) = 0$.

For uniqueness, assume that $x = p + \alpha x_0$ and $x = p' + \alpha' x_0$. Then

$$p' - p = (\alpha - \alpha') x_0.$$

If $\alpha = \alpha'$, then $p = p'$. If $\alpha \neq \alpha'$, then

$$x_0 = \frac{p' - p}{\alpha - \alpha'} \in \ker (f),$$

which contradicts the assumption that $x_0 \in X - \ker (f)$. This completes the proof of the lemma. ∎

Proof of Theorem 4.1.10. Assume that ker (f) be not closed. Then, $\exists$ a point $x_0 \notin \ker (f)$ such that x_0 is a cluster point of ker (f). Since ker (f) is a subspace of X, it follows that $\overline{\ker (f)}$ is also a subspace of X (Problem 2.26). Therefore, $\overline{\ker (f)}$ contains the linear span of x_0 and ker (f). Using Lemma 4.1.11, we get

$$X \subset \overline{\ker (f)}.$$

Hence $\overline{\ker (f)} = X.$ ∎

4.1.12 Corollary. *Let X be a normed space and let f be a linear functional on X. Then, f is bounded (continuous) if and only if ker (f) is not dense in X.*

Proof. It follows by Theorems 4.1.8 and 4.1.10. ∎

Problem 2. Let X be a normed space over the field $\mathbb{K}$ and let E be a maximal closed subspace of X. Prove that there exists a bounded linear functional f on X such that ker $(f) = E$.

[*Hint:* Take $x_0 \in X - E$. Then, prove that each $x \in X$ can be uniquely expressed as $x = p + \alpha x_0$, where $p \in E$ and $\alpha \in \mathbb{K}$. Define a mapping $f : X \to \mathbb{K}$ by $f(x) = f(p + \alpha x_0) = \alpha$. Show that f has the required properties.]

We, now, recall that if X and Y are normed spaces over the field $\mathbb{K}$, then the set $B(X, Y)$ consisting of all bounded linear operators from X into Y constitutes a normed space (Theorem 3.2.2). In particular, if the space X is a normed space over $\mathbb{K}$ and $Y = \mathbb{K}$, then the set $B(X, \mathbb{K})$ of all bounded linear functionals on X forms a normed space with the norm defined by

$$\|f\| = \sup \{ |f(x)| : x \in X, \|x\| = 1 \}.$$

Further, since $\mathbb{K}$ is complete, $B(X, \mathbb{K})$ is a Banach space. This space $B(X, \mathbb{K})$ is called the dual space. More precisely

4.1.13 Definition. Let X be a normed space over the field $\mathbb{K}$. The Banach space $B(X, \mathbb{K})$ of all bounded linear functionals on X is called the *dual space* (*conjugate space,* or *adjoint space*) of X. The dual space of X is denoted by X^*.

Remark. The dual space X^* is always complete irrespective of whether X is complete or not.

We close this section by giving some results of a slightly geometrical flavour. In fact, we also demonstrate geometrical interpretation of the norm of a non-zero bounded linear functional on a normed space.

4.1.14 Theorem. *Let $0 \neq f \in X^*$, where X is a normed space. Then, $\ker(f)$ is a closed hyperplane through the origin.*

Proof. Since f is a non-zero functional, it follows that $\ker(f) = f^{-1}(\{0\}) \neq X$. Thus, $\ker(f)$ is a proper subspace of X. Let M be a subspace of X such that

$$\ker(f) \underset{\neq}{\subseteq} M \subset X.$$

Then, $\exists\, x_0 \in M$ such that $x_0 \notin \ker(f)$. Therefore, by Lemma 4.1.11, $x \in X$ can be uniquely represented in the form

$$x = p + \alpha\, x_0,$$

where $p \in \ker(f)$ and α is some scalar. Since $p, x_0 \in M$, it follows that $x \in M$. Thus, $X \subset M$ and hence $X = M$. This shows that $\ker(f)$ is a maximal proper subspace of X and, hence is a hyperplane through the origin, Finally, $\ker(f)$ is closed because f is bounded (Theorem 4.1.8). ∎

4.1.15 Corollary. *If $0 \neq f \in X^*$, where X is a normed space, then for any fixed scalar $\alpha \neq 0$, the set*

$$S = \{x \in X : f(x) = \alpha\}$$

is a closed hyperplane.

Proof. Since f is a non-zero functional, $\exists\, x_1 \in X - \ker(f)$ such that $f(x_1) \neq 0$.

Setting $x_0 = \dfrac{\alpha\, x_1}{f(x_1)}$, we have $f(x_0) = \alpha$. Then

$$S = \{x \in X : f(x - x_0) = 0\} = x_0 + \ker(f).$$

Using Theorem 4.1.14, it follows that S is a closed hyperplane. ∎

We shall now find the distance of the origin from the hyperplane $\{x \in X : f(x) = 1\}$. In fact, we prove that the norm of a bounded linear functional f equals the reciprocal of the magnitude of the distance of the hyperplane $\{x \in X : f(x) = 1\}$ from the origin of coordinates.

4.1.16 Theorem. *Let $0 \neq f \in X^*$, where X is a normed space and let*

$$M_f = \{\, x \in X : f(x) = 1\}$$

be a closed hyperplane (Corollary 4.1.15). Then

$$d(0, M_f) = \frac{1}{\|f\|}.$$

Proof. Since f is bounded, we have

$$|f(x)| \leq \|f\| \, \|x\|, \quad \forall \, x \in X$$

and, in particular

$$\|f\| \, \|x\| \geq 1, \qquad \forall \, x \in M_f$$

$$\Rightarrow \qquad \|x\| \geq \frac{1}{\|f\|}, \qquad \forall \, x \in M_f$$

Therefore

$$d(0, M_f) = \inf \{\|x\| : x \in M_f\} \geq \frac{1}{\|f\|}.$$

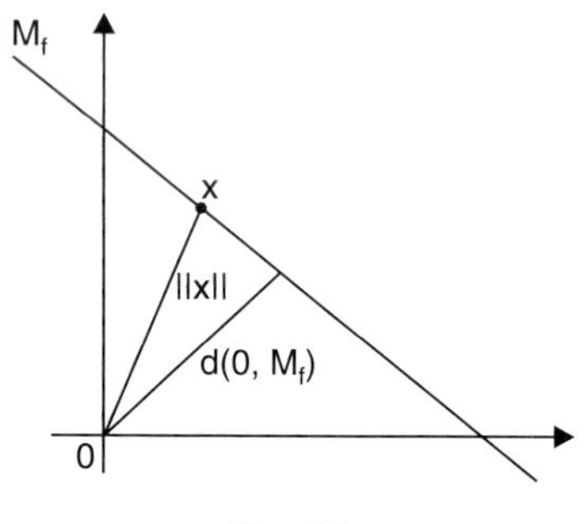

Fig. 4.1

Further

$$\|f\| = \sup \left\{ \frac{|f(x)|}{\|x\|} : x \in X, x \neq 0 \right\}$$

it follows that, for a given $K > 1$, $\exists \, 0 \neq y \in X$ such that

$$\frac{K}{\|f\|} > \left\| \frac{y}{f(y)} \right\| \geq d(0, M_f) \qquad \left(\because \ \frac{y}{f(y)} \in M_f \right)$$

But $K > 1$ being arbitrary, letting $K \to 1$, we get

$$d(0, M_f) \leq \frac{1}{\|f\|}$$

and hence the result follows. ∎

Problems

3. Let f be a bounded linear functional on a complex normed space. Is $\bar{f}$ bounded? , linear? (The bar denotes the complex conjugate.)

Solution. Let X be a complex normed space note that

$$\bar{f}(\lambda x) = \overline{f(\lambda x)} = \bar{\lambda}\,\bar{f}(x)$$

and

$$\|\bar{f}\| = \|f\|.$$

Thus $\bar{f}$ is not linear but bounded. ∎

4. Let $X = \mathbb{R}^3$ and $f(x) = x_1 + x_2 + x_3$, where $x = (x_1, x_2, x_3) \in \mathbb{R}^3$. Prove that f is a bounded linear functional. Also, find the distance of the origin from the hyperplane $x_1 + x_2 + x_3 = 1$.

Solution. f is clearly a linear functional on $\mathbb{R}^3$. Further, note that

$$|f(x)| = |x_1 + x_2 + x_3| \le \sqrt{3}\,\|x\|, \ \forall\, x \in \mathbb{R}^3.$$

This proves that f is bounded and $\|f\| \le \sqrt{3}$. Putting $x = x_0 = (1, 1, 1)$, we have

$$\|f\| \ge \frac{|f(x_0)|}{\|x_0\|} = \frac{3}{\sqrt{3}} = \sqrt{3}.$$

Thus, $\|f\| = \sqrt{3}$. By Theorem, 4.1.16, the distance d of the hyperplane $\{x \in \mathbb{R}^3 : f(x) = 1\}$ from the origin is given by

$$d = \frac{1}{\|f\|} = \frac{1}{\sqrt{3}}.$$

But

$$\{x \in \mathbb{R}^3 : f(x) = 1\} = \{(x_1, x_2, x_3) \in \mathbb{R}^3 : x_1 + x_2 + x_3 = 1\}.$$

Thus, the distance of the origin from the hyperplane $x_1 + x_2 + x_3 = 1$ is $\dfrac{1}{\sqrt{3}}$.

5. Prove that the functional defined on $C[a, b]$ by

$$g(x) = \alpha\, u(a) + \beta\, v(b),$$

$\alpha, \beta \in \mathbb{R}$ and $u, v \in C[a, b]$ is a bounded linear functional.

6. Find the norm of the linear functional f defined on $C[-1, 1]$ by

$$f(x) = \int_{-1}^{0} x(t)\, dt - \int_{0}^{1} x(t)\, dt.$$

[*Ans* : $\| f \| = 2$.]

7. On $C[a, b]$, define

$$f_1(x) = \max_{t \in [a,b]} x(t), \quad f_2(x) = \min_{t \in [a,b]} x(t).$$

(a) Are these functionals linear ?

(b) Are these functionals bounded ?

8. Let X be a normed space and let f be an additive functional on X, *i.e.*

$$f(x + y) = f(x) + f(y), \ \forall \ x, y \in X.$$

Prove that, if f is continuous on X, then f is linear.

9. Prove that on any sequence space X, a linear functional f can be defined by setting $f(x) = \xi_n$ (n fixed), where $x = \{\xi_i\} \in X$. Is f bounded if $X = l^\infty$?

10. Prove that the mapping $f : \mathbb{R}^2 \to \mathbb{R}$ defined by

$$f(\lambda, \mu) = \lambda + \mu$$

is a bounded linear functional on $(\mathbb{R}^2, \| \cdot \|_2)$.

11. Let X be a normed space and $f \in X^*$. Prove that:

(a) $I(f) = \{x \in X : f(x) = 1\}$ is a closed hyperplane, not passing through the origin, in X.

(b) The map $f \to I(f)$ gives one-to-one correspondence between non-zero bounded linear functionals and closed hyperplanes not containing zero.

4.2 The Form of Some Dual Spaces

In this section, we consider some concrete normed spaces and find out what their duals look like. The investigations of general form of the functionals on spaces and identification of their dual spaces find fruitful applications in various studies in functional analysis.

In general, $X^* \subset X^\#$, where X^* is the topological dual of X as a normed space and $X^\#$ is the algebraic dual of X as a linear space. However, we first establish the existence of non-zero bounded linear functionals on an arbitrary finite dimensional normed space, and find its dual space.

4.2.1 Theorem. *Let X be a finite dimensional normed space over the field* $\mathbb{K}$. *Then,* $X^{\#} = X^{*}$. *Furthermore, dim* $X^{*} = \dim X$.

Proof. Let $\dim X = n$ and let $\{e_1, e_2, \dots , e_n\}$ be a basis (Hamel) of X. Then, every $x \in X$ can be expressed uniquely in the form

$$(2) \qquad x = \sum_{i=1}^{n} \alpha_i e_i$$

where $\alpha_1, \alpha_2, \dots, \alpha_n$ are scalars in $\mathbb{K}$. Define the mapping $\| \cdot \|_0 : X \to \mathbb{R}$ by

$$\| x \|_0 = \sum_{i=1}^{n} |\alpha_i|.$$

It can be easily verified that $\| \cdot \|_0$ is a norm on X. Since any two norms on a finite dimensional space are equivalent (Corollary 3.4.3), it follows that $\| \cdot \|$ and $\| \cdot \|_0$ are equivalent norms on X. So, $\exists$ a real number $K > 0$ such that

$$\| x \|_0 \leq K \| x \|, \quad \forall\, x \in X \qquad \text{(Theorem 3.3.5)}$$

$$\Rightarrow \qquad \sum_{i=1}^{n} |\alpha_i| \leq K \| x \|.$$

For each $i = 1, 2, \dots, n$, consider the functional $f_i : X \to \mathbb{K}$ defined by $f_i(x) = \alpha_i$, where x and α_i are related by (2). Clearly, each f_i is a linear functional on X. Also

$$|f_i(x)| = |\alpha_i| \leq \sum_{i=1}^{n} |\alpha_i| \leq K \| x \|, \quad \forall\, x \in X.$$

Therefore, each f_i is bounded and hence the set

$$S = \{ f_1, f_2, \dots, f_n \} \subset X^{*}.$$

We, now, prove that S constitutes a basis for X^{*}. Let

$$\sum_{i=1}^{n} \lambda_i f_i = 0, \quad \lambda_i \in \mathbb{K}.$$

Then

$$\sum_{i=1}^{n} \lambda_i f_i (x) = 0, \quad \forall \, x \in X.$$

In particular, for $x = \sum_{i=1}^{n} \overline{\lambda}_i e_i$, we have

$$\sum_{i=1}^{n} |\lambda_i|^2 = 0$$

$$\Rightarrow \qquad \lambda_i = 0, \quad (i = 1, 2, ..., n).$$

This verifies that S is a linearly independent set.

Further, let $f \in X^{\#}$. Using (2) and the definition of f_i, we have

$$f(x) = \sum_{i=1}^{n} f_i (x) f(e_i)$$

$$= \left(\sum_{i=1}^{n} f(e_i) f_i \right) (x), \, \forall \, x \in X.$$

This gives

$$(3) \qquad f = \sum_{i=1}^{n} \beta_i f_i \, , \text{ where } \beta_i = f(e_i).$$

This verifies that f is bounded so that $f \in X^*$. This proves that $X^{\#} = X^*$ and also that S spans X^*. Thus, the set $S = \{f_1, f_2, ..., f_n\}$ forms a basis for X^*. Hence

$$\dim X^* = n = \dim X. \quad \blacksquare$$

Note. We have proved indirectly that a linear functional on a finite dimensional normed space is always bounded. The expression (3) gives a general form of the linear functionals defined on n-dimensional normed space.

4.2.2 Example. Dual space of $\mathbb{R}^n$ is isometrically isomorphic to $\mathbb{R}^n$.

In view of Theorem 4.2.1, it follows that $(\mathbb{R}^n)^* = (\mathbb{R}^n)^{\#}$. Therefore, it is sufficient to prove that $(\mathbb{R}^n)^{\#}$ is isometrically isomorphic to $\mathbb{R}^n$. Consider the natural basis $\{e_1, e_2, ..., e_n\}$. Then, every $x = (\alpha_1, \alpha_2, ..., \alpha_n) \in \mathbb{R}^n$ can be expressed uniquely in the form

$$x = \sum_{i=1}^{n} \alpha_i e_i \, .$$

Define the mapping $T : (\mathbb{R}^n)^{\#} \to \mathbb{R}^n$ by

$$Tf = (f(e_1), f(e_2), ..., ..., f(e_n)), f \in (\mathbb{R}^n)^{\#}.$$

Clearly, T is linear and bijective. Also, by Cauchy–Schwartz Inequality (1.5.5), we have

$$|f(x)| \leq \sum_{i=1}^{n} |\alpha_i f(e_i)|$$

$$\leq \left(\sum_{i=1}^{n} \alpha_i^2 \right)^{\frac{1}{2}} \left(\sum_{i=1}^{n} (f(e_i))^2 \right)^{\frac{1}{2}}$$

$$= \| x \| \left(\sum_{i=1}^{n} (f(e_i))^2 \right)^{\frac{1}{2}}, \quad \forall \, x \in \mathbb{R}^n$$

$$\Rightarrow \qquad \| f \| \leq \left(\sum_{i=1}^{n} (f(e_i))^2 \right)^{\frac{1}{2}} = \| Tf \|.$$

By choosing $x = x_0 \equiv (f(e_1), f(e_2), ..., f(e_n))$, we get

$$\| f \| \geq \frac{|f(x_0)|}{\| x_0 \|} = \left(\sum_{i=1}^{n} (f(e_i))^2 \right)^{\frac{1}{2}} = \| Tf \|.$$

This gives

$$\| Tf \| = \| f \|, \quad \forall f \in (\mathbb{R}^n)^{\#}.$$

Hence, $(\mathbb{R}^n)^{\#}$ and $\mathbb{R}^n$ are isometrically isomorphic spaces. ∎

Note. If the dual space of a normed space X is isometrically isomorphic to a normed space Y, we would say that the dual space of X is Y. For instance, the dual space of $\mathbb{R}^n$ is $\mathbb{R}^n$ itself.

Problem 12. Prove that the dual space of $\mathbb{C}^n$ (Example 2.2.2) is $\mathbb{C}^n$.

4.2.3 Example. Dual space of $l^p(n)$, $1 < p < \infty$, (Example 2.2.4) is $l^q(n)$, where

$$1 < q < \infty \text{ and } \frac{1}{p} + \frac{1}{q} = 1.$$

Note that $l^p(n) \equiv (l^p(n), \|\cdot\|_p)$ and $l^q(n) \equiv (l^q(n), \|\cdot\|_q)$. Consider the natural basis $\{e_1, e_2, ..., e_n\}$ for $l^p(n)$. Then, every $x = (\alpha_1, \alpha_2, ..., \alpha_n) \in l^p(n)$ can be expressed uniquely in the form.

$$x = \sum_{i=1}^{n} \alpha_i e_i .$$

Since $l^p(n)$ is a finite dimensional normed space, every linear functional on $l^p(n)$ is continuous (Theorem 4.2.1). Thus, if f is a continuous linear functional on $l^p(n)$, the unique representation for f is given by

$$f(x) = \sum_{i=1}^{n} \alpha_i f(e_i), \quad x = (\alpha_1, \alpha_2, ..., \alpha_n) \in l^p(n).$$

Clearly, $(f(e_1), f(e_2), ..., f(e_n)) \in l^q(n)$. Write

$$Tf = u \equiv (f(e_1), f(e_2), ..., f(e_n)).$$

Then, it can be easily seen that $T : (l^p(n))^* \to l^q(n)$ is a linear mapping. It is obvious that T is bijective. We, now, prove that T preserves norm.

Now, since f is bounded, it follows by Hölder Inequality (1.5.4), that

$$|f(x)| \le \sum_{i=1}^{n} |\alpha_i f(e_i)|$$

$$\le \left(\sum_{i=1}^{n} |\alpha_i|^p \right)^{\frac{1}{p}} \left(\sum_{i=1}^{n} |f(e_i)|^q \right)^{\frac{1}{q}}$$

$$= \|x\|_p \left(\sum_{i=1}^{n} |f(e_i)|^q \right)^{\frac{1}{q}}$$

$$\Rightarrow \qquad \|f\| \le \left(\sum_{i=1}^{n} |f(e_i)|^q \right)^{\frac{1}{q}} = \|u\|_q = \|Tf\|_q.$$

Further, choose the vector $x = x_0 \equiv (\lambda_1, \lambda_2, ..., \lambda_n) \in l^p(n)$, where

$$\lambda_i = \begin{cases} \dfrac{|f(e_i)|^q}{f(e_i)}, & f(e_i) \neq 0 \\ 0, & \text{otherwise.} \end{cases}$$

Then

$$\| x_0 \|_p = \left(\sum_{i=1}^{n} |\lambda_i|^p \right)^{\frac{1}{p}} = \left(\sum_{i=1}^{n} |f(e_i)|^q \right)^{\frac{1}{p}}$$

since $\dfrac{1}{p} + \dfrac{1}{q} = 1$, and

$$f(x_0) = \sum_{i=1}^{n} \lambda_i f(e_i) = \sum_{i=1}^{n} |f(e_i)|^q .$$

Therefore

$$\|f\| \geq \frac{|f(x_0)|}{\| x_0 \|_p} = \left(\sum_{i=1}^{n} |f(e_i)|^q \right)^{\frac{1}{q}}, \quad \left(\because \ \frac{1}{p} + \frac{1}{q} = 1 \right)$$

$$\Rightarrow \qquad \|f\| \geq \| u \|_q = \| Tf \|_q.$$

This gives

$$\| Tf \|_q = \|f\|, \quad \forall f \in (l^p(n))^*.$$

Thus, T is an isometric isomorphism of $(l^p(n), \| \cdot \|_p)^*$ onto $(l^q(n), \| \cdot \|_q)$ and hence $(l^p(n), \| \cdot \|_p)^*$ and $(l^q(n), \| \cdot \|_q)$ are isometrically isomorphic spaces. ∎

Remark. Note that if $p = 2$, then $q = 2$. Also, $l^2(n) \equiv \mathbb{R}^n$ or $\mathbb{C}^n$ according as the field $\mathbb{K}$ is $\mathbb{R}$ or $\mathbb{C}$. As such, the dual space of $\mathbb{R}^n$ is $\mathbb{R}^n$ and that of $\mathbb{C}^n$ is $\mathbb{C}^n$. Also, see independent proof in Example 4.2.2.

4.2.4 Example. The dual space of $(\mathbb{C}^n, \| \cdot \|_\infty)$ is the space $(\mathbb{C}^n, \| \cdot \|_1)$.

Consider the natural basis $\{e_1, e_2, ..., e_n\}$ for $\mathbb{C}^n$. Then, every continuous linear functional f on $\mathbb{C}^n$ can be expressed uniquely in the form

$$f(x) = \sum_{i=1}^{n} \alpha_i f(e_i), \quad x = (\alpha_1, \alpha_2, \ldots, \alpha_n) \in \mathbb{C}^n.$$

Clearly, $u = (f(e_1), f(e_2), \ldots, f(e_n)) \in \mathbb{C}^n$. Define the mapping $T : (\mathbb{C}^n)^* \to \mathbb{C}^n$ by

$$Tf = u = (f(e_1), f(e_2), \ldots, f(e_n)).$$

Then, T is linear and bijective. Further, since f is bounded, we have

$$|f(x)| \le \sum_{i=1}^{n} |\alpha_i f(e_i)|$$

$$\le \left(\sum_{i=1}^{n} |f(e_i)| \right) \sup_{1 \le i \le n} |\alpha_i|$$

$$= (\| u \|_1) \| x \|_\infty$$

$$\Rightarrow \qquad \| f \| \le \| u \|_1.$$

Choose an element $x = x_0 \equiv (\lambda_1, \lambda_2, \ldots, \lambda_n) \in \mathbb{C}^n$, where

$$\lambda_i = \begin{cases} \dfrac{\overline{f(e_i)}}{|f(e_i)|}, & f(e_i) \ne 0 \\ 0, & \text{otherwise.} \end{cases}$$

Then

$$\| x_0 \|_\infty \le 1 \text{ and } f(x_0) = \| u \|_1.$$

Therefore

$$\| f \| \ge \frac{|f(x_0)|}{\| x_0 \|_\infty} \ge \| u \|_1.$$

This gives

$$\| f \| = \| u \|_1 = \| Tf \|_1.$$

Thus, T is an isometric isomorphism of $(\mathbb{C}^n, \| \cdot \|_\infty)^*$ onto $(\mathbb{C}^n, \| \cdot \|_1)$. Hence, $(\mathbb{C}^n, \| \cdot \|_\infty)^*$ and $(\mathbb{C}^n, \| \cdot \|_1)$ are isometrically isomorphic spaces. ∎

Problem 13. Prove that the dual space of $(\mathbb{C}^n, \|\cdot\|_1)$ is isometrically isomorphic to $(\mathbb{C}^n, \|\cdot\|_\infty)$.

We now proceed to investigate the dual spaces of some concrete infinite dimensional normed spaces. The dual space of an infinite dimensional space X is determined not only by X but also by the norm $\|\cdot\|$ on X.

4.2.5 Examples

1. The dual space of $(c_0, \|\cdot\|_\infty)$ is $(l^1, \|\cdot\|_1)$.

   Since the sequence $\{e_i\}$ constitutes a Schauder basis for the space $(c_0, \|\cdot\|_\infty)$ (Appendix), every $x = \{\lambda_i\} \in c_0$ can be expressed uniquely in the form

$$(4) \qquad x = \sum_{i=1}^{\infty} \lambda_i e_i .$$

   Let $f \in c_0^*$ be arbitrary. Write $s_n = \sum_{i=1}^{n} \lambda_i e_i$. Then

$$f(s_n) = \sum_{i=1}^{n} \lambda_i f(e_i)$$

   and $\qquad \| x - s_n \|_\infty \to 0$, as $n \to \infty$.

   Since f is continuous, it follows that $f(s_n) \to f(x)$. Hence

$$(5) \qquad f(x) = \sum_{i=1}^{\infty} \lambda_i f(e_i), \quad x = \{\lambda_i\} \in c_0.$$

   We first show that $\{f(e_i)\} \in l^1$. For any $n \in \mathbb{N}$, consider the element $x = x_0 \equiv \{\xi_i\} \in c_0$, where

$$\xi_i = \begin{cases} \dfrac{\overline{f(e_i)}}{|f(e_i)|}, & f(e_i) \neq 0, 1 \leq i \leq n \\ 0, & f(e_i) = 0, i \geq n+1. \end{cases}$$

   Then $\qquad \| x_0 \|_\infty = \sup_{1 \leq i < \infty} |\xi_i| \leq 1$

and
$$f(x_0) = \sum_{i=1}^{\infty} \xi_i\, f(e_i) = \sum_{i=1}^{n} |f(e_i)| \le \|f\|.$$

But this is true for all $n \in \mathbb{N}$. Therefore, $\displaystyle\sum_{i=1}^{\infty} |f(e_i)| < \infty$ and hence $\{f(e_i)\} \in l^1$. Also

(6)
$$\sum_{i=1}^{\infty} |f(e_i)| \le \|f\|.$$

Now, define the mapping $T : c_0^* \to l^1$ by

$$Tf = u = \{f(e_i)\}.$$

Clearly, T is linear. From (5), it follows that $f = 0$ if $Tf = O$ so that $\ker(T) = \{0\}$. Hence, T is one-to-one.

To prove that T is onto, we suppose that $y = \{\alpha_i\} \in l^1$. Since $\{\alpha_i\}$ is a

bounded sequence, the series $\displaystyle\sum_{i=1}^{\infty} \alpha_i \lambda_i$ is absolutely convergent for each

$x = \{\lambda_i\} \in c_0$. Define the functional $g : c_0 \to \mathbb{K}$ by

$$g(x) = \sum_{i=1}^{\infty} \alpha_i \lambda_i, \quad x = \{\lambda_i\} \in c_0.$$

Clearly, g is linear. Also

$$|g(x)| \le \sum_{i=1}^{\infty} |\alpha_i \lambda_i| \le \left(\sum_{i=1}^{\infty} |\alpha_i| \right) \|x\|_{\infty}.$$

This shows that g is bounded and hence $g \in c_0^*$. Thus, T is onto.

Finally, we prove that T preserves norm. Equation (5) yields

$$|f(x)| \le \sum_{i=1}^{\infty} |\lambda_i f(e_i)|$$

$$\leq \left(\sum_{i=1}^{\infty} |f(e_i)| \right) \sup_{1 \leq i < \infty} |\lambda_i|$$

$$= \| x \|_{\infty} \left(\sum_{i=1}^{\infty} |f(e_i)| \right)$$

$$\Rightarrow \qquad \| f \| \leq \sum_{i=1}^{\infty} |f(e_i)|$$

This on combining with (6) gives

$$\| f \| = \sum_{i=1}^{\infty} |f(e_i)| = \| u \|_1 = \| Tf \|_1.$$

Thus, T is an isometric isomorphism of $(c_0, \| \cdot \|_{\infty})^*$ onto $(l^1, \| \cdot \|_1)$. Hence, $(c_0, \| \cdot \|_{\infty})^*$ and $(l^1, \| \cdot \|_1)$ are isometrically isomorphic spaces. ∎

2. The dual space of l^1 is l^{∞}.

Note that $\{e_i\}$ constitutes a (Schauder) basis for l^1 (Appendix). Proceeding as in Example 4.2.5 (1), a bounded linear functional f on l^1 can be represented uniquely in the form.

$$(7) \qquad f(x) = \sum_{i=1}^{\infty} \xi_i f(e_i), \quad x = \{\xi_i\} \in l^1.$$

Now, for each $i \in \mathbb{N}$, we have

$$|f(e_i)| \leq \| f \| \, \| e_i \|_1 = \| f \| < \infty$$

$$\Rightarrow \qquad \sup_{i \in \mathbb{N}} |f(e_i)| \leq \| f \|.$$

This proves that $u = \{ f(e_i) \} \in l^{\infty}$ and

$$(8) \qquad \| u \|_{\infty} \leq \| f \|.$$

Define the mapping $T : (l^1)^* \to l^{\infty}$ by

$$Tf = u = \{ f(e_i) \}.$$

Clearly, T is linear. From (7), it follows that $f = O$ if $Tf = 0$ so that ker $(T) = \{0\}$. Hence, T is one-to-one. To prove that T is onto, we suppose that $b = \{\beta_i\} \in l^\infty$. Define the functional $g_b : l^1 \to \mathbb{K}$ by

$$g_b(x) = \sum_{i=1}^{\infty} \xi_i \beta_i, \quad x = \{\xi_i\} \in l^1.$$

Obviously, g_b is linear and

$$|g_b(x)| \le \sum_{i=1}^{\infty} |\xi_i \beta_i| \le \left(\sup_{j \in \mathbb{N}} |\beta_j|\right) \sum_{i=1}^{\infty} |\xi_i| = \|x\|_1 \left(\sup_{j \in \mathbb{N}} |\beta_j|\right).$$

This shows that g_b is bounded and hence $g_b \in (l^1)^*$. Thus, T is onto.

Finally, we prove that T preserves norm. From equation (7), we have

$$|f(x)| = \sum_{i=1}^{\infty} |\xi_i f(e_i)| \le \left(\sup_{i \in \mathbb{N}} |f(e_i)|\right) \sum_{j=1}^{\infty} |\xi_j|$$

$$\Rightarrow \qquad \|f\| \le \sup_{i \in \mathbb{N}} |f(e_i)| = \|u\|_\infty.$$

This together with (8) yields

$$\|f\| = \|u\|_\infty = \|Tf\|_\infty.$$

Hence, T is an isometric isomorphism of $(l^1)^*$ onto l^∞. Thus, $(l^1)^* = l^\infty$. ∎

3. The dual space of l^p, $1 < p < \infty$, (Example 2.2.5) is l^q, where

$$1 < q < \infty \text{ and } \frac{1}{p} + \frac{1}{q} = 1.$$

Note that $\{e_i\}$ is a (Schauder) basis for l^p, (Appendix) and hence each $x = \{\xi_i\} \in l^p$ has a unique representation

$$(9) \qquad x = \sum_{i=1}^{\infty} \xi_i e_i.$$

Let $f \in (l^p)^*$ be arbitrary. Then, f can be expressed uniquely in the form

$$(10) \qquad f(x) = \sum_{i=1}^{\infty} \xi_i f(e_i), \ \forall \ x = \{\xi_i\} \in l^p.$$

First, we show that $\{f(e_i)\} \in l^q$. Consider $x = \{\lambda_i\} \in l^p$ with

$$\lambda_i = \begin{cases} \dfrac{|f(e_i)|^q}{f(e_i)}, & f(e_i) \neq 0, 1 \leq i \leq n \\ 0, & f(e_i) = 0, i \geq n+1. \end{cases}$$

Substituting $x = x_0$, we get

$$f(x_0) = \sum_{i=1}^{\infty} \lambda_i f(e_i) = \sum_{i=1}^{n} |f(e_i)|^q \leq \|f\| \, \|x_0\|_p$$

and

$$\|x_0\|_p = \left(\sum_{i=1}^{\infty} |\lambda_i|^p\right)^{\frac{1}{p}} = \left(\sum_{i=1}^{n} |f(e_i)|^{(q-1)p}\right)^{\frac{1}{p}}$$

$$= \left(\sum_{i=1}^{n} |f(e_i)|^q\right)^{\frac{1}{p}}.$$

This gives

$$(11) \qquad \left(\sum_{i=1}^{n} |f(e_i)|^q\right)^{\frac{1}{q}} \leq \|f\|, \quad (\because \quad 1 - \frac{1}{p} = \frac{1}{q}).$$

But this is true for all $n \in \mathbb{N}$. Therefore

$$\sum_{i=1}^{\infty} |f(e_i)|^q < \infty$$

$$\Rightarrow \qquad \{f(e_i)\} \in l^q.$$

Now, define the mapping $T : (l^p)^* \to l^q$ by

$$Tf = u = \{f(e_i)\}.$$

Clearly, T is linear. From (10), it follows that $f = O$ if $Tf = 0$ so that $\ker(T) = \{0\}$. Hence, T is one-to-one. To prove that T is onto, we suppose that $b = \{\beta_i\} \in l^q$. Define the functional $g_b : l^p \to \mathbb{K}$ by

$$g_b(x) = \sum_{i=1}^{\infty} \xi_i \beta_i, \quad x = \{\xi_i\} \in l^p.$$

Obviously, g_b is linear and

$$|g_b(x)| \le \sum_{i=1}^{\infty} |\xi_i \beta_i|$$

$$\le \left(\sum_{i=1}^{\infty} |\xi_i|^p \right)^{\frac{1}{p}} \left(\sum_{i=1}^{\infty} |\beta_i|^q \right)^{\frac{1}{q}},$$

(Hölder Inequality (1.5.8))

$$\Rightarrow \qquad |g_b(x)| \le \left(\sum_{i=1}^{\infty} |\beta_i|^q \right)^{\frac{1}{q}}, \text{ when } \|x\|_p = 1.$$

This shows that g_b is bounded and hence $g_b \in (l^p)^*$. Thus, T is onto.

Finally, we prove that T preserves norm. From equation (10) and Hölder Inequality (1.5.8), we have

$$|f(x)| \le \sum_{i=1}^{\infty} |\xi_i f(e_i)| \le \left(\sum_{i=1}^{\infty} |\xi_i|^p \right)^{\frac{1}{p}} \left(\sum_{i=1}^{\infty} |f(e_i)|^q \right)^{\frac{1}{q}}.$$

Taking the supremum overall x of norm 1, we get

$$(12) \qquad \|f\| \le \left(\sum_{i=1}^{\infty} |f(e_i)|^q \right)^{\frac{1}{q}}.$$

Inequalities (11) and (12) together yield

$$\|f\| = \|u\|_q = \|Tf\|_q.$$

Hence, T is an isometric isomorphism of $(l^p)^*$ onto l^q. Thus, $(l^p)^* = l^q$. ∎

Remarks

1. Since the space $(l^p, \|\cdot\|_p)$ is isometrically isomorphic to a dual space $(l^q, \|\cdot\|_q)^*$, where $\dfrac{1}{p} + \dfrac{1}{q} = 1$, $(l^p, \|\cdot\|_p)$ is complete for $1 < p < \infty$.

2. When $p = 2$, we have classical Hilbert space $(l^2, \| \cdot \|_2)$ which has peculiar property of being isometrically isomorphic to its own dual $(l^2, \| \cdot \|_2)^*$. Hilbert spaces will be discussed in Chapter 5.

It is possible that the duals of a Banach space and a proper subspace of it are the same.

4.2.6 Example. Consider the real Banach space $(c, \| \cdot \|_\infty)$ and it proper closed subspace $(c_0, \| \cdot \|_\infty)$, Example 2.3.4 (2). The dual spaces $(c, \| \cdot \|_\infty)^*$ and $(c_0, \| \cdot \|_\infty)^*$ are isometrically isomorphic.

Define the mapping $y \to f$ from l^1 into c^* by

$$f(x) = \mu_1 \lim_{n \to \infty} \lambda_n + \sum_{n=2}^{\infty} \mu_n \lambda_{n-1}, \ x = \{\lambda_i\} \in c \text{ and } y = \{\mu_i\} \in l^1.$$

Let $T : l^1 \to c^*$ be defined by

$$Ty = f \in c^*, \ y = \{\mu_i\} \in l^1.$$

We shall prove that T is an isometric isomorphism.

It is easy to prove that T is linear and bijective. It remains to prove that T preserves norm. Now

$$\|f\| = \sup \{ |f(x)| : x \in c, \|x\|_\infty = 1 \}$$

$$= \sup \{ |\mu_1 \lim_{n \to \infty} \lambda_n + \sum_{n=2}^{\infty} \mu_n \lambda_{n-1}| : x \in c, \|x\|_\infty = 1 \}$$

$$\leq \sup \{ |\mu_1| |\lim_{n \to \infty} \lambda_n| + \sum_{n=2}^{\infty} |\mu_n| |\lambda_{n-1}| : x \in c, \|x\|_\infty = 1 \}$$

$$= |\mu_1| |\lim_{n \to \infty} \lambda_n| + \sum_{n=2}^{\infty} |\mu_n| |\lambda_{n-1}|$$

$$= \sum_{n=1}^{\infty} |\mu_n|, \text{ on taking } \lambda_n = 1, \ \forall \ n \in \mathbb{N}$$

$$= \|y\|_1$$

$$\Rightarrow \quad \|Ty\| \leq \|y\|_1.$$

For reverse inequality, we consider for each $m \in \mathbb{N}$, a sequence $\{\delta_{nm}\}$, where

$$\delta_{mn} = \begin{cases} 0, & \mu_n = 0 \text{ and } n \leq m \\[2mm] \dfrac{\mu_n}{|\mu_n|}, & \mu_n \neq 0 \text{ and } n \leq m \\[2mm] \dfrac{\mu_1}{|\mu_1|}, & n > m. \end{cases}$$

Then, for each $m \in \mathbb{N}$, we have

$$\lim_{n \to \infty} \delta_{mn} = \frac{\mu_1}{|\mu_1|}$$

$$\Rightarrow \qquad x_0 = \{\delta_{mn}\} \in c.$$

Therefore

$$\|f\| \geq \frac{|f(x_0)|}{\|x_0\|_\infty}$$

$$= |\mu_1 \lim_{n \to \infty} \delta_{mn} + \sum_{n=1}^{\infty} \mu_{n+1} \, \delta_{mn}| \quad (\because \quad \|x_0\|_\infty = 1)$$

$$= |\mu_1| + |\mu_2| + \dots + |\mu_m| + R_m$$

where

$$|R_m| < |\mu_{m+1}| + |\mu_{m+2}| + \dots$$
$$\to 0 \qquad\qquad\qquad (\because \quad y = \{\mu_i\} \in l^1)$$

$$\Rightarrow \qquad \|f\| \geq \sum_{n=2}^{\infty} |\mu_n| = \|y\|_1$$

$$\Rightarrow \qquad \|Ty\| \geq \|y\|_1.$$

This gives

$$\|Ty\| = \|y\|_1.$$

Hence, T is an isometric isomorphism. Thus, $(l^1, \|\cdot\|_1)$ is isometrically isomorphic to a subspace of $(c, \|\cdot\|_\infty)^*$. But it can be easily seen that $(c, \|\cdot\|_\infty)^*$ is isometrically isomorphic to a subspace of $(c_0, \|\cdot\|_\infty)^*$. Further, in Example 4.2.5 (1), it is shown that $(c_0, \|\cdot\|_\infty)^*$ is isometrically isomorphic to $(l^1, \|\cdot\|_1)$. Hence, $(c_0, \|\cdot\|_\infty)^*$ and $(c, \|\cdot\|_\infty)^*$ are isometrically isomorphic. ∎

4.2.7 Example. The dual of $L^p[a, b]$ is $L^q[a, b]$ for $1 \leq p < \infty$, where p and q are conjugate exponent. The dual of $L^\infty[a, b]$ is a very big space bigger than $L^1[a, b]$ and is rarely used.

Problems

14. Let $X = (\mathbb{R}^n, \| \cdot \|_\infty)$ be the normed space (Example 2.2.3). What is the corresponding norm on the dual space X^* ? What conclusion can you draw with respect to the space X ?

[*Hint:* See Example 4.2.4 and its solution.]

15. Let $\| \cdot \|_1$ and $\| \cdot \|_2$ be two norms on a linear space X with $\| x \|_1 \leq K \| x \|_2$, $\forall\, x \in X$ and $K > 0$. Prove that $(X, \| \cdot \|_1)^* \subset (X, \| \cdot \|_2)^*$.

16. Let X be a normed space and $\dim X = \infty$. Prove that the dual space X^* is not identical with the algebraic dual space $X^\#$.

4.3 Hahn-Banach Theorem and its Consequences

In this section, we discuss the existence of non-zero bounded linear functionals on an arbitrary non-zero normed space. In fact, while dealing with normed spaces, one often requires to construct linear functionals with certain properties. This can be achieved in two steps: first, one defines a linear functional on a subspace of the normed space where it is easy to verify the desired properties; second, one appeals to an extension theorem which says that any such functional can be extended to the whole space while retaining the desired properties. The Hahn-Banach Theorem is basically an extension theorem for linear functionals. In this theorem, we consider a bounded linear functional f defined on a subspace M of a given normed space X and then we extend this f from M to the entire space X in such a way that certain basic properties of f continue to hold good for the extended functional.

We first derive an extension theorem, called the *Hahn-Banach Lemma,* for linear functionals on an arbitrary real linear space, we then apply this lemma to prove the fundamental extension theorem, called the *Hahn-Banach Theorem,* for bounded linear functionals. For real Banach spaces, the Hahn-Banach Theorem was first proved by H. Hahn in 1927 and then rediscovered in the present general form by S. Banach in 1929. The generalizations that include complex linear spaces were obtained by H.F. Bohnenblust and A. Sobczyk in 1938.

For proving Hahn-Banach Lemma, it is now customary to use Zorn's Lemma. Zorn's Lemma is an established and routine tool of mathematics and the proofs of such extension theorems provide concrete examples of the way in which it is used. First, we define a sublinear functional on a linear space.

4.3.1 Definition. Let X be a linear space over the field $\mathbb{K}$. A mapping $p : X \to \mathbb{R}$ is said to be a *sublinear functional* on X if

   (i) $p(x + y) \leq p(x) + p(y)$, $\forall\ x, y \in X$ (Subadditive)

   (ii) $p(\alpha x) = \alpha p(x)$, $\forall\ x \in X$ and $\alpha \in \mathbb{R}$ with $\alpha \geq 0$. (Positive homogeneous)

Remark. A norm on a linear space X is obviously a sublinear functional on X (Example 4.1.7).

4.3.2 Theorem (Hahn-Banach Lemma). *Let X be a real linear space and let p be a sublinear functional on X. If f is a real linear functional on a subspace M of X satisfying*

$$f(x) \leq p(x), \quad \forall\ x \in M$$

then, $\exists$ a real linear functional F on X such the $F/M = f$ and

$$F(x) \leq p(x), \quad \forall\ x \in X.$$

Proof. We shall give the proof of the lemma in two steps.

   Step (i). Let $\mathscr{P}$ denote the set of all linear extensions (M_α, h_α) of (M, f) which satisfy the condition

$$h_\alpha(x) \leq p(x), \quad \forall\ x \in M_\alpha$$

that is

$$\mathscr{P} = \{(M_\alpha, h_\alpha) : M_\alpha \text{ is a linear subspace of } X \text{ containing } M,$$
$$h_\alpha\big|_M = f \text{ and } h_\alpha(x) \leq p(x), \forall\ x \in M_\alpha\}.$$

Clearly, $\mathscr{P} \neq \phi$ since $(M, f) \in \mathscr{P}$. We define a relation "$\leq$" on $\mathscr{P}$ as follows :

$$(M_\alpha, h_\alpha) \leq (M_\beta, h_\beta) \quad \Leftrightarrow \quad M_\alpha \subset M_\beta$$

and $h_\beta\big|_{M_\alpha} = h_\alpha$, i.e., by definition, h_β is an extension of h_α.

   It is easy to see that the relation "$\leq$" is reflexive, antisymmetric and transitive and hence $(\mathscr{P}, \leq)$ becomes a partially ordered set. Let $\mathscr{Y}$ be any totally ordered subset (*i.e.* chain) of $\mathscr{P}$ and let

$$M' = \beta \cup \{ M_\beta : (M_\beta, h_\beta) \in \mathscr{Y} \}.$$

Note that M' is a subspace because of the total ordering of $\mathscr{Y}$. Now, define $h' : M' \to \mathbb{R}$ by

$$h'(x) = h_\beta(x), \quad \forall\ x \in M_\beta.$$

Then, h' is clearly a linear functional on M' and $h'|_M = f$. We claim that (M', h') is an upper bound of $\mathscr{Y}$ because if

$$M_\beta \subset M' \text{ and } h'|_{M_\beta} = h_\beta$$

then

$$(M_\beta, h_\beta) \le (M', h').$$

Hence, by Zorn's Lemma 1.3.5, it follows that $\mathscr{P}$ has a maximal element (M_0, h_0), say.

We need to show that $(M_0, h_0) \equiv (X, F)$. But, if we can show that $M_0 = X$, then obviously $F = h_0$ and then the proof of the theorem is complete.

Step (ii). We assume that $M_0 \ne X$. Then $\exists$ a vector $x_0 \in X - M_0$. Now, consider the linear subspace $M_1 = M_0 + [x_0]$ spanned by M_0 and x_0. Then, each point $y \in M_1$ can be expressed uniquely in the form

$$y = x + \lambda x_0,$$

where $x \in M_0$ and $\lambda \in \mathbb{R}$. Define $h_1 : M_1 \to \mathbb{R}$ by

$$h_1 (x + \lambda x_0) = h_0 (x) + \lambda C, \quad x \in M_0$$

where C is any real constant. Then, h_1 is obviously a linear functional on M_1. Also $h_1|_M = h_0$ since for $\lambda = 0$, we have

$$h_1 (x) = h_0 (x), \quad \forall x \in M_0.$$

Thus, it is now clear that M_1 is a linear subspace of X containing M and $h_1|_M = f$.

We, now, wish to choose the constant C in such a way that

$$h_1 (y) \le p (y), \quad \forall y \in M_1$$

or, equivalently, the inequality

(13) $$\lambda C \le p (x + \lambda x_0) - h_0 (x)$$

holds for all $\lambda \in \mathbb{R}$ and $x \in M_0$.

Observe that, for any two points $x_1, x_2 \in M_0$, we have

$$h_0 (x_2) - h_0 (x_1) = h_0 (x_2 - x_1)$$
$$\le p (x_2 - x_1)$$

$$= p\left(x_2 + x_0 - (x_1 + x_0)\right)$$

$$\leq p\left(x_2 + x_0\right) + p\left(- x_1 - x_0\right)$$

$$\Rightarrow \quad - p\left(- x_1 - x_0\right) - h_0\left(x_1\right) \leq p\left(x_2 + x_0\right) - h_0\left(x_2\right).$$

Setting

$$\begin{cases} A = \sup_{x_1 \in M_0} \{- p\left(- x_1 - x_0\right) - h_0\left(x_1\right)\} \\ B = \inf_{x_1 \in M_0} \{p\left(x_2 + x_0\right) - h_0\left(x_2\right)\} \end{cases}$$

it follows that $A \leq B$. Therefore, $\exists$ a real number C such that

$$A \leq C \leq B.$$

This yields

(14) $$\qquad - p\left(- x_1 - x_0\right) - h_0\left(x_1\right) \leq C, \quad \forall\, x_1 \in M_0.$$

(15) $$\qquad C \leq p\left(x_2 + x_0\right) - h_0\left(x_2\right), \qquad \forall\, x_2 \in M_0.$$

We, now, come back to (13) and prove it in three cases.

Case (i). Let $\lambda < 0$. Putting $x_1 = \dfrac{x}{\lambda}$ in (14), we get

$$- p\left[(- \frac{1}{\lambda})\,(x + \lambda\, x_0)\right] - \frac{1}{\lambda}\, h_0\left(x\right) \leq C$$

$$\Rightarrow \qquad \frac{1}{\lambda}\, p\left(x + \lambda x_0\right) - \frac{1}{\lambda}\, h_0\left(x\right) \leq C.$$

Case (ii). Let $\lambda = 0$. Then, $y = x$ and there is nothing to prove.

Case (iii). Let $\lambda > 0$. Putting $x_2 = \dfrac{x}{\lambda}$ in (15), we get

$$C \leq \frac{1}{\lambda}\, p\left(x + \lambda x_0\right) - \frac{1}{\lambda}\, h_0\left(x\right).$$

Multiplication by λ on both sides yields (13).

We thus conclude that $(M_1, h_1) \in \mathscr{P}$ and

$$(M_0, h_0) \leq (M_1, h_1), \quad M_0 \neq M_1.$$

But this contradicts the maximality of (M_0, h_0). Hence, the supposition that $M_0 \neq X$ is false.

This completes the proof. ∎

Before presenting our main theorem, we recall that associated with each complex linear space $(X, \mathbb{C})$, there is a real linear space $(X, \mathbb{R})$. Let us denote this associated real linear space by $X_{\mathbb{R}}$. Thus, to any given normed space $(X, \| \cdot \|)$ over $\mathbb{C}$, we clearly have the associated real normed space $(X_{\mathbb{R}}, \| \cdot \|)$ over $\mathbb{R}$.

4.3.3 Theorem (Hahn-Banach). *Let X be a normed space over the field $\mathbb{K}$ and M be a subspace of X. Then, for every bounded linear functional f on M, $\exists$ a bounded linear functional g on X such that*

$$g\big|_M = f \text{ and } \| g \| = \| f \|.$$

Proof. Case (i). Suppose that X is a real normed space. Set

$$p(x) = \| f \| \, \| x \|, \quad x \in X.$$

Then, p is a sublinear functional on X and

$$f(x) \leq p(x), \qquad \forall \, x \in M.$$

Hence, by Hahn-Banach Lemma (Theorem 4.3.2), it follows that there exists a real linear functional g on X such that

$$g\big|_M = f \text{ and } g(x) \leq \| f \| \, \| x \|, \quad \forall \, x \in X.$$

Now, we show that $\| g \| = \| f \|$. Note that for any $x \in X$, we have

$$g(x) = \pm \, | \, g(x) \, |.$$

Then

$$| \, g(x) \, | = \pm \, g(x) = g(\pm x)$$
$$\leq p(\pm x)$$
$$= \| f \| \, \| \pm x \|$$

$$\Rightarrow \qquad | \, g(x) \, | \leq \| f \| \, \| x \|, \quad \forall \, x \in X$$

$\Rightarrow$ g is bounded and $\| g \| \leq \| f \|$.

For the reverse inequality, take $x = x_1 \in M, x_1 \neq 0$. Then

$$\| g \| \geq \frac{| \, g(x_1) \, |}{\| x_1 \|} = \frac{| \, f(x_1) \, |}{\| x_1 \|}$$

$$\Rightarrow \qquad \| g \| \geq \| f \|.$$

This completes the proof for the case when X is a real normed space.

Case (ii). Suppose that X is a complex normed space. Then, M is a complex linear subspace and $f \in M^*$ is a complex valued functional on M. We write

$$f(x) = f_1(x) + i f_2(x), \quad x \in M$$

where f_1, f_2 are real valued linear functionals on the associated real linear subspace $M_{\mathbb{R}}$. (Note that $M_{\mathbb{R}} \equiv M$ except that in M we restrict multiplication by scalars to real numbers).

Now, for every $x \in M$, we have

$$i\,[\,f_1(x) + i f_2(x)\,] = i f(x)$$

$$= f(ix) = f_1(ix) + i f_2(ix)$$

Equating the real parts on both sides, we obtain

$$f_2(x) = - f_1(ix), \quad \forall\, x \in M.$$

Hence, $f \in M^*$ can be expressed as

(16)
$$f(x) = f_1(x) - i f_1(ix), \quad x \in M.$$

Also, we have

$$f_1(x) = \operatorname{Re} f(x) \le |f(x)| \le \|f\|\,\|x\|, \quad \forall\, x \in M_{\mathbb{R}}$$

so that f_1 is a bounded, real valued linear functional on $M_{\mathbb{R}}$. Thus, considering the associated real normed space $X_{\mathbb{R}}$, it follows from Case (i) of this theorem, that $\exists$ a real bounded linear functional g_1 on $X_{\mathbb{R}}$ such that

(17)
$$g_1\big|_{M_{\mathbb{R}}} = f_1 \text{ and } \|g_1\| = \|f_1\|.$$

Define $g : X \to \mathbb{C}$ by

$$g(x) = g_1(x) - i g_1(ix), \quad x \in X.$$

Then, for all $x_1, x_2, x \in X$ and $\lambda \in \mathbb{R}$, we have

$$\begin{cases} g(x_1 + x_2) = g(x_1) + g(x_2) \\ g(\lambda x) = \lambda g(x). \end{cases}$$

Also

$$g\,(ix) = g_1\,(ix) - ig_1\,(-x)$$

$$= i\,(g_1\,(x) - ig_1(ix))$$

$$= ig\,(x), \quad \forall\, x \in X.$$

Hence, g is a (complex) linear functional on complex normed space X. Further, g is an extension of f since, for all $x \in M$, we have

$$g\,(x) = g_1\,(x) - ig_1\,(ix)$$

$$= f_1\,(x) - i\,f_1\,(ix) \qquad\qquad (\because\; g_1|_M = g_1|_{M_{\mathbb{R}}} = f_1)$$

$$= f\,(x) \quad \text{(from (16))}.$$

It also implies that $\|f\| \le \|g\|$. Thus, it only remains to prove that g is bounded and $\|g\| \le \|f\|$.

Let $x \in X$ and let θ be a real number with

$$g\,(x) = e^{i\theta}\,|\,g\,(x)\,|.$$

Then

$$|\,g\,(x)\,| = e^{-i\theta}\,g\,(x)$$

$$= \mathrm{Re}\,\left(e^{-i\theta}\,g\,(x)\right)$$

$$= \mathrm{Re}\,\left(g\left(e^{-i\theta}\,x\right)\right)$$

$$\le |\,g_1\,(e^{-i\theta}\,x)\,|$$

$$\le \|g_1\|\,\|e^{-i\theta}\,x\| \qquad\qquad (\because\; g_1 \text{ is bounded})$$

$$= \|g_1\|\,\|x\|$$

$$\le \|f\|\,\|x\| \qquad\qquad (\because\; \|g_1\| = \|f_1\| \le \|f\|).$$

This shows that g is bounded and that $\|g\| \le \|f\|$. This completes the proof of the case when X is a complex normed space. ∎

The Hahn-Banach Theorem provides guarantee that the dual space of any Banach space (or normed space) consists of sufficiently many bounded linear functionals. This property is understood in the sense of the following theorem, on which most of the applications depend; and, in fact, this plays a great role in the ideas developed in functional analysis.

4.3.4 Theorem. *Let X be a normed space over the field $\mathbb{K}$ and let $0 \neq x_0 \in X$. Then, $\exists$ a bounded linear functional g on X such that*

(i) $\qquad\qquad g(x_0) = \| x_0 \|$

(ii) $\qquad\qquad \| g \| = 1.$

Proof. Consider the set

$$M = \{x : x = \alpha\, x_0,\ \alpha \in \mathbb{K}\}.$$

Then, M is clearly a subspace of X spanned by x_0. Define $f : M \to \mathbb{K}$ by

$$f(x) = f(\alpha\, x_0) = \alpha \| x_0 \|, \quad \alpha \in \mathbb{K}.$$

Note that f is a linear functional on M with the property that $f(x_0) = \| x_0 \|$. Further, for any $x \in M$, we have

$$|f(x)| = |f(\alpha\, x_0) = \alpha\, | \| x_0 \| = \| \alpha\, x_0 \| = \| x \|$$

$\Rightarrow$ $\quad f$ is bounded and $\| f \| = 1.$

By the Hahn-Banach Theorem, $\exists$ a bounded linear functional g and X such that

$$g\big|_M = f \text{ and } \| g \| = \| f \|.$$

Hence, $g(x_0) = f(x_0) = \| x_0 \|$ and $\| g \| = 1$. This completes the proof. $\blacksquare$

4.3.5 Corollary. *Let X be a normed space over the field $\mathbb{K}$ and let x_1 and x_2 be distinct points of X. Then, $\exists$ a $g \in X^*$ such that $g(x_1) \neq g(x_2)$.*

(In other words, the dual space X^* separates points in X).

Proof. Write $x = x_1 - x_2$. Then, $0 \neq x \in X$ and, therefore, it follows from Theorem 4.3.4 that $\exists$ a $g \in X^*$ such that

$$g(x) = \| x \|$$

$\Rightarrow \qquad\qquad g(x_1 - x_2) = \| x_1 - x_2 \| \neq 0$

$\Rightarrow \qquad\qquad g(x_1) \neq g(x_2).$ $\blacksquare$

The following is an immediate consequence of Theorem 4.3.4.

4.3.6 Corollary. *If $X \neq \{0\}$ is a normed space, then there are always non-trivial bounded linear functionals on X, i.e., $X \neq \{0\} \Rightarrow X^* \neq \{O\}$.*

4.3.7 Corollary. *Let X be a normed space. If $f(x) = 0,\ \forall\, f \in X^*$, then $x = 0$.*

Proof. Let, if possible, $x \neq 0$. Then, by Corollary 4.3.6, $\exists$ a $g \in X^*$ such that $g(x) \neq 0$. This is a contradiction to the hypothesis. $\blacksquare$

We may restate Corollary 4.3.7 as:

4.3.7′ Corollary. *If all bounded linear functionals on a normed space X vanish on a given vector in X, then the vector must be zero.*

4.3.8 Theorem. *Let X be a normed space over the field $\mathbb{K}$ and $x \in X$. Then*

$$\| x \| = \sup \{ \frac{| f (x) |}{\| f \|} : f \in X^{*}, f \neq 0 \}.$$

Proof. If $x = 0$, there is nothing to prove. Let $x \neq 0$ be any vector in X. By Theorem 4.3.4, $\exists$ a $g \in X^{*}$ such that $g(x) = \| x \|$ and $\| g \| = 1$. Therefore

$$\sup \{ \frac{| f (x) |}{\| f \|} : f \in X^{*}, f \neq 0 \} \geq \frac{| g (x) |}{\| g \|} = \| x \|.$$

For the reverse inequality, noting that

$$| f (x) | \leq \| f \| \, \| x \|, \ \forall f \in X^{*}$$

we obtain

$$\sup \left\{ \frac{| f (x) |}{\| f \|} : f \in X^{*}, f \neq 0 \right\} \leq \| x \|. \ \blacksquare$$

We now rephrase Theorem 4.3.4 in a geometrical language. Before we do so, we give the concept of supporting hyperplanes in analogy with the theory of convex bodies of a k-dimensional Euclidean space. Let X be a normed space and let $0 \neq f \in X^{*}$. The hyperplane $\{ x \in X : f(x) = r \| f \| \}$ is called a *support* of the sphere $\| x \| \leq r$.

4.3.9 Theorem (Geometrical Interpretation of Theorem 4.3.4). *Let X be a normed space over the field $\mathbb{K}$ and let x_0 by any point on the surface of the sphere $\| x \| \leq r$, i.e., such that $\| x_0 \| = r$. Then, $\exists$ a supporting hyperplane to the sphere $\| x \| \leq r$ at the point x_0.*

Proof. The equation of the supporting hyperplane for the sphere $\| x \| \leq r$ is of the form

$$\{ x \in X : f (x) = r \| f \| \}, \text{ for some } f \in X^{*}, f \neq 0.$$

But, in view of Theorem 4.3.4, $\exists f_0 \in X^{*}$ such that

$$f_0 (x_0) = \| x_0 \| = r \text{ and } \| f_0 \| = 1.$$

Therefore

$$H = \{x \in X : f_0(x) = r \, \| f_0 \| \} = \{x \in X : f_0(x) = r\}$$

is a supporting hyperplane to the sphere $\| x \| \leq r$. Since $f_0(x_0) = \| x_0 \| = r$, it follows that the supporting hyperplane H passes through x_0. ∎

Remark. Theorem 4.3.9 is a generalization of the theorem proved by Minkowski for the n-dimensional spaces.

We now proceed to present another implication of the Hahn-Banach Theorem. We shall establish that there are always sufficient bounded linear functionals on a normed space which separate points from proper subspaces.

4.3.10 Theorem. *Let X be a normed space over the field $\mathbb{K}$, M a subspace of X and let $x_0 \in X$ be such that $d(x_0, M) = d > 0$, Then, $\exists$ a $g \in X^*$ such that*

  (i) $g(x_0) = 1$
  (ii) $g(M) = 0$, (*i.e.*, $g(m) = 0, \quad \forall \; m \in M$)

  (iii) $\| g \| = \dfrac{1}{d}.$

Proof. Let

$$M_0 = M + [x_0]$$

be the space spanned by M and x_0. Since $d > 0$, we have $x_0 \notin M$. Therefore, each point $x \in M_0$ can be expressed uniquely in the form

$$x = m + \alpha \, x_0, \quad m \in M \text{ and } \alpha \in \mathbb{K}.$$

Consider the functional $f : M_0 \to \mathbb{K}$ defined by

$$f(m + \alpha \, x_0) = \alpha.$$

Note that f is a linear functional on M_0 satisfying $f(M) = 0$ and $f(x_0) = 1$.

Also, we have

$$| f(x) | = | f(m + \alpha \, x_0) | = | \alpha |, \quad \forall \; x \in M_0.$$

Therefore, if $\alpha \neq 0$, then

$$\| x \| = \| m + \alpha \, x_0 \|$$

$$= | \alpha | \, \left\| \frac{m}{\alpha} + x_0 \right\|$$

$$\geq |\alpha| d$$

$$= d |f(x)|, \quad x \in M_0$$

If $\alpha = 0$, then

$$\| x \| = \| m \| \geq 0 = d |f(x)|, \quad x \in M_0.$$

Hence, in either case, it follows that

$$|f(x)| \leq \frac{\| x \|}{d}, \quad \forall x \in M_0$$

$$\Rightarrow \quad f \text{ is bounded on } M_0 \text{ and } \| f \| \leq \frac{1}{d}.$$

To prove $\| f \| \geq \dfrac{1}{d}$, we consider a sequence $\{m_k\} \subset M$ such that $\| x_0 - m_k \| \to d$. Then

$$1 = f(x_0 - m_k) \leq \| f \| \, \| x_0 - m_k \| \to \| f \| \, d$$

$$\Rightarrow \qquad \| f \| \geq \frac{1}{d}.$$

Thus, we have established that $\exists$ an $f \in M_0^*$ such that

$$f(M) = 0, f(x_0) = 1 \text{ and } \| f \| = \frac{1}{d}.$$

Applying the Hahn-Banach theorem, we obtain a functional $g \in X^*$ such that

$$g\big|_{M_0} = f \text{ and } \| g \| = \| f \| = \frac{1}{d}.$$

But

$$g\big|_{M_0} = f \quad \Rightarrow \quad \begin{cases} g(M) = f(M) = 0 \\ g(x_0) = f(x_0) = 1. \end{cases}$$

Hence, the proof of the theorem is complete. ∎

Theorem 4.3.4 can also be derived as a corollary of Theorem 4.3.10.

Alternative Proof of Theorem 4.3.4

Take $M = \{0\}$ and $d = \| x_0 \|$ in Theorem 4.3.10. Then $\exists$ a functional $g_0 \in X^*$ such that

$$\| g_0 \| = \frac{1}{d} = \frac{1}{\| x_0 \|} \text{ and } g_0 (x_0) = 1.$$

Setting $g = \| x_0 \| \, g_0$, we obtain

$$g (x_0) = \| x_0 \| \text{ and } \| g \| = 1. \ \blacksquare$$

4.3.11 Corollary. *Let X be a normed space over the field $\mathbb{K}$, M a linear subspace of X and let $x_0 \in X$ be such that $d (x_0, M) = d > 0$. Then, $\exists$ a $g \in X^*$ such that*

(i) $g (x_0) = d$

(ii) $g (M) = 0$

(iii) $\| g \| = 1$.

Proof. By Theorem 4.3.10, $\exists$ a functional $g_0 \in X^*$ such that $g_0 (x_0) = 1$,

$g_0 (M) = 0$ and $\| g_0 \| = \dfrac{1}{d}$.

Define the bounded linear functional g on X by $g = d \, g_0$. Then

$$g (x_0) = d \, g_0 (x_0) = d,$$

$$g (M) = d \, g_0 (M) = 0,$$

and $\qquad\qquad\qquad \| g \| = d \, \| g_0 \| = \dfrac{d}{d} = 1. \ \blacksquare$

Problem 17. Let X be a normed space and x_1, x_2 be two linearly independent vectors in X, Prove that $\exists$ a functional $f \in X^*$ such that $f (x_1) = 0$ and $f (x_2) = 1$. Determine $\| f \|$.

[*Hint:* Take $M = span \, \{x_1\}$, $x_0 = x_2$ and apply Corollary 4.3.11].

4.3.12 Corollary. *Let X be a normed space over the field $\mathbb{K}$, M a closed linear subspace of X and let $x_0 \in X - M$. If d is the distance from x_0 to M, then $\exists$ a $g \in X^*$ such that*

(i) $g (x_0) = 1$

(ii) $g(M) = 0$

(iii) $\| g \| = \dfrac{1}{d}$.

Proof. Recall that

$$d(x_0, M) = 0 \quad \Leftrightarrow \quad x_0 \in \overline{M} \qquad \text{(Theorem 1.7.8 (2))}.$$

But $M = \overline{M}$ since M is closed and $x_0 \notin M$. Therefore

$$d \equiv d(x_0, M) > 0.$$

Now, the result follows from Theorem 4.3.10. ∎

As an immediate consequence of Corollary 4.3.12, we have

4.3.13 Corollary. *Let X be a normed space over the field $\mathbb{K}$, M a closed linear subspace of X and let $x_0 \in X - M$. Then, $\exists$ a $g \in X^*$ such that*

(i) $g(x_0) \neq 0$

(ii) $g(M) = 0$.

Problem 18. Obtain the result of Corollary 4.3.13 by using the map in Problem 2.39.

4.3.14 Corollary. *Let X be a normed space over the field $\mathbb{K}$ and M be a subspace of X. If M is not dense in X, then $\exists$ a $0 \neq g \in X^*$ such that $g(M) = 0$.*

Proof. Since $\overline{M} \neq X$, $\exists$ a point $x_0 \in X$ such that $d(x_0, M) = d > 0$. Now, the result follows in view of Theorem 4.3.10. ∎

The Hahn-Banach Theorem has several applications in other branches of mathematics. For instance, the familiar Liouville's Theorem in complex analysis[*] can be generalized by using the Hahn-Banach Theorem. We recall that usually Liouville's Theorem asserts that a bounded integral function must be a constant. For generalized form of this theorem, we need the following concepts:

4.3.15 Definition. Let X be a complex Banach space and D a domain in the complex plane $\mathbb{C}$.

(a) The function $T : D \to X$ is said to be *analytic* in D if

---

[*] For various concepts and results in complex analysis, we may refer to any standard book on the topic, for instance [1].

$$\lim_{h \to 0} \left(\frac{T(z+h) - Tz}{h} \right)$$

exists for every $z \in D$ and $z + h \in D$. This limit is taken in the sense of the norm on X.

(b) The function $T : \mathbb{C} \to X$ is said to be an *integral function* if T is analytic on $\mathbb{C}$.

(c) The function $T : \mathbb{C} \to X$ is said to be *bounded* if $\exists$ a real number $K > 0$ such that

$$\| Tz \| \leq K, \quad \forall z \in \mathbb{C}.$$

4.3.16 Theorem. *Let X be a complex Banach space. If $T : \mathbb{C} \to X$ is a bounded integral function, then T is a constant.*

Proof. Let f be a bounded linear functional on X. Consider the function $f \circ T : \mathbb{C} \to \mathbb{C}$ defined by

$$(f \circ T)(z) = f(Tz), \quad z \in \mathbb{C}.$$

Note that $f \circ T$ is a complex valued function of a complex variable. Further, it is easy to verify that $f \circ T$ is analytic on $\mathbb{C}$.

Since f is a bounded linear functional on X, we have

$$| (f \circ T)(z) | = |f(Tz)|$$

$$\leq \| f \| \, \| Tz \|, \quad \forall z \in \mathbb{C}$$

$$\leq \mathbb{K} \, \| f \|,$$

for some constant $K > 0$, since T is bounded. Thus, $f \circ T$ is bounded on $\mathbb{C}$. Applying classical Liouville's Theorem to $f \circ T$, we obtain

$$f(T z_1) = f(T z_2), \quad z_1, z_2 \in \mathbb{C}$$

$$\Rightarrow \quad f(T z_1 - T z_2) = 0, \quad z_1, z_2 \in \mathbb{C}.$$

But f being an arbitrary bounded linear functional on X, we have

$$f(Tz_1 - Tz_2) = 0, \quad \forall f \in X^*.$$

By Corollary 4.3.7, it follows that

$$T z_1 = T z_2, \quad \forall z_1, z_2 \in \mathbb{C}.$$

This proves that T is a constant function. $\blacksquare$

The Hahn-Banach Theorem and its consequences can be used to reveal much between the properties of a normed space and its dual space. We discuss separability of spaces to explain the applications of the Hahn-Banach Theorem and its immediate implications given in the preceding theorems and their corollaries. In fact, our next theorem relates the separability of the dual space to the separability of the original space.

4.3.17 Theorem. *Let X be a normed space and X^* be its dual. Then, X^* is separable $\Rightarrow X$ is separable.*

Proof. Since X^* is separable, $\exists$ a countable set

$$S = \{\, f_n \in X^* : n \in \mathbb{N}\,\}$$

such that S is dense in X^*, *i.e.*, $\overline{S} = X^*$.

For each $n \in \mathbb{N}$, choose $x_n \in X$ such that

$$\| x_n \| = 1 \text{ and } | f_n(x_n) | \ge \frac{1}{2} \| f_n \|.$$

Let M be the closed subspace of X generated by the sequence $\{x_n\}$, *i.e.*,

$$M = \overline{\text{span}} \,\{x_n \in X : n \in \mathbb{N}\}.$$

Suppose $M \ne X$. Then, $\exists$ a point $x_0 \in X - M$. As such, by Corollary 4.3.13, $\exists$ a functional $0 \ne g \in X^*$ such that

$$g(x_0) \ne 0 \text{ and } g(M) = 0.$$

Thus

$$\begin{cases} g(x_n) = 0,\, n \in \mathbb{N} \\ \dfrac{1}{2} \| f_n \| \le | f_n(x_n) | = | (f_n - g)(x_n) | \le \| f_n - g \| \end{cases}$$

Therefore

$$\| g \| \le \| f_n - g \| + \| f_n \|$$
$$\le 3 \| f_n - g \|, \quad \forall\, n \in \mathbb{N}.$$

But $\overline{S} = X^*$, it follows that $g = 0$, which contradicts the assumption that $M \ne X$. Hence, $M = X$ and thus X is separable. ∎

Remark. The converse of the above theorem is not true.

4.3.18 Example. The sequence space l^1 is separable while its dual space $(l^1)^*$ is not separable, since $(l^1)^*$ is isometrically isomorphic to l^∞, and l^∞ is not separable.

Problems

19. Let M be the smallest closed subspace of the normed space l^∞ containing all sequences of the form

$$\{\lambda_1, \lambda_2 - \lambda_1, \lambda_3 - \lambda_2, \ldots\}, \text{ where } \{\lambda_i\} \in l^\infty.$$

Prove that $e \equiv \{1, 1, 1, \ldots\} \notin M$ and that $\exists$ a bounded linear functional f^* on l^∞ such that $\|f^*\| = 1$, $f^*(e) = 1$ and $f^*(M) = 0$.

20. If x_0 in a normed space X is such that $|f(x_0)| \le K$, $\forall f \in X^*$ with $\|f\| = 1$, prove that $\|x_0\| \le K$.

[*Hint:* If $\|x_0\| > K$, then $x_0 \neq 0$ and by Theorem 4.3.4, $\exists$ a $g \in X^*$ such that $\|g\| = 1$ and $g(x_0) = \|x_0\| > K$].

21. Let M be a closed subspace of a normed space X such that $f(M) = 0 \implies f(X) = 0$, $\forall f \in X^*$. Prove that $M = X$.

22. A subset M of a normed space X is said to be *total* in X if $\overline{\text{span } M} = X$. Prove that $M \subset X$ is total in X if and only if

$$f(M) = 0 \implies f(X) = 0, \quad \forall f \in X^*.$$

23. Let M be a subspace of a normed space X. Write

$$S = \{ f \in X^* : f(x) = 0, \forall x \in M \}.$$

Prove that S is a closed linear subspace of X^*. Further, prove that the dual space S^* is isometrically isomorphic to the quotient space X^*/S.

24. Let X be a normed space over $\mathbb{C}$ and let $T : D \to X$ be a function, where D is a domain in $\mathbb{C}$. Prove that, if T is analytic in D and $f \in X^*$, then $f \circ T$ is analytic in D.

25. Let L be a simple closed path in $\mathbb{C}$ and let X be a Banach space. If $T : L \to X$ is continuous on L, prove that $\int_L T(z)\, dz$ exists.

[*Hint.* The integral is defined as in complex analysis except that the limit is taken in the norm. The completeness of X is used in the proof.]

26. Let X be a Banach space and L be a simple closed path in $\mathbb{C}$. If $T : L \to X$ is analytic within L and continuous on L, prove the generalized Cauchy Theorem:

$$\int_L T(z)\, dz = 0.$$

[*Hint:* Use the classical Cauchy Theorem and the Hahn-Banach Theorem.]

4.4 Embedding and Reflexivity of Normed Spaces

In Section 4.2, we defined the dual space of a normed space to be the set consisting of all bounded linear functionals on the space. However, in this section we are primarily concerned with 'what sort of bounded linear functionals exist on the dual space', *i.e.,* we wish to investigate the dual space of the dual space. We shall call such a space as *second dual space.* In some cases, the second dual space of a normed space, under a specific mapping–called the natural embedding, is isometrically isomorphic to the original space. Such normed spaces are known as *reflexive spaces.* Finally, we shall establish some nice properties of reflexive Banach spaces.

Recall that given a normed space $X \neq \{0\}$, the dual space X^* is a normed space with norm $\| . \| : X^* \to \mathbb{R}$ defined by

$$\| f \| = \sup \{ | f(x) | : x \in X, \| x \| = 1 \}.$$

Further, note that the dual space X^* is always complete and hence is a Banach space. Also, in view of Corollary 4.3.6, $X^* \neq \{o\}$ and, therefore, as a normed space the dual space X^* has its own (topological) dual space $(X^*)^*$, denoted by X^{**} and is called the *second dual* (*second conjugate* or *bidual*) space of X, which is again a Banach space under the norm.

$$\| \varphi \| = \sup \{ | \varphi(f) | : f \in X^*, \| f \| \leq 1 \}, \varphi \in X^{**}.$$

We shall usually denote the elements of X by x, elements of X^* by f and elements of X^{**} by φ, so that f operates on the element x and φ operates on the element f.

The importance of X^{**} lies in the fact that each vector $x \in X$ gives rise to a unique functional φ_x (say) in X^{**} having the same norm. More precisely, we have

4.4.1 Theorem. *Let $X \neq \{0\}$ be a normed space over the field $\mathbb{K}$. Given $x \in X$, let*

$$(18) \qquad \varphi_x(f) = f(x), \quad \forall f \in X^*.$$

Then, φ_x is a bounded linear functional on X^*, i.e., $\varphi_x \in X^{**}$. Further, the mapping $x \to \varphi_x$ is an isometric isomorphism of X onto the subspace $\hat{X} = \{\varphi_x : x \in X\}$ of X^{**}.

Proof. The functional φ_x is linear since

$$\varphi_x(\alpha f + \beta g) = (\alpha f + \beta g)(x)$$

$$= \alpha f(x) + \beta g(x)$$
$$= \alpha \varphi_x(f) + \beta \varphi_x(g), \quad \forall f, g \in X^* \text{ and } \alpha, \beta \in \mathbb{K}.$$

Also, φ_x is bounded since

$$|\varphi_x(f)| = |f(x)| \leq \|f\| \, \|x\|, \quad \forall f \in X^*.$$

Consequently, $\varphi_x \in X^{**}$. Moreover, such a φ_x is determined uniquely (Verify!).

Thus, to every $x \in X$, $\exists$ a unique element $\varphi_x \in X^{**}$ given by (18). This defines a function $\pi : X \to X^{**}$ given by $\pi(x) = \varphi_x$.

Observe that:

(*i*) π *is linear.*

For $x, y \in X$ and $\alpha, \beta \in \mathbb{K}$, we have

$$(\pi(\alpha x + \beta y))(f) = \varphi_{\alpha x + \beta y}(f)$$
$$= f(\alpha x + \beta y)$$
$$= (\alpha \varphi_x + \beta \varphi_y)(f)$$
$$= (\alpha \pi(x) + \beta \pi(y))(f), \quad \forall f \in X^*$$
$$\Rightarrow \qquad \pi(\alpha x + \beta y) = \alpha \pi(x) + \beta \pi(y)$$

(*ii*) π *preserves norm.*

For each $x \in X$, we have

$$\|\pi(x)\| = \|\varphi_x\| = \sup \left\{ \frac{|\varphi_x(f)|}{\|f\|} : f \in X^*, f \neq 0 \right\}$$

$$= \sup \left\{ \frac{|f(x)|}{\|f\|} : f \in X^*, f \neq 0 \right\}$$

$$= \|x\| \quad \text{(Theorem 4.3.8)}.$$

(*iii*) π *is injective.*

Let $x, y \in X$. Then

$$x - y \neq 0 \quad \Rightarrow \quad \|x - y\| \neq 0$$
$$\Rightarrow \quad \|\pi(x - y)\| \neq 0$$
$$\Rightarrow \quad \|\pi(x) - \pi(y)\| \neq 0$$
$$\Rightarrow \quad \pi(x) \neq \pi(y).$$

We thus conclude that π is an isometric isomorphism of X onto the subspace $\hat{X}\,(=\pi(X))$ of X^{**}. This completes the proof. ∎

4.4.2 Definition. Let X be a normed space over the field $\mathbb{K}$. The isometric isomorphism $\pi : X \to X^{**}$ defined by $\pi(x) = \varphi_x$ is called the *natural embedding* (or, the *canonical mapping*) of X into its second dual space X^{**}. The functional $\varphi_x \in X^{**}$ is called the functional induced by the vector x. We refer to the functionals of this kind as *induced functionals*.

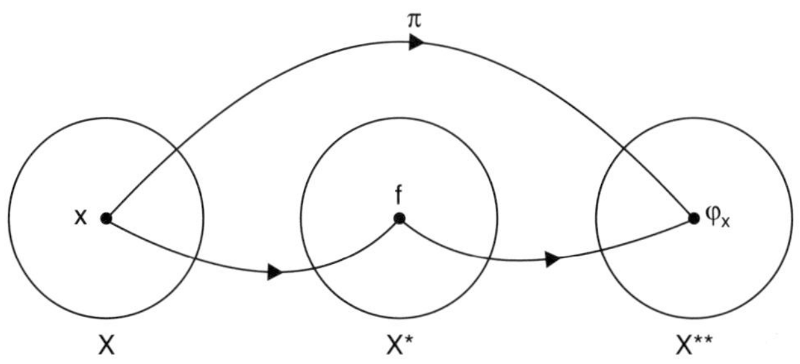

Fig. 4.2

Note. The isometric isomorphism $\pi : X \to X^{**}$ defined by $\pi(x) = \varphi_x$ is called natural embedding of X into X^{**} because it allows us to regard X as a part of X^{**} without altering any of its structures as a normed space. We write $X \subset X^{**}$ to mean that X is embedded in X^{**} ; in fact, $\pi(X) \subset X^{**}$.

4.4.3 Definition. A normed space X is said to be *reflexive* if the natural embedding π maps the space X onto its second dual space X^{**}, *i.e.*, $\pi(X) = X^{**}$.

Remark. If X is a reflexive normed space, then X is isometrically isomorphic to X^{**} under the natural embedding. In this regard, we have the following significant observations:

(a) Since X^{**} is always complete, the space X must be complete. Hence, completeness is a necessary condition for a normed space to be reflexive. However, this condition need not be sufficient (Example 4.4.4(3). Thus, it is clear that if X is not a Banach space, then we must have $\pi(X) \neq X^{**}$ and hence X is non-reflexive.

(b) It is not sufficient for reflexivity that X be isometrically isomorphic to X^{**} under mappings different from the natural embeddings.

4.4.4 Examples

1. $\mathbb{R}^n$ is reflexive, since $(\mathbb{R}^n)^* = \mathbb{R}^n$ (Example 4.2.2) and $(\mathbb{R}^n)^{**} = \mathbb{R}^n$.

2. Every finite dimensional normed space is reflexive.

Let X be an n-dimensional normed space. Then $\dim X = \dim X^* = \dim X^{**} = n$. Further, since the natural embedding $\pi : X \to X^{**}$ defined by $\pi(x) = \varphi_x$ is injective and X is n-dimensional, it follows that $\pi(X)$ is n-dimensional. But $\pi(X)$ being a subspace of the n-dimensional space X^{**}, we conclude that $\pi(X) = X^{**}$.

3. The Banach space $(c_0, \| \cdot \|_\infty)$ (Problem 2.11) is not reflexive.

Note that

$$(c_0 \, \| \cdot \|_\infty)^{**} = (l^1, \| \cdot \|_1)^* \qquad \text{(Example 4.2.5 (1))}$$
$$= (l^\infty, \| \cdot \|_\infty) \qquad \text{(Example 4.2.5 (2))}.$$

Since, c_0 is separable and l^∞ is non-separable, there cannot exist any isometric isomorphism of c_0 onto l^∞. Hence, c_0 is not reflexive.

In general, for any normed space X (reflexive or not), we prove the following:

4.4.5 Theorem. *A normed space is isometrically isomorphic to a dense subspace of a Banach space.*

Proof. Let X be a normed space. If $\pi : X \to X^{**}$ be the natural embedding, then X and $\pi(X)$ are isometrically isomorphic spaces. But $\pi(X)$ is a dense subspace of $\overline{\pi(X)}$, and $\overline{\pi(X)}$ is a closed subspace of the Banach space X^{**}, it follows that $\overline{\pi(X)}$, itself is a Banach space. Hence, X is isometrically isomorphic to the dense subspace $\pi(X)$ of the Banach space $\overline{\pi(X)}$. ∎

We next discuss the connection between reflexivity and separability of normed spaces. Recall that separability of the dual space X^* implies the separability of the original space X (Theorem 4.3.17) but the converse may not hold (Example 4.3.18).

4.4.6 Theorem. *Let X be a separable normed space. If the dual space X^* is non-separable, then X is non-reflexive.*

Proof. Let, if possible, X be reflexive. Then, X^{**} is isometrically isomorphic to X under the natural embedding. Since X is separable, it follows that X^{**} is separable. By Theorem 4.3.17, the space X^* is separable. This contradicts the hypothesis. ∎

4.4.7 Example. The normed space $(l^1, \| \cdot \|_1)$ is not reflexive.

The space l^1 is separable (Problem 2.34 (c)) and $(l^1)^* = l^\infty$ is not separable (Problem 2.34 (d)). Hence, by Theorem 4.46, the space l^1 is not reflexive. ∎

We now use the natural embedding of X into X^{**} for deriving results concerning X and X^* in the following theorem.

4.4.8 Theorem. *Let $\{x_n\}$ be a sequence in a normed space X. Suppose that*

$$\sup_{1 \le n < \infty} |f(x_n)| < \infty, \ \forall f \in X^*. \ Then \ \sup_{1 \le n < \infty} \| x_n \| < \infty.$$

Proof. Consider the natural embedding $x \to \varphi_x$, $x \in X$. Since $\{x_n\}$ is a sequence of vectors in X, $\{\varphi_{x_n}\}$ is a sequence of functionals in X^{**}. Also

$$|\varphi_{x_n}(f)| = |f(x_n)| \ \le \ \sup_{1 \le n < \infty} |f(x_n)|.$$

Therefore, under the hypothesis, $\{\varphi_{x_n}(f)\}$ is bounded for each $f \in X^*$. Applying the Principle of Uniform Boundedness (Theorem 3.7.3) to the family $\{\varphi_{x_n}\}$, we conclude that the set $\{\|\varphi_{x_n}\|\}$ and hence $\{\|x_n\|\}$ is bounded. This proves the result. ∎

4.4.9 Theorem. *Let X and Y be isometrically isomorphic normed spaces over the field $\mathbb{K}$. If Y is reflexive, then X is reflexive.*

Proof. Let $T : X \to Y$ be an isometric isomorphism of X onto Y. Consider the map $T_0 : X^* \to Y^*$ defined by

$$(T_0 f)(y) = f(T^{-1}y), \quad (y \in Y, f \in X^*) \ (\text{Figure 4.3})$$

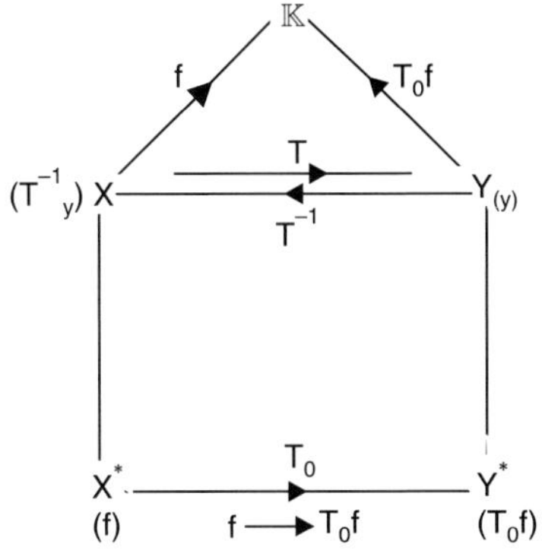

Fig. 4.3

Claim. T_0 is an isometric isomorphism of X^* onto Y^*.

The map T_0 is linear since, for $f, g \in X^*$ and $\alpha, \beta \in \mathbb{K}$, we have

$$(T_0 \, (\alpha f + \beta \, g)) \, (y) = (\alpha f + \beta g) \, (T^{-1} \, y)$$

$$= \alpha \, (f \, (T^{-1} \, y)) + \beta \, (\, g \, (T^{-1} \, y))$$

$$= (\alpha \, (T_0 f) + \beta \, (T_0 \, g)) \, (y), \; \forall \; y \in Y$$

$$\Rightarrow \quad T_0 \, (\alpha f + \beta \, g) = \alpha \, (T_0 f) + \beta \, (T_0 \, g).$$

Further, T_0 is bounded and preserves norm since

$$\| T_0 f \| = \sup \{ \, | \, (T_0 f) \, (y) \, | : y \in Y, \| y \| = 1 \, \}$$

$$= \sup \{ \, | f \, (T^{-1} y) \, | : y \in Y, \| y \| = 1 \}$$

$$\leq \| f \| \sup \{ \, \| T^{-1} y \, \| : y \in Y, \| y \| = 1 \}$$

$$\leq \| f \| \quad (\because \; T^{-1} \text{ is bounded and preserves the norm}).$$

The reverse inequality $\| f \| \leq \| T_0 f \|$ can be easily seen.

Finally, it can be easily seen that T_0 is bijective. This establishes our claim.

Next, define $T_1 : X^{**} \to Y^{**}$ by

$$(T_1 \, \varphi) \, (g) = \varphi \, (T_0^{-1} \, g), \quad \varphi \in X^{**}, g \in Y^*.$$

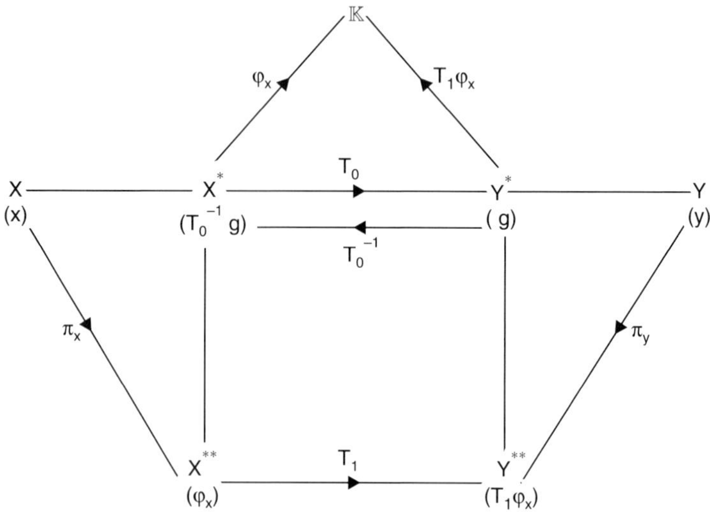

Fig. 4.4

Like T_0, it can be easily verified that T_1 is an isometric isomorphism of X^{**} onto Y^{**}.

Let $\pi_X : X \to X^{**}$ and $\pi_Y : Y \to Y^{**}$ be the natural embeddings of X into X^{**} and Y into Y^{**}, respectively. Let $\varphi_0 \in X^{**}$ be an arbitrary but fixed element. Set

$$x_0 = T^{-1} \, \pi_Y^{-1} \, T_1 \, \varphi_0.$$

It is clear (Fig. 4.4) that $x_0 \in X$. Further, for $f \in X^*$, we have

$$f(x_0) = f(T^{-1}(\pi_Y^{-1} \, T_1 \, \varphi_0))$$

$$= (T_0 f)(\pi_Y^{-1}(T_1 \, \varphi_0))$$

$$= (T_1 \, \varphi_0)(T_0 f)$$

$$= \varphi_0(T_0^{-1}(T_0 f))$$

$$= \varphi_0(f)$$

$$\Rightarrow \qquad \pi_X(x_0) = \varphi_0.$$

But φ_0 is an arbitrary element in X^{**}. Hence, $\pi(X) = X^{**}$. This proves that X is reflexive. ∎

4.4.10 Theorem. *A closed subspace of a reflexive Banach space is reflexive.*

Proof. Let X be a reflexive Banach space and let Y be a closed subspace of X. Let $T : X^* \to Y^*$ be an operator defined by

$$(Tf)(y) = f(y), \quad y \in Y.$$

Then

$$\| Tf \| = \sup \left\{ \frac{|f(y)|}{\|y\|} : y \in Y, y \neq 0 \right\}$$

$$= \| f \|.$$

Let $y \to \psi_y$ be the natural embedding of Y into Y^{**}. Define $T_1 : Y^{**} \to X^{**}$ by

$$(T_1 \psi_y)(f) = \psi_y(Tf), \ f \in X^*.$$

Note that $T_1 \, \psi_y \in X^{**}$ (Verify !) and hence T_1 is well defined.

Since X is reflexive, the natural embedding $\pi_X : X \to X^{**}$ is such th

$= X^{**}$. Therefore, $T_1 \, \psi_y \in X^{**}$ implies that $\pi_X^{-1}(T_1 \, \psi_y) \in X$. Write $x = \pi_X^{-1}(T_1 \, \psi_y)$ so that $\pi_X(x) = T_1 \, \psi_y$.

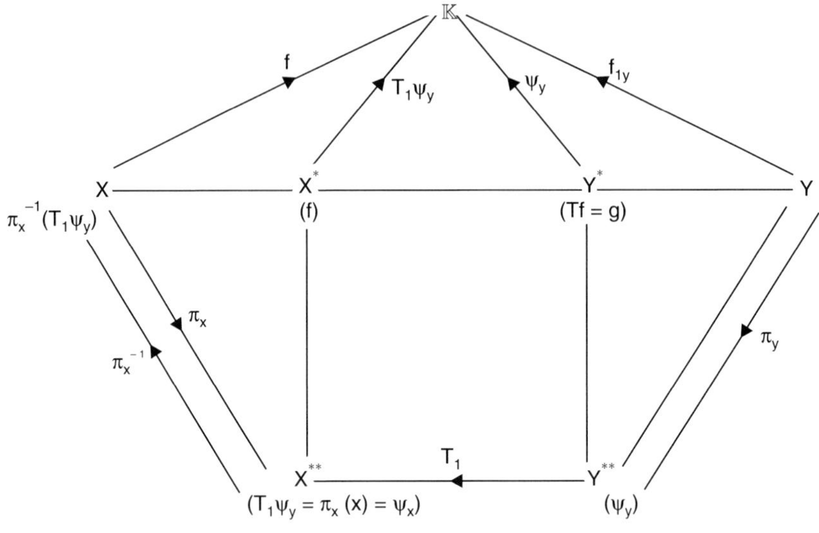

Fig. 4.5

We need to prove that $x \in Y$. Let, if possible, $x \in X - Y$. Then, by Corollary 4.3.13, $\exists$ a functional $f \in X^*$ such that $f(x) \neq 0$ and $f(Y) = 0$. Consequently, $Tf = 0$ and as such $\psi_y(Tf) = 0$. This leads to $\varphi_x(f) = 0$ and hence $f(x) = 0$ which is a contradiction. We thus conclude that $x = \pi_X^{-1}(T_1\,\psi_y) \in Y$. This verifies that $\pi_X^{-1}(T_1(Y^{**})) \subset Y$.

Now, let $\psi \in Y^{**}$. Set $x_0 = \pi_X^{-1}(T_1\,\psi)$ so that $x_0 \in Y$. Let $g \in Y^*$. Then, $\exists f \in X^*$ such that $f|_Y = g$, $g = Tf$. Therefore

$$\begin{aligned}
\psi(g) &= (T_1\,\psi)(f) \\
&= (\pi_X(x_0))(f) \\
&= \varphi_{x_0}(f) \\
&= f(x_0) \\
&= g(x_0) && (\because \quad x_0 \in Y)
\end{aligned}$$

This proves that $\pi_Y(x_0) = \psi$, where π_Y is the natural embedding of Y into Y^{**}. Hence $\pi_Y(Y) = Y^{**}$. This proves that Y is reflexive. ∎

4.4.11. Theorem. A *Banach space is reflexive if and only if its dual space is reflexive.*

Proof. Let X be reflexive. Then, X and X^{**} are isometrically isomorphic under the natural embedding π and $\pi(X) = X^{**}$. Also, X^{**} is reflexive (Theorem 4.4.9). Note that

$$(X^*)^{**} = ((X^*)^*)^* = (X^{**})^* = X^{***}.$$

Let $\tilde{\pi}$ be the natural embedding of X^* into X^{***}. We shall now show that $\tilde{\pi}(X^*) = X^{***}$. Consider $\theta \in X^{***}$. Define the functional f on X by

$$f(x) = \theta(\varphi_x), \quad \varphi_x \in X^{**}$$

where $\varphi_x = \pi(x)$. Then, f is clearly a linear functional on X. Also, f is bounded since

$$|f(x)| = |\theta(\varphi_x)| \leq \|\theta\| \|\varphi_x\| = \|\theta\| \|x\|.$$

Thus, $f \in X^*$. Observe that the result follows, if we prove that $\tilde{\pi}(f) = \theta$. We have

$$\theta(\varphi_x) = f(x) = \varphi_x(f) = (\tilde{\pi}(f))(\varphi_x).$$

But each $\varphi \in X^{**}$ is φ_x for some $x \in X$. Therefore

$$\theta(\varphi) = (\tilde{\pi}(f))(\varphi), \quad \forall \varphi \in X^{**}.$$

Hence, we conclude that X^* is reflexive.

Conversely, assume that X^* is reflexive. Then, from the first part of this theorem it follows that X^{**} is reflexive. Since $\pi(X)$ is a closed linear subspace of X^{**}, by Theorem 4.4.10, it follows that $\pi(X)$ is reflexive. Hence, X is reflexive since π is isometric isomorphism of X onto $\pi(X)$ (Theorem 4.4.9). ∎

Problems

27. Let M be a subspace of a reflexive space X. Prove that:

   (a) M is reflexive.

   (b) The quotient space X/M is reflexive.

28. Prove that it is possible to have a non-reflexive normed space with its dual space reflexive.

29. Prove that a Banach space X is reflexive if and only if every $f \in X^*$ attains its supremum on the unit sphere of X, *i.e.,* for every $f \in X^*$, $\exists$ an $x \in X$, $\|x\| = 1$ such that $f(x) = \|f\|$.

30. Determine whether the following normed spaces are reflexive:

   (a) $(\Phi, \|\cdot\|_\infty)$.

   (b) $(c, \|\cdot\|_\infty)$.

   (c) $(l^p, \|\cdot\|_p)$, $1 < p < \infty$.

   (d) $(l^\infty, \|\cdot\|_\infty)$.

(e) $(C\,[a,\,b],\,\|\cdot\|_\infty)$.

(f) $(L^p\,[a,\,b],\,\|\cdot\|_p)$, $1 \le p \le \infty$.

31. A Banach space X is called *uniformly convex* if for every $\in\, > 0$, there is a $\delta > 0$ such that for $x,\,y \in X$ with $\|\,x\,\| = 1$, $\|\,y\,\| = 1$ and $\|\,\dfrac{x+y}{2}\,\| > 1 - \delta$, we have $\|\,x - y\,\| < \in$. Prove that:

(a) Every uniformly convex Banach space is reflexive.

(b) Give an example of a reflexive Banach space which is not uniformly convex.

4.5 Adjoint of Bounded Linear Operators

The concept of a bounded linear operator from a normed space into a normed space was introduced and studied in depth in Chapter 3. In the present chapter, we have been dealing with bounded linear functionals on normed spaces. Now, we look in a special setting–the framework of normed spaces, bounded linear functionals and bounded linear operators. In fact, we use the Hahn-Banach Theorem (via Theorem 4.3.4) to define the adjoint (or conjugate) of a bounded linear operator. The adjoint of a bounded linear operator is defined on the dual of the space which contains the range of the bounded linear operator. The adjoint of a linear operator is an old friend and is one of the most beautiful concept from linear algebra.

Let $T : X \to Y$ be in $B\,(X,\,Y)$, where X and Y are normed spaces over the field $\mathbb{K}$ and $g \in Y^*$. Then, the mapping $f : X \to \mathbb{K}$ given by

$$(19) \qquad f(x) = (g o T)\,(x) = g\,(T\,x)$$

is in X^* and $\|\,f\,\| \le \|\,g\,\|\,\|\,T\,\|$. Indeed, we have

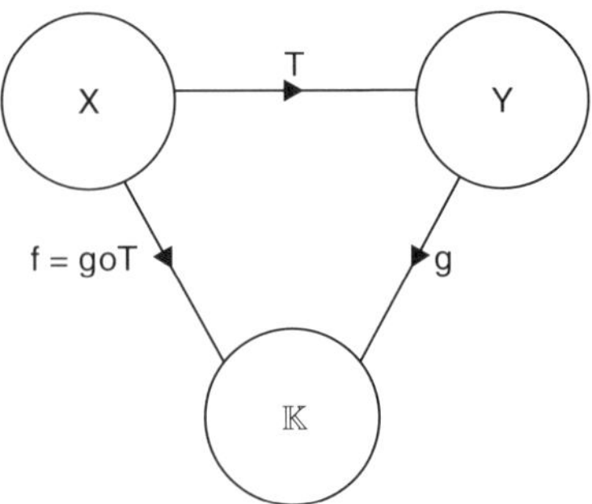

Fig. 4.6

$$f(\alpha x + \beta y) = g(T(\alpha x + \beta y))$$
$$= g(\alpha Tx + \beta Ty) \qquad (\because \quad T \text{ is linear})$$
$$= \alpha g(Tx) + \beta g(Ty)$$
$$= \alpha f(x) + \beta f(y), \qquad \forall x, y \in X \text{ and } \alpha, \beta \in \mathbb{K}$$

and

$$|f(x)| = |g(Tx)| \le (\|g\| \|T\|) \|x\|, \quad \forall x \in X.$$

Hence, f is a bounded linear functional on X and

$$\|f\| \le \|g\| \|T\|.$$

Thus, we see that for each $g \in Y^*$, $\exists$ a (unique) functional $f \in X^*$, defined by (19). This gives rise to a mapping $T^\times : Y^* \to X^*$ defined by $T^\times g = f$. The mapping $T^\times$ is called the adjoint of bounded linear operator T. More precisely.

4.5.1 Definition. Let X and Y be normed spaces over the field $\mathbb{K}$ and $T : X \to Y$ a bounded linear operator. The mapping $T^\times : Y^* \to X^*$ defined by

$$T^\times g = g \, o \, T,$$

that is

(20) $$(T^\times g)(x) = (g \, o \, T)(x) = g(Tx), \; \forall \, g \in Y^* \text{ and } x \in X$$

is called the *adjoint* (or *conjugate*) of bounded linear operator T.

Notes

1. If a given bounded linear operator T is defined from X into Y, then the adjoint $T^\times$ is defined from Y^* into X^*.

2. The equation (20) is very useful because all the properties of adjoint $T^\times$ are derived from (20).

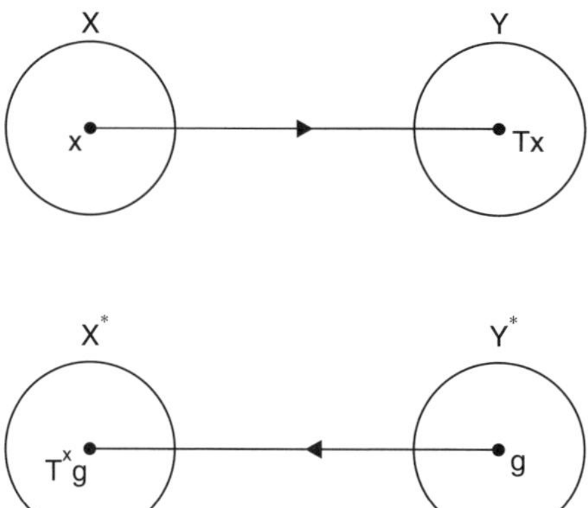

Fig. 4.7

Now, our interest is to see what properties the adjoint $T^\times$ of T inherits from T and to investigate the properties of the map $T \to T^\times$. We see that $T^\times$ is a bounded linear operator and has the same norm as the operator T itself.

4.5.2 Theorem. *Let X and Y be normed spaces over the field $\mathbb{K}$ and $T : X \to Y$ a bounded linear operator. Then:*

    (a) *The adjoint $T^\times$ is a bounded linear operator Y^* into X^*, i.e.,*

$$T^\times \in B\,(Y^*, X^*).$$

    (b) $\| T^\times \| = \| T \|.$

    (c) *The mapping $T \to T^\times$ is an isometric isomorphism of $B\,(X,\ Y)$ into $B\,(Y^*, X^*).$*

Proof. (a) Let $g, h \in Y^*$ and $\alpha, \beta \in \mathbb{K}$. Then

$$
\begin{aligned}
(T^\times (\alpha\, g + \beta\, h))\,(x) &= (\alpha\, g + \beta\, h)\,(T\,x) \\
&= \alpha\, g\,(Tx) + \beta\, h\,(T\,x) \\
&= \alpha\,(T^\times g)\,(x) + \beta\,(T^\times h)\,(x) \\
&= (\alpha\, T^\times g + \beta\, T^\times h)\,(x), \quad \forall\, x \in X
\end{aligned}
$$

$$\Rightarrow \qquad T^\times (\alpha\, g + \beta\, h) = \alpha\, T^\times g + \beta\, T^\times h.$$

This shows that $T^\times$ is a linear operator.

    Further, for all $g \in Y^*$, we have

$$
\begin{aligned}
\| T^\times g \| &= \sup\,\{|\,(T^\times g)\,(x)\,| : x \in X,\, \| x \| = 1\} \\
&= \sup\,\{|\, g\,(Tx)\,| : x \in X,\, \| x \| = 1\} \\
&\leq \| g \|\, \sup\,\{\| Tx \| : x \in X,\, \| x \| = 1\} \\
&= \| g \|\, \| T \|.
\end{aligned}
$$

This verifies that $T^\times$ is bounded and

$$(21) \qquad\qquad \| T^\times \| \leq \| T \|.$$

    (b) By a consequence of the Hahn-Banach Theorem (Theorem 4.3.4), any non-zero vector $Tx \in Y$ yields a $g \in Y^*$ such that

$$\| g \| = 1 \text{ and } g\,(Tx) = \| Tx \|.$$

Therefore

$$
\begin{aligned}
\| Tx \| &= |\, g\,(Tx)\,| \\
&= |\,(T^\times g)\,(x)\,| \\
&\leq \| T^\times g \|\, \| x \|
\end{aligned}
$$

$$\leq \| T^{\times} \| \, \| g \| \, \| x \|$$

$$= \| T^{\times} \| \, \| x \|$$

$$\Rightarrow \qquad \| T \| \leq \| T^{\times} \|$$

which on combining with (21) completes the proof of (b).

(c) Let $\varphi : B(X, Y) \to B(Y^{*}, X^{*})$ be the mapping defined by $\varphi(T) = T^{\times}$. Let $T, S \in B(X, Y)$. Then, for $g \in Y^{*}$, we have

$$(\varphi(T + S)(g))(x) = ((T + S)^{\times}(g))(x)$$

$$= g((T + S)(x))$$

$$= g(Tx + Sx)$$

$$= (T^{\times} g)(x) + (S^{\times} g)(x)$$

$$= ((T^{\times} + S^{\times})(g))(x)$$

$$= ((\varphi(T) + \varphi(S))(g))(x), \quad \forall\, x \in X$$

$$\Rightarrow \qquad \varphi(T + S) = \varphi(T) + \varphi(S).$$

Further, let $\alpha \in \mathbb{K}$. Then, for $g \in Y^{*}$, we have

$$(\varphi(\alpha T)(g))(x) = ((\alpha T)^{\times}(g))(x)$$

$$= g(\alpha T)(x)$$

$$= \alpha(g(Tx))$$

$$= \alpha((T^{\times} g)(x))$$

$$= (\alpha\, \varphi(T)(g))(x), \, \forall\, x \in X$$

$$\Rightarrow \qquad \varphi(\alpha T) = \alpha\, \varphi(T).$$

Hence, the map φ is linear.

Finally, in view of (b), the map φ is injective and preserves the norm. ∎

4.5.3 Corollary. *Let T be a bounded linear operator on a normed space X. Then, the adjoint* $T^{\times} : X^{*} \to X^{*}$ *defined by*

$$(T^{\times} g)(x) = g(Tx), \qquad g \in X^{*} \text{ and } x \in X$$

is a bounded linear operator on X^{*}.

(In other words, $T \in B(X) \;\Rightarrow\; T^{\times} \in B(X^{*})$).

4.5.4 Theorem. *Let X, Y, Z be normed spaces over the field* $\mathbb{K}$. *Then:*

(a) *If* $T \in B(X, Y)$ *and* $S \in B(Y, Z)$, *then*

$$(S \circ T)^{\times} = T^{\times} \circ S^{\times}.$$

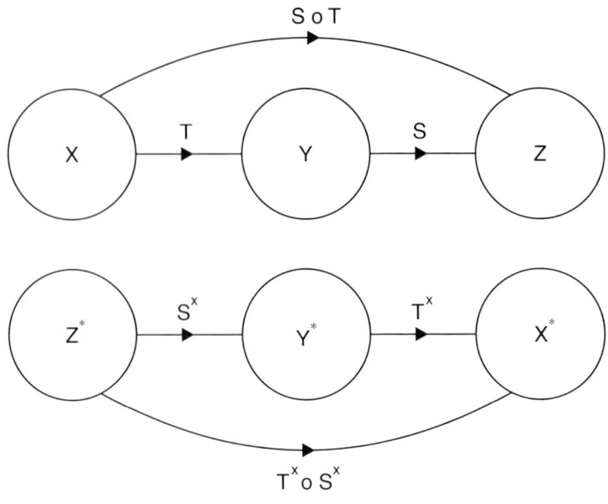

Fig. 4.8

(b) *The adjoint of the identity I_X in $B(X)$ is the identity I_X^* in $B(X^*)$.*

(c) *If $T \in B(X, Y)$ and T^{-1} exists, then*

$$(T^x)^{-1} = (T^{-1})^x.$$

Proof. (a) Let $h \in Z^*$. Then

$$((S \ o \ T)^x (h)) (x) = h((S \ o \ T)(x))$$

$$= h(S(Tx))$$

$$= (S^x h)(Tx)$$

$$= ((T^x \ o \ S^x)(h))(x), \ \forall \ x \in X$$

$$\Rightarrow \quad (S \ o \ T)^x (h) = (T^x \ o \ S^x)(h), \qquad \forall \ h \in Z^*$$

Hence

$$(S \ o \ T)^x = T^x o \ S^x.$$

(b) Let $f \in X^*$. Then

$$(I_X^x f)(x) = f(I_X x)$$

$$= f(x), \quad \forall \ x \in X$$

$$\Rightarrow \qquad I_X^x f = f, \qquad \forall f \in X^*.$$

Hence $I_X^x = I_{X^*}$ the identity in $B(X^*)$.

(c) Under the hypothesis, $T^{-1} \in B(Y, X)$. Also, $T^x \in B(Y^*, X^*)$ and $(T^{-1})^x \in B(X^*, Y^*)$. Further, note that

$$\begin{cases} (T^{-1})^\times \, T^\times = (T \, T^{-1})^\times = I_X^\times = I_{X^*} \\ T^\times \, (T^{-1})^\times = (T^{-1} \, T)^\times = I_X^\times = I_{X^*} \end{cases}$$

$$\Rightarrow \qquad (T^{-1})^\times \, T^\times = I_{X^*} = T^\times \, (T^{-1})^\times.$$

Hence

$$(T^{-1})^\times = (T^\times)^{-1}. \ \blacksquare$$

4.5.5 Corollary. *The mapping $T \to T^\times$ is an isometric isomorphism of normed algebra $B\,(X)$ into the normed algebra $B\,(X^*)$ which reverses products and preserves the identity operator.*

Let $T \in B\,(X, Y)$. Then, $T^\times \in B\,(Y^*, X^*)$. So we can define $(T^\times)^\times \in B\,(X^{**}, Y^{**})$. We write $(T^\times)^\times = T^{\times\times}$ and we call $T^{\times\times}$ as the *second adjoint* of T. Formally, $T^{\times\times} : X^{**} \to Y^{**}$ is defined by

$$(T^{\times\times} \varphi)\,(g) = (\varphi_0 T^\times)\,(g)$$
$$= \varphi\,(T^\times g), \quad \forall \, \varphi \in X^{**} \text{ and } g \in Y^*.$$

We give below the relationship between $T : X \to Y$ and $T^{\times\times} : X^{**} \to Y^{**}$.

4.5.6 Theorem. *Let X and Y be normed spaces over the field $\mathbb{K}$ and let $T \in B\,(X, Y)$. Let $\hat{X}, \hat{Y}$ be the images under the natural embeddings of X, Y into X^{**}, Y^{**}, respectively. Define $\hat{T} \in B\,(\hat{X}, \hat{Y})$ by $\hat{T}\,\hat{x} = \hat{y}$, where $y = T\,x$. Then, the second adjoint $T^{\times\times} : X^{**} \to Y^{**}$ of T is an extension of $\hat{T} : \hat{X} \to \hat{Y}$. In particular, if X is reflexive, then $T^{\times\times} = \hat{T}$.*

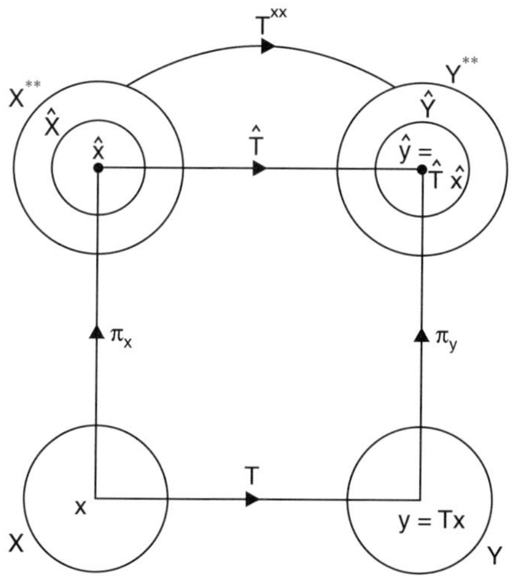

Fig. 4.9

Proof. Note that $\hat{X}$ is a subspace of X^{**}. Let $x \in X$. Then

$$(T^{\times\times} \hat{x}) = \hat{x} \, (T^{\times} g)$$

$$= (T^{\times} g) \, (x)$$

$$= g \, (T x)$$

$$= g \, (y)$$

$$= \hat{y} \, (g)$$

$$= (\hat{T} \hat{x}) \, (g), \quad \forall \, g \in Y^*$$

$$\Rightarrow \qquad T^{\times\times} \hat{x} = \hat{T} \hat{x}, \qquad \forall \, \hat{x} \in \hat{X}$$

$$\Rightarrow \qquad T^{\times\times}|\hat{X} = \hat{T}.$$

It is obvious that $\| \hat{T} \| = \| T \|$. Hence, we conclude that $T^{\times\times}$ is an extension of $\hat{T}$.

Finally, if X is reflexive, then $\hat{X} = X^{**}$ and hence $T^{\times\times} = \hat{T}$. This completes the proof. $\blacksquare$

4.5.7 Corollary. $\| T^{\times\times} \| = \| T^{\times} \| = \| T \|$.

Remark. If X is considered as a part of X^{**} by means of the natural embedding and, similarly, Y as a part of Y^{**}, then one can infer that $T^{\times\times}$ is an extension of T itself. If X is reflexive, then $T^{\times\times} = T$.

It is of interest to investigate dual properties between a bounded linear operator and its conjugate when one of these mappings is a topological isomorphism (Definition 3.3.2).

4.5.8 Theorem. *Let X and Y be normed spaces over the field $\mathbb{K}$ and let $T \in B$ (X, Y). Then:*

(a) *If T is a topological isomorphism of X onto Y, then $T^{\times}$ is a topological isomorphism of Y^* onto X^* and $(T^{\times})^{-1} = (T^{-1})^{\times}$.*

(b) *The converse of (a) holds if X is a Banach space.*

Proof. (a) Let T be a topological isomorphism of X onto Y. Then, T and T^{-1} are both continuous and

$$T \circ T^{-1} = I_Y \text{ and } T^{-1} \circ T = I_X,$$

where I_X and I_Y are the identity operators on X and Y, respectively. By Theorem 4.5.4, we have

$$(T^{-1})^\times o \, T^\times = (T \, o \, T^{-1})^\times = I_Y^\times = I_{Y^*}$$

and

$$T^\times o \, (T^{-1})^\times = (T^{-1} \, o \, T)^\times = I_X^\times = I_{X^*} \, .$$

It follows that $(T^\times)^{-1}$ exists and

$$(T^\times)^{-1} = (T^{-1})^\times.$$

Since T^{-1} is continuous, $(T^\times)^{-1}$ is also continuous. We thus conclude that $T^\times$ is a topological isomorphism of Y^* onto X^*.

(b) We assume that X is a Banach space and that $T^\times$ is a topological isomorphism of Y^* onto X^*. Then, from (a), it follows that $T^{\times\times}$ is a topological isomorphism of X^{**} onto Y^{**}. Therefore, $\exists \, K_1, K_2 > 0$ such that

$$(22) \qquad K_1 \, \| \, \varphi \, \| \leq \| \, T^{\times\times} \, \varphi \, \| \leq K_2 \, \| \, \varphi \, \|, \quad \forall \, \varphi \in X^{**}. \qquad \text{(Theorem 3.3.3)}$$

Let $\hat{X}, \hat{Y}$ be the images under the natural embeddings of X, Y into X^{**}, Y^{**} respectively. Define $\hat{T} : \hat{X} \to \hat{Y}$ by $\hat{T}\hat{x} = \hat{y}$, where $y = Tx$. Then, by Theorem 4.5.6, $T^{\times\times}$ is an extension of $\hat{T}$. Therefore, in view of (22) together with the fact that the natural embeddings are isometry, we have

$$K_1 \, \| \, x \, \| \leq \| \, Tx \, \| \leq K_2 \, \| \, x \, \|, \quad \forall \, x \in X.$$

Consequently, T is a topological isomorphism of X onto $T\,(X)$. The proof is complete if we show that $T\,(X) = Y$.

Since X is complete , $T(X)$ is complete. Hence, $T(X)$ is closed in Y (Theorem 2.3.6). Suppose $T\,(X)$ is proper closed subspace of Y. Let $0 \neq y \in Y - T\,(X)$. Then, by Corollary 4.3.13, $\exists$ a non-zero functional $g \in Y^*$ such that

$$\begin{cases} g(Tx) = 0, & \forall \, x \in X \\ g(y) \neq 0. \end{cases}$$

The first assertion gives

$$(T^\times g)\,(x) = o, \quad \forall \, x \in X$$
$$\Rightarrow \qquad T^\times g = o.$$

But $T^\times$ is one-to-one and so $g = 0$. This contradicts the second assertion. This proves that $T(X) = Y$. Hence, T is a topological isomorphism of X onto Y. ∎

Problems

32. Let X and Y be normed spaces. A mapping $T : X \to Y$ is said to be *norm increasing* if

$$\| Tx \|_Y \leq \| x \|_X, \; \forall \, x \in X.$$

Prove that if T is onto, then $T^\times$ is bijective and the converse holds if X is a Banach space.

[*Hint:* Use Theorem 4.5.8]

33. Let $T : X \to X$ be a bounded linear operator, where X is a Banach space. Prove that T has inverse T^{-1} if and only if the adjoint operator $T^\times$ has inverse $(T^\times)^{-1}$ and that in this case

$$(T^\times)^{-1} = (T^{-1})^\times.$$

4.6 WEAK CONVERGENCE

By exploiting a certain kind of interplay between a normed space and its dual space, we wish to introduce an interesting notion of "weak convergence" of a sequence of vectors in a normed space. This concept demonstrates a fundamental principle of functional analysis which in turn states that the investigation of normed spaces is generally linked with that of their dual spaces. Weak convergence has various applications in the calculus of variations, general theory of differential equations, and, in fact, plays an important role in many problems of analysis. The famous "Principle of Uniform Boundedness" (Theorem 3.7.3) will find its major applications in the theory of weak convergence.

We now introduce the notion of weak convergence for sequences of vectors in a normed space X in terms of bounded linear functionals on X.

4.6.1 Definition. A sequence $\{x_n\}$ in a normed space X is said to be *weakly convergent* if $\exists$ an element $x \in X$ such that

$$\lim_{n \to \infty} f(x_n) = f(x), \quad \forall f \in X^*.$$

The vector x is called the *weak limit* of the sequence $\{x_n\}$, and we say that $\{x_n\}$ *converges weakly* to x which we denote by $x_n \xrightarrow{w} x$.

Note that, for each $f \in X^*$, $\{f(x_n)\}$ is a sequence of scalars in $\mathbb{K}$. Therefore, a sequence $\{x_n\}$ in X is weakly convergent if and only if the sequence $\{f(x_n)\}$ of scalars is convergent (in the ordinary sense) for each $f \in X^*$.

We now state and prove some basic properties on weak convergence.

4.6.2 Theorem. *Let $\{x_n\}$ be a weakly convergent sequence in a normed space X. Then:*

(a) *The weak limit of $\{x_n\}$ is unique.*

(b) *$\{ \| x_n \| \}$ is a bounded sequence in $\mathbb{R}$.*

(c) *Every subsequence of $\{x_n\}$ converges weakly to the weak limit of $\{x_n\}$.*

Proof. (a) Suppose that $\{x_n\}$ converges weakly to x as well as to y. Then

$$f(x) = \lim_{n \to \infty} f(x_n) = f(y), \quad \forall f \in X^*$$

$$\Rightarrow \qquad f(x - y) = 0, \qquad\qquad \forall f \in X^*.$$

Hence $x = y$.

(b) Since $x_n \xrightarrow{w} x$, we have

$$\lim_{n \to \infty} f(x_n) = f(x), \ \forall f \in X^*.$$

Therefore, for each $f \in X^*$, $\{ f(x_n) \}$ is a convergent sequence of numbers and so $\{ f(x_n) \}$ is bounded. Consequently, $\exists$ a constant K_f (depending of f) such that

$$|f(x_n)| \leq K_f, \qquad \forall n \in \mathbb{N}.$$

Let $x \to \varphi_x$ be the natural imbedding of X into X^{**}. Then, for each $n \in \mathbb{N}$, $\| \varphi_{x_n} \| = \| x_n \|$ and

$$| \varphi_{x_n}(f) | = |f(x_n)| \leq K_f, \forall n \in \mathbb{N}.$$

Thus, $\{ \varphi_{x_n}(f) \}$ is bounded for each $f \in X^*$. But the dual space X^* being a Banach space, by the Principle of Uniform Boundedness, it follows that $\{ \| \varphi_{x_n} \| \}$ is bounded and hence $\{ \| x_n \| \}$ is bounded.

(c) By the definition of weak convergence, the sequence $\{ f(x_n) \}$ converges to $f(x)$ for each $f \in X^*$. But $\{ f(x_n) \}$ being a sequence of numbers, it follows that every subsequence of $\{ f(x_n) \}$ converges and has the same limit as the sequence $\{ f(x_n) \}$. ∎

4.6.3 Theorem. *Let $\{x_n\}$ and $\{y_n\}$ be sequences in a normed space X.*

(a) *If $x_n \xrightarrow{w} x$ and $y_n \xrightarrow{w} y$, then $x_n + y_n \xrightarrow{w} x + y$.*

(b) *if $x_n \xrightarrow{w} x$ and α is any scalar, then $\alpha\, x_n \xrightarrow{w} \alpha\, x$.*

Proof. If follows from the linearity of the functionals on X. ∎

In definition 2.1.3 (*iii*), we introduced the concept of convergence of a sequence of vectors in a normed space X. This was an ordinary notion of convergence or what may be called convergence in the norm. To distinguish the new concept of weak convergence from the ordinary convergence of vectors in a normed space, we refer to the ordinary convergence as the strong convergence. Thus, we can restate Definition 2.1.3 (*iii*) as follows.

4.6.4 Definition. A sequence $\{x_n\}$ in a normed space X is said to be *strongly convergent* (or *convergent in the norm*) if $\exists$ a vector $x \in X$ such that

$$\lim_{n \to \infty} \| x_n - x \| = 0.$$

The vector x is called the *strong limit* or simply the limit of the sequence $\{x_n\}$ and we say that $\{x_n\}$ *converges strongly* to x which we denote by

$$x_n \to x \text{ or } \lim_{n \to \infty} x_n = x.$$

We now explore the relationship between strong and weak convergences.

4.6.5 Theorem. *Let $\{x_n\}$ be a sequence in a normed space X. Then $x_n \to x$ in $X \Rightarrow x_n \xrightarrow{w} x$ in X.*

Proof. For every $f \in X^*$, we have

$$|f(x_n) - f(x)| = |f(x_n - x)| \qquad (\because f \text{ is linear})$$

$$\leq \| f \| \, \| x_n - x \| \qquad (\because f \text{ is bounded})$$

$$\to 0, \qquad (\because x_n \to x)$$

$$\Rightarrow \quad \lim_{n \to \infty} f(x_n) = f(x), \quad \forall f \in X^*.$$

Hence $x_n \xrightarrow{w} x$.

Remark. The converse of the Theorem 4.6.5 need not be true.

4.6.6 Examples

1. Take $X = l^p$, $1 < p < \infty$. Consider the sequence $\{x^k\} \subset X$ with $x^k = \{ \xi_i^k \}_i$, where

$$\xi_i^k = \begin{cases} 1, & i = k \\ 0, & i \neq k \end{cases}$$

Then, for $f = \{ f_i \} \in l^q \ (= X^*)$, we have

$$f(x^k) = f_k.$$

But, since $f \in l^q$, $\displaystyle\sum_{k=1}^{\infty} |f_k|^q < \infty$, so that

$$f_k \to 0 \text{ on } k \to \infty.$$

Thus $x^k \overset{w}{\to} 0$. On the other hand

$$\| x^k - 0 \|_p = 1, \quad \forall \, k$$

which verifies that $x^k \nrightarrow 0$, strongly.

2†. Consider the sequence $\{e_n\}$ of unit vectors in the Hilbert space l^2. It is enough to see that every $f \in (l^2)^*$ has the Riesz representation (Theorem 6.1.1)

$$f(x) = <x, a>,$$

for some fixed vector $a = \{a_n\} \in l^2$. Therefore

$$f(e_n) = <e_n, a> = a_n.$$

Since $a_n \to 0$ as $n \to \infty$, we have

$$\lim_{n \to \infty} f(e_n) = 0 = f(0).$$

This is true for all $f \in (l^2)^*$. Consequently, $e_n \overset{w}{\to} 0$. However, $e_n \to 0$ strongly since

$$\| e_n - e_m \|^2 = <e_n - e_m, e_n - e_m> = 2, \quad (m \neq n). \ \blacksquare$$

$\dagger$In this example and Example 4.7.8 (2) that follows, we use certain results and concepts from Hilbert spaces, which will be discussed in Chapter 6. If the reader feels like may postpone the details of these examples till then.

One is interested to know under what conditions the converse of Theorem 4.6.5 holds.

4.6.7 Theorem. *In a finite dimensional normed space, weak convergence implies strong convergence.*

Proof. We assume that dim $X = m$. Then, $\exists$ a basis $\{e_1, e_2, ..., e_m\}$ for X. Let $\{x_n\}$ be a sequence in X such that $x_n \xrightarrow{w} x$. We can write

$$x_n = \lambda_1^{(n)} e_1 + ... + \lambda_m^{(n)} e_m , \ (n = 1, 2, ...)$$

and
$$x = \lambda_1 e_1 + ... + \lambda_m e_m.$$

Consider the functionals $\{f_1, f_2, ..., f_m\}$ in X^*, where

$$f_i(e_j) = \begin{cases} 1, & i = j \\ 0, & i \neq j, \end{cases} \ 1 \leq i, j \leq m.$$

Note that

$$f_i(x_n) = f_i\left(\sum_{j=1}^{m} \lambda_j^{(n)} e_j\right) = \sum_{j=1}^{m} \lambda_j^{(n)} f_i(e_j) = \lambda_i^{(n)}, \quad (1 \leq i \leq m)$$

and similarly

$$f_i(x) = \lambda_i, \ (1 \leq i \leq m).$$

By the definition of weak convergence

$$\lim_{n \to \infty} f(x_n) = f(x), \quad \forall f \in X^*$$

and, in particular, we have

$$\lim_{n \to \infty} f_i(x_n) = f_i(x), \quad (1 \leq i \leq m)$$

$$\Rightarrow \qquad \lim_{n \to \infty} \lambda_i^{(n)} = \lambda_i, \qquad (1 \leq i \leq m).$$

Therefore

$$\| x_n - x \| = \Big\| \sum_{i=1}^{m} (\lambda_i^{(n)} - \lambda_i)\, e_i \Big\| \le \sum_{i=1}^{m} |\lambda_i^{(n)} - \lambda_i| \, \| e_i \|,$$

and hence $x_n \to x$ in X. $\blacksquare$

Note. In ordinary calculus, we deal only with finite dimensional spaces. Thus, in such cases, there is no distinction between strong and weak convergence. As such weak convergence does not play any specific role in ordinary calculus.

Remark. There also exist some infinite dimensional spaces in which strong and weak convergence are equivalent. The example of such an infinite dimensional space is the space l^1 (Example 2.2.5 with $p = 1$), for details refer to Example 4.7.10 that follows.

4.6.8 Definition. A subset M of a normed space X is said to be a *fundamental* (or *total*) set if the *span M* is dense in X, i.e., $\overline{span\ M} = X$.

4.6.9 Theorem. *A sequence $\{x_n\}$ in a normed space X converges weakly to $x \in X$ if and only if*

(a) *the sequence $\{\| x_n \|\}$ is bounded; and*

(b) $\displaystyle\lim_{n \to \infty} f(x_n) = f(x), \quad \forall f \in M,$

   where the subset M of X^* is fundamental.

Proof. Let $x_n \xrightarrow{w} x$. Then, (a) follows by Theorem 4.6.2 (b). Since $M \subset X^*$, (b) follows from the weak convergence of $\{x_n\}$.

   Conversely, in view of (a), $\exists$ a constant K such that

$$\| x_n \| \le K, \quad \forall\, n \in \mathbb{N}$$

and $$\| x \| \le K.$$

Since $\overline{span\ M} = X^*$, for each $f \in X^*$, $\exists$ a sequence $\{ f_m \}$ in *span M* such that

$$\lim_{m \to \infty} f_m = f.$$

Therefore, for a given $\varepsilon > 0$, $\exists f_m \in$ *span* M $(m = 1, 2, ...)$ such that

$$\| f_m - f \| < \frac{\varepsilon}{3 K} .$$

Furthermore, by hypothesis (b), $\exists$ a positive integer N such that

$$| f_m (x_n) - f_m (x) | < \frac{\varepsilon}{3}, \quad \forall\, n > N.$$

Now, for $n > N$, we have

$$| f (x_n) - f (x) | \leq | f (x_n) - f_m (x_n) | + | f_m (x_n) - f_m (x) | +$$
$$| f_m (x) - f (x) |$$

$$< \| f_m - f \| \, \| x_n \| + \frac{\varepsilon}{3} + \| f_m - f \| \, \| x \|$$

$$< \frac{\varepsilon}{3} + \frac{\varepsilon}{3} + \frac{\varepsilon}{3} = \varepsilon$$

$$\Rightarrow \qquad \lim_{n \to \infty} f(x_n) = f(x), \quad \forall\, f \in X^*.$$

Hence, $x_n \overset{w}{\to} x$. ∎

We now look at weak convergence in a concrete space.

4.6.10 Corollary (Weak Convergence in l^p). *Consider the normed space* $(l^p, \| \cdot \|_p)$, $1 < p < \infty$ *(Example 2.2.5). Let*

$$x_n = \left\{ \xi_i^{(n)} \right\}_{i=1}^{\infty} \in l^p \text{ and } x = \{ \xi_i \}_{i=1}^{\infty} \in l^p.$$

Then, $x_n \overset{w}{\to} x$ in l^p if and only if

(i) *the sequence of real numbers $\{ \| x_n \| \}$ is bounded; and*

(ii) $\lim_{n \to \infty} \xi_i^{(n)} = \xi_i, \forall\, i \in \mathbb{N}$

Proof. Consider the unit vector basis $\{ e_n \}$ for l^q, where $\dfrac{1}{p} + \dfrac{1}{q} = 1$. But $(l^p)^*$

being isometrically isomorphic to l^q (Example 4.2.5 (3)), it follows that there is a corresponding set of sequence $\{ f_n \}$ in $(l^p)^*$, where f_n corresponds to e_n, for each n, such that any $f \in (l^p)^*$ can be uniquely expressed as

$$f = \sum_{n=1}^{\infty} \lambda_n f_n, \quad \lambda_n = f (e_n).$$

Let

$$M = \{\, f_1, f_2, \ldots, f_n, \ldots \}.$$

Then, clearly

$$\overline{span\ M} = (l^P)^*.$$

Hence, by Theorem 4.6.9, the result follows. ∎

Recall that the strong convergence of a sequence in $(C\,[a,\,b],\,\|\cdot\|_\infty)$ is exactly the same as the uniform convergence of the sequence, regarded as the sequence of functions, Example 2.2.7 and Remark 1 that follows. Note Remark 2 thereto also.

Problem 34. Show that the weak convergence of a sequence in $(C\,[a,\,b],\,\|\cdot\|_\infty)$ is equivalent to the pointwise convergence and the uniform boundedness of the sequence.

An important result that gives a test for weak convergence in reflexive spaces in the following.

4.6.11 Theorem (Eberlein-Shmulyan). *A Banach space X is reflexive if and only if every strongly bounded sequence in X contains a subsequence which converges weakly to an element in X.*

We now introduce some concepts related to weak convergence.

4.6.12 Definition. Let X be a normed space.

(a) A sequence $\{x_n\}$ in X is said to be a *weak Cauchy sequence* if $\{f(x_n)\}$ is a Cauchy sequence for all $f \in X^*$.

(b) The space X is said to be *weakly complete* if every weak Cauchy sequence in X has a weak limit in X.

(c) A subset M of X is said to be *weakly closed* if the weak limit of every weakly convergent sequence in M is also in M.

(d) A subset M of X is said to be *weakly sequentially compact* if every sequence $\{x_n\}$ in M contains a subsequence converging weakly to a point in M.

4.6.13 Theorem. Let X be a normed space. Then:

(a) A weak Cauchy sequence in X is bounded.

(b) If $\{x_n\} \subset X$ converges weakly to $x \in X$, then

$$\|x\| \le \varliminf_{n \to \infty} \|x_n\|.$$

(c) If X is strongly complete, it need not be weakly complete.

Proof. (a) Let $\{x_n\}$ be a weak Cauchy sequence in X, Then, $\{f(x_n)\}$ is a Cauchy sequence in $\mathbb{K}$, for all $f \in X^*$. Therefore, $\lim\limits_{n \to \infty} f(x_n)$ exists for each $f \in X^*$. This implies that $f(x_n) \xrightarrow{w} x$. Hence, in view of Theorem 4.6.2 (b), $\{\|x_n\|\}$ is bounded.

(b) Since $x_n \xrightarrow{w} x$ in X, we have

$$\lim_{n \to \infty} f(x_n) = f(x), \ \forall f \in X^*.$$

Now

$$|f(x)| = \lim_{n \to \infty} |f(x_n)| \le \|f\| \ \varliminf_{n \to \infty} \|x_n\|$$

$$\Rightarrow \qquad \|x\| = \sup \{|f(x)| : f \in X^*, \|f\| = 1\} \le \varliminf_{n \to \infty} \|x_n\|.$$

(c) Consider the Banach space $X = (c_0, \|\cdot\|_\infty)$. We show that X is not weakly complete. If $x = \{\xi_i\} \in c_0$ and $y = \{\eta_j\} \in l^1$, then

$$f(x) = \sum_{i=1}^{\infty} \xi_i \, \eta_i$$

$\Rightarrow \ \ f(e_k) = \eta_k$, $\{e_n\}$ being the unit vectors in c_0.
Therefore

$$\lim_{k \to \infty} f(e_k) = \lim_{k \to \infty} \eta_k = 0.$$

Thus, $\{e_k\}$ is a weakly Cauchy sequence in c_0.

Let, if possible, $e_k \xrightarrow{w} x_0$ in X for some $x_0 = \{\xi_i^0\} \in c_0$. Then

$$f(e_k - x_0) \to 0 \text{ as } k \to \infty, \quad \forall f \in l^1.$$

Taking $f = e_k$, we find that

$$|1 - \xi_n^0| = 0, \quad \forall\, n \geq 1$$

$$\Rightarrow \qquad \xi_n^0 = 1, \qquad \forall\, n \geq 1.$$

Thus, $x_0 = (1, 1, 1, 1, ...) \notin c_0$

Hence $\{e_k\}$ does not converge weakly in c_0

4.6.14 Theorem. *Let M be a weakly sequentially compact subset of a normed space X. Then:*

(a) *M is bounded.*

(b) *M is weakly closed.*

Proof. (a) Suppose that M is unbounded. Since M is weakly sequentially compact, for every sequence $\{x_n\} \subset M$, $\exists$ a subsequence $\{x_{n_i}\}$ converging weakly to a point in M. By Theorem 4.6.2 (b), $\{\| x_{n_i} \|\}$ is bounded. This is a contradiction to the assumption. Hence the result follows.

(b) Let $\{x_n\}$ be a sequence in M such that $x_n \xrightarrow{w} x$. Then, $\exists$ a subsequence $\{x_{n_i}\}$ converging weakly to a point $y \in M$ since M is weakly sequentially compact. By Theorem 4.6.2 (a), $x = y$ and hence $x \in M$. Thus, M is weakly closed. $\blacksquare$

4.6.15 Theorem. *Let X and Y be normed spaces, $T \in B(X, Y)$ and $\{x_n\}$ be a sequence in X. If $x_n \xrightarrow{w} x$, prove that $Tx_n \xrightarrow{w} Tx$.*

Proof. Note that $\{Tx_n\}$ is a sequence in Y. Let $g \in Y^*$ be arbitrary. Then

$$\begin{cases} g\,(Tx_n) = (g \circ T)\,(x_n) = f\,(x_n), & (g \circ T = f \in X^*) \\ g\,(Tx) = (g \circ T)\,(x) = f\,(x), & f \in X^*. \end{cases}$$

$$\Rightarrow \qquad \lim_{n \to \infty} g\,(Tx_n) = g\,(Tx). \qquad (\because \lim_{n \to \infty} f\,(x_n) = f\,(x), \forall f \in X^*)$$

But $g \in Y^*$ being arbitrary, it follows that

$$Tx_n \xrightarrow{w} Tx. \quad \blacksquare$$

Problems

35. Let A be a set in a normed space X such that every non-empty subset of A contains a weak Cauchy sequence. Prove that A is bounded.

Solution. Let A be unbounded. Then, $\exists$ an unbounded sequence $\{x_n\}$ in A such that $\lim_{n \to \infty} \| x_n \| = \infty$. In particular, for every subsequence $\{x_{n_i}\}$ of $\{x_n\}$, we have

$$\lim_{i \to \infty} \| x_{n_i} \| = \infty.$$

Therefore, $\{x_n\}$ has no weak Cauchy subsequence. This contradicts the hypothesis and hence the result follows.

36. Let $\{x_n\}$ be a sequence in a normed space X such that $x_n \xrightarrow{w} x$ in X. Prove that there is a sequence $\{y_m\}$ of linear combinations of elements of $\{x_n\}$ which converges strongly to x.

[*Hint:* Use Hahn-Banach Theorem.]

37. Prove that every closed subspace of a normed space is weakly closed.

38. Let $\{x_n\}$ be a weakly convergent sequence in a normed space X. Prove that its weak limit x belongs to $\overline{M}$, where M is the linear subspace spanned by $\{x_n\}$.

[*Hint:* Use Hahn-Banach Theorem.]

39. Prove that the converse of Theorem 4.6.12 holds if X is reflexive.

40. Prove that every reflexive normed space is weakly complete.

4.7 WEAK* CONVERGENCE

As discussed for the normed space X, one can talk about strong convergence and weak convergence of sequences in the dual space X^*, that is, a sequence $\{f_n\} \subset X^*$ is

(a) strongly convergent to f in X^* if

$$\| f_n - f \|_{X^*} \to 0, \text{ as } n \to \infty,$$

(b) weakly convergent to f in X^* if

$$\varphi(f_n) \to \varphi(f), \text{ as } n \to \infty, \quad \forall \, \varphi \in X^{**}.$$

Observe that while considering the weak convergence of a sequence in X^*, we make use of the elements of X^{**}, the dual space of X^*. It is natural to think about a type of convergence by going back to the pre-dual space of X^*; namely, the space X. The convergence so obtained is termed as weak* convergence. More precisely,

4.7.1 Definition. Let X^* be the dual space of a normed space X. A sequence $\{f_n\}$ in X^* is said to be weak* *convergent* if $\exists$ an $f \in X^*$ such that

$$\lim_{n \to \infty} f_n(x) = f(x), \quad \forall \, x \in X.$$

The limit f is a called the weak* *limit* of the sequence $\{f_n\}$ which we denote by

$f_n \xrightarrow{w^*} f.$

Remark. The weak convergence in X^* implies weak* convergence but converse need not be true.

4.7.2 Example. Consider the Banach space $X = (c_0, \| \cdot \|_\infty)$ so that $X^* = (l^1, \| \cdot \|_1)$. Let $f_n : c_0 \to \mathbb{K}$ be defined by

$$f_n(x) = \xi_n, \quad x = \{\xi_n\} \in c_0.$$

Then $f_n(x) \to 0$ as $n \to \infty$ for each $x \in X$ so that the sequence $\{f_n\}$ is weak* convergent in X^* (Verify !). But $\|f_n\| = 1, n \in \mathbb{N}$. Therefore $\{f_n\}$ is not strongly convergent in X^* and hence does not converge weakly in X^*, (Example 4.7.10).

One is interested to know under what conditions the converse also holds. Before we workout the converse, we would like to have a very significant observation in this regard. Note that $\{f_n\} \subset X^*$ is weak* convergent if

$$f_n(x) \to f(x), \quad \forall \, x \in X$$

and that X is a subspace of X^{**}. Thus, the elements of X define bounded linear functionals on X^* and so the weak* convergence is like the weak convergence, except the fact that only a subspace of all possible bounded linear functionals on X^* (*i.e.,* the element in X^{**}) is used. Of course, if X is reflexive, then weak* convergence on X^* is precisely the weak convergence on X^*.

4.7.3 Theorem. *Let X be a reflexive normed space. Then, weak* convergence in X^* implies weak convergence in X^*.*

Proof. Let $\{f_n\}$ be a sequence in X^* such that $f_n \xrightarrow{w} f$, i.e.

$$\lim_{n \to \infty} f_n(x) = f(x), \quad \forall\, x \in X$$

$$\Rightarrow \qquad \lim_{n \to \infty} \varphi_x(f_n) = \varphi_x(f), \ \forall\, \varphi_x \in \pi(X),$$

where $\pi : X \to X^{**}$ is the natural imbedding with $\pi(x) = \varphi_x$. Since X is reflexive, $\pi(X) = X^{**}$. Hence $f_n \xrightarrow{w} f$ in. $\blacksquare$

4.7.4 Corollary. *If X is reflexive, then weak convergence and weak* convergence in X^* are equivalent.*

Note. In case X^* is the dual of a non-reflexive space X, then weak convergence and weak* convergence are different. It is preferable to use weak* convergence in X^* instead of weak convergence.

4.7.5 Theorem. *Let X be a Banach space and let $\{f_n\} \subset X^*$ be a sequence. Then, $\{f_n\}$ is weak* convergent if and only if*

(a) *the sequence $\{\|f_n\|\}$ is bounded; and*

(b) *the sequence $\{f_n(x)\}$ is Cauchy, for each $x \in M$, where M is a fundamental subset of X.*

Proof. Let $f_n \xrightarrow{w} f$ in X^*. Then

$$\lim_{n \to \infty} f_n(x) = f(x), \quad \forall\, x \in X$$

$\Rightarrow \quad \{f_n(x)\}$ is bounded for each $x \in X$.

But X being complete, Principle of Uniform Boundedness when applied to bounded linear functionals yields that $\{\|f_n\|\}$ is bounded. This proves (a). Note that (b) is trivial since $\{f_n(x)\}$ is a convergent sequence of numbers for each $x \in X$, in particular, for $x \in M$.

Conversely, suppose that (a) and (b) hold. Since the sequence $\{\|f_n\|\}$ is bounded, $\exists$ a constant K such that

$$\|f_n\| \leq K, \quad \forall\, n \in \mathbb{N}.$$

Let $\varepsilon > 0$ be given. Since $\overline{span\ M} = X$, it follows that for each $x \in X$, $\exists$ a $y \in span\ M$ such that

$$\| x - y \| < \frac{\varepsilon}{3\,K}.$$

For $y \in span\ M$, (b) implies that the sequence $\{ f_n\,(y) \}$ is Cauchy. Hence, $\exists$ an integer $N > 0$ such that

$$|f_n\,(y) - f_m\,(y)\,| < \frac{\varepsilon}{3}, \quad \forall\ m,\,n \geq N.$$

Now, for an arbitrary $x \in X$, we have

$$|f_n\,(x) - f_m\,(x)\,| \leq |f_n\,(x) - f_n\,(y)\,| + |f_n\,(y) - f_m\,(y)\,| + |f_m\,(y) - f_m\,(x)\,|$$

$$< \|f_n\|\,\|x - y\| + \frac{\varepsilon}{3} + \|f_m\|\,\|x - y\|$$

$$< \frac{\varepsilon}{3} + \frac{\varepsilon}{3} + \frac{\varepsilon}{3} = \varepsilon, \quad \forall\ m,\,n \geq N.$$

This shows that $\{ f_n\,(x) \}$ is a Cauchy sequence in $\mathbb{R}$. But $\mathbb{R}$ being complete, $\{ f_n\,(x) \}$ converges to $f\,(x)$, (say), in $\mathbb{R}$. Further, x is an arbitrary element of X, it follows that

$$\lim_{n \to \infty}\ f_n\,(x) = f\,(x), \quad \forall\ x \in X.$$

Thus $f_n \xrightarrow{w^*} f.$ ∎

Problems

41. Let $\{e_n\}$ be a sequence of unit vectors in l^p, $1 \leq p \leq \infty$. Prove that:

(a) $e_n \xrightarrow{w}\!\!\!\!\!/\ \ 0$ in l^1.

(b) $e_n \xrightarrow{w} 0$ in l^p, $1 < p < \infty$.

(c) $e_n \xrightarrow{w^*} 0$ in l^∞.

42. Let $\{f_n\}$ be a sequence of functions in $L^p\,(\mathbb{R})$, $1 \le p \le \infty$, which is uniformly bounded in $L^p(\mathbb{R})$. Prove that:

    (a) If $f_n = X_{[n,\,n+1]}$, then $f_n \overset{w}{\nrightarrow} 0$ in $L^1\,(\mathbb{R})$.

    (b) If $f_n|_{[-n,\,n]} = 0$, then $f_n \overset{w}{\rightarrow} 0$ in $L^p\,(\mathbb{R})$, $1 < p < \infty$.

    (c) $f_n \overset{w^*}{\rightarrow} 0$ in $L^\infty\,(\mathbb{R})$.

43. Consider the following statements:

    (a) $f_n \overset{w^*}{\rightarrow} f$ in $L^\infty\,[a,\,b]$.

    (b) $f_n \overset{w}{\rightarrow} f$ in $L^p\,[a,\,b]$.

    (c) $f_n \overset{w}{\rightarrow} f$ in $L^q\,[a,\,b]$.

    Show that (a) $\Rightarrow$ (b) $\Rightarrow$ (c) whenever $1 \le q < p < \infty$.

44. Show that $e^{2\pi inx} \overset{w^*}{\rightarrow} 0$ in $L^\infty\,[0,\,1]$, *i.e.*,

$$\lim_{n \to \infty} \int_0^1 g\,(x)\,e^{2\pi inx}\,dx = 0, \quad \forall\, g \in L^1\,[0,\,1].$$

Weak* convergence has useful applications in the theory of divergent sequences and series, numerical integration and differentiation, interpolation etc.

We now give a theorem which plays an important role in various applications of the concept of weak* convergence.

4.7.6 Theorem. *Let X be a separable Banach space and let M be a bounded subset of X^*. Then, every sequence of bounded linear functionals in M contains a subsequence which is weak* convergent to a bounded linear functional in X^*.*

Proof. Choose a countable dense subset $A = \{x_1,\, x_2,\, ...\}$ of X. Let $\{f_n\}$ be a sequence in M. Then

$$\{ f_1(x_1), f_2(x_1), \ldots \}$$

is a bounded sequence of scalars. Hence, $\exists$ a subsequence $\{ f_1^{(1)}, f_2^{(1)}, \ldots \}$ of $\{ f_n \}$ such that the sequence $\{ f_1^{(1)}(x_1), f_2^{(1)}(x_1), \ldots \}$ converges. Further, we can select a subsequence $\{ f_1^{(2)}, f_2^{(2)}, \ldots \}$ from the sequence $\{ f_n^{(1)} \}$ such that the sequence $\{ f_1^{(2)}(x_2), f_2^{(2)}(x_2), \ldots \}$ converges. Proceeding in this way, we obtain sequences such that each one is a subsequence of the preceding one. Now, consider the diagonal subsequence

$$\{ f_1^{(1)}, f_2^{(2)}, f_3^{(3)}, \ldots \} \, .$$

Then $\{ f_1^{(1)}(x_n), f_2^{(2)}(x_n), \ldots \}$ converges for each $n \in \mathbb{N}$ and therefore in view of A being dense in X, $\{ f_1^{(1)}(x), f_2^{(2)}(x), \ldots \}$ also converges for an arbitrary $x \in X$. Hence the diagonal subsequence $\{ f_1^{(1)}, f_2^{(2)}, \ldots \}$ is weak* convergent to a functional in X^*. $\blacksquare$

One may note that weak* convergence of sequences in X^* can be renamed as pointwise convergence of sequence of functionals defined on X. Towards the uniform convergence of these functionals, we have the following important result.

4.7.7 Theorem. *Let X be a Banach space. A sequence $\{ f_n \} \subset X^*$ converges strongly to f in X^* if and only if it (as a sequence of functions defined on X) converges uniformly to f in the unit disc $\{ x \in X : \| x \| \le 1 \}$.*

Proof. It follows by using the fact that

$$| f_n(x) - f(x) | \le \| f_n - f \| \, \| x \|. \quad \blacksquare$$

The notions like weak* Cauchy sequence, weak* complete, weak* closed, weak* compact etc. can be introduced on the lines of Definition 4.6.11, replacing weakly convergent sequence in X by weak* convergent sequence in X^*. In particular, a set $\Gamma \subset X^*$ is said to be weak* compact if every sequence $f_n \subset \Gamma$ contains a subsequence which is weak* convergent to an element in Γ.

One of the reasons for importance of weak* convergence is the following compactness result, which we state without proof, the proof being out of the present scope of study.

4.7.8 Theorem (Banach-Alaoglu). *Let X^* be the dual of a Banach space X. Then, the closed unit ball in X^* is weak* compact.*

4.7.9 Corollary. If $f_n \xrightarrow{w^*} f$ in X^*, then

$$\| f \|_{X^*} \leq \liminf_{n \to \infty} \| f_n \|_{X^*}.$$

Proof. Let $\lambda = \liminf\limits_{n \to \infty} \| f_n \|$ and let $\varepsilon > 0$ be arbitrary. Then, $\exists$ a subsequence

$\{ f_{n_k} \}$ of $\{ f_n \}$ with $\| f_{n_k} \| < \lambda + \varepsilon$. The ball of radius $\lambda + \varepsilon$ being weak* compact, it is weak* closed. Thus $\| f \| < \lambda + \varepsilon$. Since $\varepsilon > 0$ is arbitrary, the result follows. ∎

We are now ready to exhibit an example of an infinite dimensional Banach space in which weak convergence implies strong convergence of sequences.

4.7.10 Example. Consider the Banach space l^1 (Example 2.2.5 with $p = 1$). Let $x^{(n)} \xrightarrow{w} 0$ is l^1. Write $x^{(n)} = \{ \xi_i^{(n)} \}_i$. Let S_1 be the closed unit ball in l^∞ $\left(= (l^1)^* \right)$ and let $\varepsilon > 0$ be given. For $k \in \mathbb{N}$, define

$$F_k = \left\{ f \in S_1 : | f(x^{(n)}) | \leq \frac{\varepsilon}{3}, \forall \, n \geq k \right\}.$$

Then F_k is weak* closed in l^∞ for each fixed k. Also

$$S_1 = \bigcup_{k=1}^{\infty} F_k.$$

By Banach-Alaoglu Theorem, S_1 is weak* compact. Therefore, S_1 is a complete metric space with respect to the weak* subspace topology.

By Baire's Category Theorem (Theorem 1.7.32′), F_{k_0} has non-empty weak* interior for some positive integer k_0. Hence, there exists $f = \{ \alpha_n \} \in S_1$ such that

$$G = \{ g \in S_1 : g = \{ \beta_n \}, | \alpha_n - \beta_n | < r, 1 \leq n \leq m \} \subset F_{k_0}$$

for some $r > 0$ and some integer m. Since $\xi^{(n)} \xrightarrow{w} 0$ in l^1, we can choose a positive integer n_0 such that

$$\sum_{i=1}^{m} \xi_i^{(n)} < \frac{\varepsilon}{3}, \quad \forall \, n \geq n_0$$

Let n be fixed but arbitrary positive integer such that $n \geq \max \{k_0, n_0\}$. Define $g = \{\beta_k\} \in S_1$, where

$$\beta_k = \begin{cases} \alpha_k, & 1 \leq k \leq m \\ \operatorname{sgn} \xi_k^{(n)}, & k > n \end{cases}.$$

Then $g \in G$ and therefore $g \in F_{n_0}$. Thus

$$\left| \sum_{k=1}^{m} | \xi_k^{(n)} | \alpha_k + \sum_{k=m+1}^{\infty} | \xi_k^{(n)} | \right| \leq \frac{\varepsilon}{3}, \quad \forall\, n \geq n_0$$

giving

$$\sum_{k=m+1}^{\infty} | \xi_k^{(n)} | \leq \sum_{k=1}^{m} | \xi_k^{(n)} | + \frac{\varepsilon}{3} < \frac{2\varepsilon}{3}.$$

Consequently, we have

$$\| x^{(n)} \| = \sum_{k=1}^{\infty} | \xi_k^{(n)} | < \varepsilon.$$

Hence $x^{(n)} \to 0$ in l^1. ∎

As a consequence of Banach-Alaoglu Theorem, one can establish that a subspace of the Banach space of continuous functions on a compact Hansdorff space is the most general form of a normed space.

4.7.11 Theorem. *Every normed space X is isometrically isomorphic to a subspace of $C(K)$, the Banach space of continuous functions defined on K, where K is the closed unit ball $S[0, 1]$ in X^* endowed with the weak* topology. In particular, every normed space is isometrically isomorphic to a subspace of $C(K)$ for some compact Hansdorff space K.*

Proof. For $x \in X$, let φ_x be the restriction of $\varphi \in X^{**}$ to k, where φ is defined by $\varphi(f) = f(x), f \in X^*$. By the definition of the weak* topology, $\varphi_x \in C(K)$; in fact, the weak* topology on X^* is precisely the weakest topology on X^* in which every function φ is continuous. Clearly, the map $\pi : X \to C(K)$ given by $\pi(x) = \varphi_x$ is linear. Also, by Hahn-Banach Theorem, we note that

$$\| \varphi_x \| = \sup \{| \varphi_x(f) | : f \in K\}$$

$$= \sup \{ \, | \, f(x) \, | : f \in S\,[0,\,1] \}$$

$$= \| \, x \, \|.$$

Hence the map π is linearly isometry. ∎

Problem 45. Let M be a closed subspace of a Banach space X. Define

$$M^{\perp} = \{ \, f \in E^{*} : f(x) = 0, \ \forall \, x \in M \, \}.$$

Prove that:

(a) The collection $M^{\perp}$ is a weak* closed subspace of E^{*}.

     ($M^{\perp}$ is called the *orthogonal complement* of M)

(b) Every weak* closed subspace $G \subset E^{*}$ is the orthogonal complement of the subspace $G_{\perp} \subset E$, where

$$G_{\perp} = \{ x \in E : f(x) = 0, \ \forall \, f \in G \}.$$

(c) The quotient space $(E/M)^{*}$ is isometric to the space $M^{\perp}$.

(d) The space M^{*}, conjugate to the space M, is isometric to the space $E^{*}/M^{\perp}$.

Finally, towards the close of this section, we discuss the possible type of convergence for a sequence of operators in $B\,(X,\,Y)$ which again, in a particular situation when $Y = \mathbb{K}$, leads to the concept of weak* convergence as discussed above.

4.7.12 Definition. Let X and Y be normed spaces over the scalar field $\mathbb{K}$. A sequence $\{T_{n}\} \subset B\,(X,\,Y)$ is said to be:

(a) *Uniformly operator convergent,* if $\exists$ a $T \in B\,(X,\,Y)$ such that

$$\lim_{n \to \infty} \| \, T_{n} - T \, \| = 0.$$

     The limit T is called the *uniform operator limit* of the sequence $\{T_{n}\}$.

(b) *Strongly operator convergent,* if $\exists$ a $T \in B\,(X,\,Y)$ such that

$$\lim_{n \to \infty} \| \, T_{n}x - Tx \, \| = 0, \quad \forall \, x \in X.$$

     The limit T is called the *strong operator limit* of the sequence $\{T_{n}\}$.

(c) *Weakly operator convergent,* if $\exists$ a $T \in B\,(X,\,Y)$ such that

$$\lim_{n \to \infty} | \, f\,(T_{n}x) - f(Tx) \, | = 0, \quad \forall \, x \in X \text{ and } \forall \, f \in Y^{*}.$$

It is easy to verify that (a) $\Rightarrow$ (b) $\Rightarrow$ (c). But the converse implication are not true in general.

4.7.13 Examples

1. In the space l^2, we consider a sequence $\{T_n\}$, where $T_n : l^2 \to l^2$ is defined by

$$T_n x = (0, ..., 0, \xi_{n+1}, \xi_{n+2}, ...), \quad x = \{\xi_i\} \in l^2.$$

This operator T_n is linear and bounded. Clearly, $\{T_n\}$ is strongly operator convergent to O since $T_n x \to O = Ox$. However, $\{T_n\}$ is not uniformly operator convergent since $\| T_n - O \| = \| T_n \| = 1$.

2. In the space l^2, we consider a sequence $\{T_n\}$, where $T_n : l^2 \to l^2$ is defined by

$$T_n x = (0, ..., 0, \xi_1, \xi_2, ...), \quad x = \{\xi_i\} \in l^2.$$

This operator T_n is linear and bounded. We show that $\{T_n\}$ is weakly operator convergent to O but not strongly.

Every bounded linear functional f on l^2 has a Riesz representation

$$f(x) = <x, z> = \sum_{i=1}^{\infty} \xi_i \, \overline{\eta}_i \, , z = \{\eta_i\} \in l^2.$$

Hence, setting $i = n + p$ and using the definition of T_n, we have

$$f(T_n x) = < T_n x, z > = \sum_{i=n+1}^{\infty} \xi_{i-n} \overline{\eta}_i = \sum_{p=1}^{\infty} \xi_p \overline{\eta}_{n+p} .$$

By Cauchy-Schwartz Inequality, we have

$$|f(T_n x)|^2 = |< T_n x, z >|^2 \le \sum_{i=1}^{\infty} |\xi_i|^2 \; \sum_{j=n+1}^{\infty} |\eta_j|^2 .$$

The last series is the remainder of a convergent series. Hence the right hand side tends to 0 as $n \to \infty$. Thus $f(T_n x) \to 0 = f(O\,x)$. Consequently, $\{T_n\}$ is weakly operator convergent to O. However, $\{T_n\}$ is not strongly operator convergent since for $x = (1, 0, 0,...)$, we have

$$\| T_m x - T_n x \| = \sqrt{1^2 + 1^2} = \sqrt{2} \quad (m \neq n).$$

Remark. In view of Theorem 4.6.7, if Y is finite dimensional and in particular when $Y = \mathbb{K}$, the strong operator convergence and weak operator convergence become equivalent. Thus, we are left with two types of convergence in this case; namely, (a) the uniform operator convergence, in fact, the strong convergence (or norm convergence) (Definition 2.1.3 (c)) and (d) (equivalent (c)) – the strong operator convergence, in fact, the weak* convergence (Definition 4.7.1).

The Concept and Specific Geometry of Hilbert Spaces

5

CHAPTER

In the preceding chapters, we studied normed and Banach spaces. These spaces enjoy linear properties as well as metric properties. Although the norm on a linear space generalizes the elementary concept of the length of a vector, but the main geometric concept, missing in abstract normed and Banach spaces, is the angle between two vectors. In fact, these spaces are still too general to yield a really rich theory of operators. In this chapter, we study linear spaces having an inner product, a generalization of the usual dot product on finite dimensional linear spaces. The concept of linner product in a linear space leads to an inner product space and a complete inner product space which is called a Hilbert space. An inner product space (Hilbert space) is a special type of normed space (Banach space) which possesses an additional structure–an inner product.

The theory of Hilbert spaces does not deal with angles in general. But the presence of the additional algebraic structure of inner product greatly enriches the geometric properties of the space. Most significantly, it enables us to introduce a notion of perpendicularity for two vectors and the geometry corresponds in several fundamental respects with Euclidean geometry. In particular, this leads to the important projection theorems, a generalized theory of Fourier series, a most sensible definition of a self-adjoint operator and a powerful body of the theory based on this new concept. This chapter is mainly concerned with the geometric properties of the inner product spaces and Hilbert spaces.

The foundation of the theory of Hilbert spaces was laid down in 1912 by the work of a great German mathematician, D. Hilbert (1862–1943), on integral equations. However, an axiomatic basis of the theory was provided by the famous mathematician, J. Von Neumann (1903–1957). Since then this branch of mathematics has become one of the most interesting and powerful subject. Moreover, Hilbert spaces are the simplest type of infinite dimensional Banach spaces to play a remarkable role in functional analysis.

233

5.1 Definitions and Basic Properties of Inner Product Spaces and Hilbert Spaces

The readers may be familiar with the usual dot product (or scalar product)

$$x \cdot y = \xi_1 \eta_1 + \xi_2 \eta_2 + \xi_3 \eta_3$$

of two vectors $x = (\xi_1, \xi_2, \xi_3)$ and $y = (\eta_1, \eta_2, \eta_3)$ in three dimensional Euclidean space $\mathbb{R}^3$ which is known to satisfy the following properties:

(i) $x \cdot x \geq 0, \ \forall \ x \in \mathbb{R}^3$

   and $x \cdot x = 0 \Leftrightarrow x = 0$

(ii) $x \cdot y = y \cdot x,$ $\qquad\qquad \forall \ x, y \in \mathbb{R}^3$

(iii) $(x + y) \cdot z = x \cdot z + y \cdot z, \quad \forall \ x, y, z \in \mathbb{R}^3$

(iv) $(\alpha x) \cdot y = \alpha \ (x \cdot y), \qquad \forall \ x, y \in \mathbb{R}^3$ and $\forall \ \alpha \in \mathbb{R}.$

On the other hand, the dot product of two vectors $x = (\xi_1, \xi_2, \xi_3)$ and $y = (\eta_1, \eta_2, \eta_3)$ in three dimensional unitary space $\mathbb{C}^3$ can be defined by

$$x \cdot y = \xi_1 \overline{\eta_1} + \xi_2 \overline{\eta_2} + \xi_3 \overline{\eta_3} .$$

It is readily seen that it satisfies all the above properties except (*ii*), which when modified to

(ii)$'$ $x \cdot y = \overline{y \cdot x}$ is also satisfied.

Note that in $\mathbb{R}^3$, (*ii*)$'$, is equivalent to (*ii*). We are now being motivated to use the properties (*i*), (*ii*)$'$, (*iii*) and (*iv*) of the standard dot product in $\mathbb{K}^3$ ($\mathbb{R}^3$ or $\mathbb{C}^3$) as our foundation for generalizing the motion of the dot product to any real or complex linear space. Here X is an arbitrary linear space over $\mathbb{K}$ (not necessarily finite dimensional).

5.1.1 Definition. Let X be a linear space over the field $\mathbb{K}$. A function $< \cdot , \cdot > : X \times X \to \mathbb{K}$ that assigns to each ordered pair (x, y) of vectors in X a scalar $< x, y >$ is said to be an *inner product* on X if it satisfies the following conditions:

(i) $\langle x, x \rangle \geq 0, \ \forall \ x \in X$ (positive definite)

   and $< x, x > = 0 \Leftrightarrow x = 0$

(ii) $< x, y > = \overline{< y, x >}, \quad \forall \quad x, y \in X$ (conjugate symmetry)

(iii) $< \alpha, x, y > = \alpha < x, y >,$ $\forall \, x, y \in X$ and $\forall \, \alpha \in \mathbb{K}$

(iv) $< x + y, z > = < x, z > + < y, z >, \quad \forall \, x, y, z \in X.$

The scalar $< x, y >$ is called the inner product of x and y. A linear space X equipped with an inner product $< \cdot, \cdot >$ defined on X is called an *inner product space* (or *pre-Hilbert space*).

5.1.2 Observations

(i) If X is a real space, then $< x, y > = < y, x >$.

(ii) $< \alpha\, x + \beta\, y, z > = \alpha < x, z > + \beta < y, z >$.

(iii) $< x, \alpha\, y > = \overline{< \alpha\, y, x >} = \overline{\alpha} \, \overline{< y, x >} = \overline{\alpha} < x, y >$.

(iv) $< x, \alpha\, y + \beta\, z > = \overline{\alpha} < x, y > + \overline{\beta} < x, z >$.

The above observation (*ii*) demonstrates that the inner product is linear in the first argument, while (*iv*) shows that the inner product is conjugate linear in the second argument. Consequently, the inner product is sesquilinear which means

$1\dfrac{1}{2}$ times linear (Definition 1.6.32).

5.1.3. Example. For vectors $x = (\xi_1, \xi_2, ..., \xi_n)$ and $y = (\eta_1, \eta_2, ..., \eta_n)$ in $\mathbb{R}^n$, define

$$< x, y > = \sum_{i=1}^{n} \xi_i \, \eta_i \, .$$

One can readily verify that this function satisfies the conditions of Definition 5.1.1. Hence $(\mathbb{R}^n, < \cdot, \cdot >)$ is an inner product space. Note that this space is real.

Note. For $n = 3$, the space $\mathbb{R}^n$ gives the standard inner product space $\mathbb{R}^3$.

5.1.4. Example. For vectors $x = (\xi_1, \xi_2)$ and $y = (\eta_1, \eta_2)$ in $\mathbb{R}^2$, define

$$< x, y > = \xi_1\eta_1 - \xi_2\eta_1 - \xi_1\eta_2 + 3\,\xi_2\eta_2.$$

The space $(\mathbb{R}^2, < \cdot, \cdot >)$ is a real inner product space. The inner product $< \cdot, \cdot >$ is, of course, not the standard inner product on $\mathbb{R}^2$.

Note. Example 5.1.4 shows that we may have many different inner products on a linear space since we also have the standard inner product on $\mathbb{R}^2$ obtained from Example 5.1.3 for $n = 2$.

5.1.5 Example. Consider the linear space $\mathbb{C}^n$ $(= l^2(n))$. For vectors $x = (\xi_1, \xi_2, ..., \xi_n)$ and $y = (\eta_1, \eta_2, ..., \eta_n)$ in $\mathbb{C}^n$, define

$$< x, y > = \sum_{i=1}^{n} \xi_i \, \overline{\eta}_i .$$

The space $(\mathbb{C}^n, < \cdot, \cdot >)$ is a complex inner product space.

Note. Sometimes we do not specify which system of vectors is to be used. It is because we wish to allow both the possibilities unless the contrary is clearly stated. Further, no distinction in notation is made.

5.1.6 Examples

1. Consider the linear space Φ (Example 2.3.9 (1)). For vectors $x = (\xi_1, \xi_2, ...)$ and $y = (\eta_1, \eta_2, ...)$ in Φ, define

$$< x, y > = \sum_{i} \xi_i \overline{\eta}_i .$$

   The space $(\Phi, < \cdot, \cdot >)$ is an inner product space.

2. Consider the linear space l^2. For vectors $x = \{\xi_i\}$ and $y = \{\eta_i\}$ in l^2, define

$$< x, y > = \sum_{i=1}^{\infty} \xi_i \, \overline{\eta}_i .$$

   The space $(l^2, < \cdot, \cdot >)$ is an inner product space.

3. Consider the linear space $C[a, b]$, the set of all continuous real valued (or complex valued) functions on $[a, b]$. For vectors $x, y \in C[a, b]$, define

$$< x, y > = \int_a^b \overline{x(t)} \, y(t) \, dt .$$

   The space $(C[a, b], < \cdot, \cdot >)$ is an inner product space. ∎

   An inner product on X generates a norm on X given by

$$\| x \|^2 = < x, x >$$

   and a metric on X given by

$$d(x, y) = \| x - y \| = (< x - y, x - y >)^{1/2}.$$

Indeed, note that

(i) $\| x \| \geq 0$ since $< x, x > \geq 0$.

(ii) $\| x \| = 0 \Leftrightarrow x = 0$ since $< x, x > = 0 \Leftrightarrow x = 0$.

(iii) $\quad \| \alpha x \|^2 = < \alpha x, \alpha x >$

$$= \alpha \, \overline{\alpha} \, < x, x >$$

$$= | \alpha |^2 \| x \|^2$$

$\Rightarrow \quad \| \alpha x \| = | \alpha | \| x \|.$

Finally, to establish the triangle inequality

(iv) $\| x + y \| \leq \| x \| + \| y \|,$

we establish a fundamental inequality which shows that $\big| < x, y > \big|$ is dominated by the product of $\| x \|$ and $\| y \|$.

5.1.7 Lemma (Schwartz Inequality). *If x, y are any two vectors in an inner product space, then*

(1) $$\big| < x, y > \big| \leq \| x \| \| y \|.$$

The equality occurs if and only if $\{x, y\}$ is a linearly dependent set.

Proof. If $y = 0$, then

$$< x, y > = < x, 0 > = \overline{< 0, x >} = 0 \overline{< x, x >} = 0$$

and the conclusion is clear. Let $y \neq 0$. Then, for any scalar λ, we have

$$0 \leq \| x - \lambda y \|^2 = < x - \lambda y, x - \lambda y >$$

$$= < x, x > - \overline{\lambda} \, < x, y > - \lambda < y, x > + \lambda \, \overline{\lambda} \, < y, y >$$

$$= \| x \|^2 - \overline{\lambda} \, < x, y > - \lambda \, [< y, x > - \overline{\lambda} \, \| y \|^2].$$

In particular, choose λ such that

$$\overline{\lambda} = < y, x > / \| y \|^2.$$

Consequently,

$$0 \leq \| x \|^2 - \frac{| < x, y > |^2}{\| y \|^2}.$$

This verifies the inequality.

From the derivation of Schwartz Inequality, it is quite clear that the sign of equality occurs, if and only if $y = 0$ or $\| x - \lambda y \|^2 = 0$, *i.e.*, $x - \lambda y = 0$, so that $x = \lambda y$. This shows the linear dependence of the set $\{x, y\}$.

5.1.8 Corollary. *If $X \neq \{0\}$ is an inner product space, then*

$$\| x \| = \sup \{ |<x, y>| : \| y \| = 1 \}.$$

Proof. For $x = 0$, there is nothing to prove. Let $x \neq 0$. Then

$$\| x \| = <x, \frac{x}{\| x \|}>$$

$$\leq \sup \{ |<x, y>| : \| y \| = 1 \}, \text{ where } y = \frac{x}{\| x \|}$$

$$\leq \sup \{ \| x \| \| y \| : \| y \| = 1 \} \qquad \text{(Schwartz Inequality)}$$

$$= \| x \|. \quad \blacksquare$$

Now, in order to establish the triangle inequality in (iv) preceding Lemma 5.1.7, note that

$$\| x + y \|^2 = <x + y, x + y>$$

$$= <x, x> + <y, y> + <x, y> + <y, x>$$

$$= \| x \|^2 + \| y \|^2 + <x, y> + \overline{<x, y>}$$

$$= \| x \|^2 + \| y \|^2 + 2 \operatorname{Re} <x, y>$$

$$\leq \| x \|^2 + \| y \|^2 + 2 |<x, y>|$$

$$\leq \| x \|^2 + \| y \|^2 + 2 \| x \| \| y \| \qquad \text{(Schwartz Inequality)}$$

$$= (\| x \| + \| y \|)^2$$

$$\Rightarrow \qquad \| x + y \| \leq \| x \| + \| y \|.$$

Also note that in the derivation of the triangle inequality the sign of equality occurs if and only if

$$<x, y> + \overline{<y, x>} = 2 \| x \| \| y \|$$

$$\Leftrightarrow \qquad \operatorname{Re} <x, y> = \| x \| \| y \| \geq |<x, y>|. \quad \blacksquare$$

Summarily, we have proved that $\| \cdot \| : X \to \mathbb{R}$ satisfies all the conditions of being a norm on X. Hence, the inner product space X is a normed space too under the norm defined with the help of the inner product $< \cdot , \cdot >$ on X. Such a norm on X is called *the norm induced by the inner product* on X. We then have the notions of distance, convergence, completeness, separability and density in inner product spaces as defined in normed spaces and metric spaces.

In view of what we have stated above, an inner product space X is complete as a normed space (or as a metric space) if every Cauchy sequence $\{x_n\}$ in X converges, *i.e.*, if $\| x_n - x_m \| \to 0$, then $\exists$ a vector $x \in X$ such that $\| x_n - x \| \to 0$.

5.1.9 Definition. An inner product space $(X, < \cdot , \cdot >)$ over the field $\mathbb{K}$ is said to be a *Hilbert space* if $(X, < \cdot , \cdot >)$ is complete with respect to the norm $\| . \|$ given by $\| x \| = (< x, x >)^{1/2}$. The Hilbert space is called a *real* or *complex Hilbert space* according as the field $\mathbb{K} = \mathbb{R}$ or $\mathbb{C}$.

Clearly, a Hilbert space is a special type of Banach space. On the other hand, a Banach space $(X, \| \cdot \|)$ will be a Hilbert space if $\exists$ an inner product $< \cdot , \cdot >$ on X such that $< x, x > = \| x \|^2, \ \forall \ x \in X$. However, we shall shortly give an example to show that a Banach space, in general, may not be a Hilbert space.

Notation. In the sequel the letter H, with or without suffix, will generally denote a Hilbert space with the inner product $< \cdot , \cdot >$ defined on H.

Before proceeding to give various examples and counter examples of Hilbert spaces, we give a necessary condition which the norm induced by the inner product on X must satisfy.

The norm induced by an inner product on a linear space X satisfies the *parallelogram equality:*

(2) $\qquad \| x + y \|^2 + \| x - y \|^2 = 2 \ (\| x \|^2 + \| y \|^2), \quad \forall \ x, y \in X.$

Indeed, we have

$$\| x + y \|^2 = < x + y, x + y >$$

$$= < x, x > + < x, y > + < y, x > + < y, y >$$

(3) $$= \| x \|^2 + \| y \|^2 + < x, y > + < y, x >$$

which on changing y to $- y$ gives

(4) $$\| x - y \|^2 = \| x \|^2 + \| y \|^2 - < x, y > - < y, x >.$$

Adding (3) and (4), the parallelogram equality follows.

Note. The name 'parallelogram equality' is suggested by elementary geometry, where we know that "In a parallelogram, the sum of the squares of the diagonals equals the twice of the sum of the squares of its sides". If we consider the norm as the length of a vector, then this result may be interpreted as parallelogram equality in $\mathbb{R}^2$.

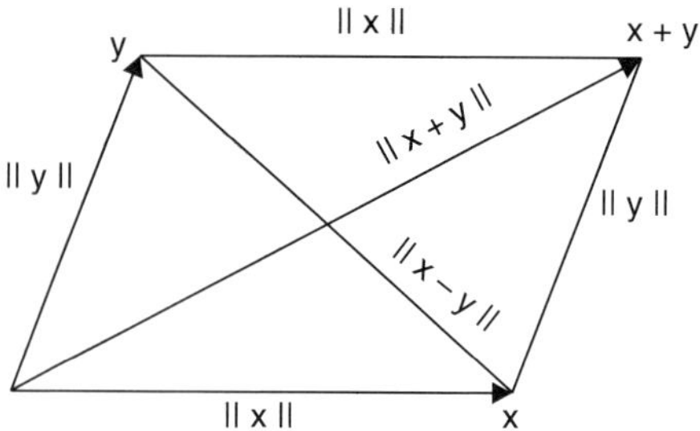

Fig. 5.1. Illustration of parallelogram equality in $\mathbb{R}^2$

We thus observe that if a norm does not satisfy the parallelogram equality, then it can not be generated by an inner product, but such norms do exist as we shall see soon.

We now give below examples and counter examples of Hilbert spaces.

5.1.10 Examples

1. The inner product space $(\mathbb{R}^n, < \cdot\,, \cdot >)$ (Example 5.1.3) equipped with the induced norm given by

$$\| x \| = (< x, x >)^{1/2} = \left(\sum_{i=1}^{n} \xi^2 \right)^{1/2}, \quad x = (\xi_1, \xi_2, ..., \xi_n) \in \mathbb{R}^n$$

   is a Hilbert space. Indeed, in Example 2.2.1, it is proved that $\mathbb{R}^n$ is complete with respect to the above norm.

2. The inner product space $(\mathbb{C}^n, < \cdot\,, \cdot >)$ (Example 5.1.5) equipped with the induced norm given by

$$\| x \| = (< x, x >)^{1/2} = \left(\sum_{i=1}^{n} |\xi_i|^2 \right)^{1/2}, \quad x = (\xi_1, \xi_2, ..., \xi_n) \in \mathbb{C}^n$$

   is a Hilbert space.

3. The inner product space $(\Phi, < \cdot\,, \cdot >)$ (Example 5.1.6 (1)) equipped with the induced norm given by

$$\| x \|_2 = (<x, x>)^{1/2} = \left(\sum_i |\xi_i|^2 \right)^{1/2}, \quad x = (\xi_1, \xi_2, ...) \in \Phi$$

is not a Hilbert space. Indeed, in Example 2.3.9 (3), it has been observed that $(\Phi, \| . \|_2)$ is an incomplete normed space.

4. The inner product space $(l^2, <\cdot, \cdot>)$ (Example 5.1.6 (2)) equipped with the induced norm given by

$$\| x \| = (<x, x>)^{1/2} = \left(\sum_{i=1}^{\infty} |\xi_i|^2 \right)^{1/2}, \quad x = \{\xi_i\} \in l^2$$

is a Hilbert space. Indeed, in Example 2.2.5, it is proved that the space l^p, $1 \le p < \infty$, and in particular, l^2 is complete with respect to the above norm.

5. The linear space l^p, $1 \le p < \infty$, $p \ne 2$ equipped with the norm given by

$$\| x \|_p = \left(\sum_{i=1}^{\infty} |\xi_i|^p \right)^{1/p}, x = \{\xi_i\} \in l^p$$

is not an inner product space and hence not a Hilbert space.

Consider $x = (-1, -1, 0, 0, ...)$ and $y = (-1, 1, 0, 0, ...)$ in l^p. Then

$$\begin{cases} \| x \|_p = 2^{1/p}, \| y \|_p = 2^{1/p} \\ \| x + y \|_p = 2, \| x - y \|_p = 2 \end{cases}$$

$$\Rightarrow \quad \begin{cases} \| x + y \|_p^2 + \| x - y \|_p^2 = 8 \\ 2 (\| x \|_p^2 + \| y \|_p^2) = 4 \times 4^{1/p}. \end{cases}$$

Hence, $\| \cdot \|_p$ does not satisfy parallelogram equality if $p \ne 2$. Thus, l^p $(p \ne 2)$ is not an inner product space, and hence not a Hilbert space. Although, it is a Banach (Example 2.2.5).

6. The linear space $C[a, b]$ equipped with the norm given by

$$\| x \|_\infty = \sup_{t \in [a,b]} |x(t)|, \quad x \in C[a, b]$$

is not an inner product space and hence not a Hilbert space.

Consider $x, y \in C\,[a, b]$ defined by

$$\begin{cases} x(t) = 1 \\ y(t) = \dfrac{t-a}{b-a} \end{cases}$$

$$\Rightarrow \quad \begin{cases} x(t) + y(t) = 1 + \dfrac{t-a}{b-a} \\[2mm] x(t) - y(t) = 1 - \dfrac{t-a}{b-a} \end{cases}$$

$$\Rightarrow \qquad \| x + y \|_\infty^2 + \| x - y \|_\infty^2 = 2^2 + 1^2 = 5.$$

But

$$2\,(\| x \|_\infty^2 + \| y \|_\infty^2) = 2(1 + 1) = 4.$$

Therefore, for $x, y \in C\,[a, b]$, we have

$$\| x + y \|_\infty^2 + \| x - y \|_\infty^2 \neq 2\,(\| x \|_\infty^2 + \| y \|_\infty^2).$$

Hence, $\| . \|_\infty$ can not be the norm induced by any inner product on $C[a, b]$. Thus, $C[a, b]$ is not a Hilbert space. Although, $C[a, b]$ is a Banach space (Example 2.2.7).

7. The inner product space $C[a, b]$ (Example 5.1.6 (3)) equipped with the induced norm is not a Hilbert space.

8. The linear space $L^2\,[a, b]$ equipped with the induced norm given by

$$\| x \|_2 = (<x, x>)^{1/2} = \left(\int_a^b |x(t)|^2\, dt \right)^{1/2}, \quad x \in L^2\,[a, b]$$

is a Hilbert space. See Example 2.2.8 with $p = 2$.

9. The linear space $L^p\,[a, b]$, $1 \le p < \infty$, $p \neq 2$. equipped with the norm given by

$$\| x \|_p = \left(\int_a^b |x(t)|^p\, dt \right)^{1/p}, \quad x \in L^p\,[a, b]$$

is not an inner product space and hence not a Hilbert space.

10. The linear space $L^\infty\,[a, b]$ equipped with the norm given by

$$\| x \|_\infty = \operatorname*{ess\,sup}_{t \in [a, b]}\, |x\,(t)|, \qquad x \in L^\infty\,[a, b]$$

is not an inner product space and hence not a Hilbert space. ∎

The norm of a vector in an inner product space as observed is expressed in terms of the inner product : $\| x \| = \sqrt{<x, x>}$. There is also a useful formula, the *polarization identity,* which expresses the inner product in terms of its induced norm given by

$$
(5) \qquad <x, y> = \begin{cases} \dfrac{1}{4}[(\| x + y \|^2 - \| x - y \|^2) + i\,(\| x + iy \|^2 \\ \quad - \| x - iy \|^2)], \qquad\qquad \mathbb{K} = \mathbb{C} \\ \dfrac{1}{4}(\| x + y \|^2 - \| x - y \|^2), \qquad \mathbb{K} = \mathbb{R}. \end{cases}
$$

Indeed, replacing successively y by $-y$, and iy by $-iy$ in (3), we obtain

$$
\begin{cases} \| x - y \|^2 = \| x \|^2 - <x, y> - <y, x> + \| y \|^2 \\ \| x + iy \|^2 = \| x \|^2 - i<x, y> + i<y, x> + \| y \|^2 \\ \| x - iy \|^2 = \| x \|^2 + i<x, y> - i<y, x> + \| y \|^2 \end{cases}
$$

These equations, respectively yield

(6) $\quad -\| x - y \|^2 = -\| x \|^2 + <x, y> + <y, x> - \| y \|^2.$

(7) $\quad i\| x + iy \|^2 = i\| x \|^2 + <x, y> - <y, x> + i\| y \|^2.$

(8) $\quad -i\| x - iy \|^2 = -i\| x \|^2 + <x, y> - <y, x> - i\| y \|^2.$

Adding (3), (6), (7) and (8), we obtain the polarization identity (5).

We now characterize the norms which are induced by the inner products.

5.1.11 Theorem. *A norm $\| \cdot \|$ on a linear space X is induced by an inner product $< \cdot , \cdot >$ on it if and only if the norm satisfies the parallelogram equality. If it is so, the inner product $< \cdot , \cdot >$ is given by the polarization identity.*

Proof. Suppose the norm $\| \cdot \|$ is induced by the inner product $< \cdot , \cdot >$. Then, the parallelogram equality holds true. Furthermore, the inner product can be recaptured from the norm by the polarization identity.

Conversely, we assume that $\| \cdot \|$ obeys the parallelogram equality and $< \cdot , \cdot >$ is defined by the polarization identity as given in (5). It only remains to prove that $< \cdot , \cdot >$ is an inner product and generates the norm $\| \cdot \|$ on X. Note that the real case is simpler and follows from the complex case. So we give the proof in the complex case as below:

(i) For all $x \in X$, we have

$$<x, x> = \frac{1}{4}[\|2x\|^2 - \|0\|^2 + i(\|(1+i)x\|^2 - \|(1-i)x\|^2)]$$

$$= \frac{1}{4}[4\|x\|^2 + 2i(\|x\|^2 - \|x\|^2)]$$

$$= \|x\|^2.$$

Therefore, inner product $< \cdot , \cdot >$ generates the norm $\| \cdot \|$.

(ii) For all $x, y \in X$, we have

$$\overline{<y, x>} = \frac{1}{4}[\|y+x\|^2 - \|y-x\|^2 - i(\|y+ix\|^2 - \|y-ix\|^2)]$$

$$= \frac{1}{4}[\|x+y\|^2 - \|x-y\|^2 + i(\|x+iy\|^2 - \|x-iy\|^2)]$$

$$= <x, y>.$$

(iii) Let $u, v, w \in X$. By parallelogram equality, we obtain

$$\begin{cases} \|(u+v)+w\|^2 + \|(u+v)-w\|^2 = 2(\|u+v\|^2 + \|w\|^2) \\ \|(u-v)+w\|^2 + \|(u-v)-w\|^2 = 2(\|u-v\|^2 + \|w\|^2) \end{cases}$$

On subtracting, we get

$$(\|(u+w)+v\|^2 - \|(u+w)-v\|^2) +$$

$$(\|(u-w)+v\|^2 - \|(u-w)-v\|^2)$$

$$= 2(\|u+v\|^2 - \|u-v\|^2).$$

Using the polarization identity, we get

$$\begin{cases} \mathrm{Re} <u+w, v> + \mathrm{Re} <u-w, v> = 2\,\mathrm{Re} <u, v> \\ \mathrm{Im} <u+w, v> + \mathrm{Im} <u-w, v> = 2\,\mathrm{Im} <u, v>. \end{cases}$$

This implies

$$(9) \qquad <u+w, v> + <u-w, v> = 2 <u, v>.$$

Setting $u + w = x$, $u - w = y$ and $v = z$, we obtain

$$<x, z> + <y, z> = 2 < \frac{x+y}{2}, z >.$$

But (9) with $w = u$ gives

$$< 2\, u, v > = 2 < u, v >, \quad \forall\, u, v \in X.$$

Hence

$$< x, z > + < y, z > = < x + y, z >.$$

This proves the condition (*iv*) of Definition 5.1.1.

(iv)   We need to prove condition (*iii*) of Definition 5.1.1; namely

$$< \lambda\, x, y > = \lambda < x, y >$$

for every complex scalar λ and $\forall\ x, y \in X$. We shall prove it in stages.

Stage 1. Let $\lambda = m$, a positive integer. Then

$$< m, x, y > = < (m - 1)x + x, y >$$

$$= < (m - 1)x, y > + < x, y >$$

$$= < (m - 2)x, y > + 2 < x, y >$$

$$\cdots\cdots\cdots\cdots\cdots\cdots\cdots\cdots$$

$$\cdots\cdots\cdots\cdots\cdots\cdots\cdots\cdots$$

$$= m < x, y >.$$

Also, for any positive integer n, we have

$$n < \frac{x}{n}, y > = < n \left(\frac{x}{n} \right), y > = < x, y >$$

$$\Rightarrow \qquad < \frac{x}{n}, y > = \frac{1}{n} < x, y >.$$

Stage 2. Let $\lambda = r = \dfrac{m}{n}$ be a rational number. Then

$$r < x, y > = \frac{m}{n} < x, y > = m < \frac{x}{n}, y > = < \frac{m}{n} x, y >$$

$$= < rx, y >.$$

Stage 3. Let λ be any real number. Then, $\exists$ a sequence $\{r_n\}$ of rational numbers such that $r_n \to \lambda$ so that

$$r_n < x, y > \to \lambda < x, y >.$$

But

$$r_n < x, y > = < r_n x, y > \quad \text{and} \quad \| r_n x + y \| \to \| \lambda x + y \|.$$

Therefore

$$r_n < x, y > \; \to \; < \lambda \, x, y >.$$

We thus conclude that

$$\lambda < x, y > \; = \; < \lambda \, x, y > \text{ for any real } \lambda.$$

Stage 4. Let $\lambda = i$. Then, the polarization identity yields

$$< i \, x, y > \; = \; \frac{1}{4} [\| \, ix + y \, \|^2 - \| \, ix - y \, \|^2 + i(\| \, i(x + y) \, \|^2 - \| \, i(x - y) \, \|^2]$$

$$= \frac{i}{4} [\| \, x + y \, \|^2 - \| \, x - y \, \|^2 + i(\| \, x + iy \, \|^2 - \| \, x - iy \, \|^2)]$$

$$= i < x, y >$$

Stage 5. Finally, let $\lambda = a + ib$ be any complex number. Then

$$\lambda < x, y > \; = \; a < x, y > + i \, b < x, y >$$

$$= < a \, x, y > + < i \, b \, x \, , y >$$

$$= < (a + ib) \, x, y >$$

$$= < \lambda \, x, y >.$$

Hence

$$< \lambda \, x, y > \; = \; \lambda < x, y >, \quad \forall \, \lambda \in \mathbb{K} \text{ amd } \forall \, x, y \in X.$$

Thus we have shown that $< \cdot \, , \, \cdot >$ is the inner product inducing the norm $\| \cdot \|$ on X. ∎

Remark. Hilbert spaces (inner product spaces) are precisely those Banach spaces (normed spaces) in which parallelogram equality holds.

Note. An inner product space (a Hilbert space) will be equipped from now onwards with its induced norm, called the *canonical norm*.

Finally, we discuss the notion of a subspace of an inner product space and a Hilbert space.

5.1.12 Definition. A non-empty subset Y of an inner product space X is said to be a *subspace* of X if

(i) Y is a (linear) subspace of X considered as a linear space, and

(ii) Y is equipped with the inner product $< \cdot \, , \, \cdot >_Y$ induced by the inner product $< \cdot \, , \, \cdot >$ on X, *i.e.*,

$$< x, y >_Y \; = \; < x, y >, \quad \forall \, x, y \in Y.$$

Note. It can be easily seen that a subspace Y of an inner product space X is an inner product space. Furthermore, the induced norm $\| \cdot \|_Y$ on Y coincides with the induced norm $\| \cdot \|$ on X.

5.1.13 Definition. A subspace Y of an inner product space X is called a *closed subspace* of X if Y is closed in X considered as a normed space under the induced norm.

5.1.14 Definition. Let H be a Hilbert space. A subspace (closed subspace) Y of H regarded as an inner product space is said to be a *subspace* (*closed subspace*) of the Hilbert space H.

Remark. A subspace Y of a Hilbert space H need not be a Hilbert space because Y may not be complete as a normed space.

It is well known that every subspace of a finite dimensional inner product space is closed. More generally, every finite dimensional subspace of an inner product space is closed. This is not true, in general, as the following example shows.

5.1.15 Example. Consider the Hilbert space l^2 (Example 5.1.10 (4)) and let M be the subset of all finite sequences in l^2 given by

$$M = \{x = \{\xi_i\} \in l^2 : \xi_i = 0 \text{ for } i > N, N \text{ is some positive integer}\}.$$

Obviously, M is a proper subspave of l^2. But M is dense in l^2. Therefore

$$\overline{M} = l^2 \neq M.$$

Hence, M is not closed in l^2. ∎

Since a Hilbert space is a special type of Banach space, Corollary 2.3.8 and Problem 2.24 yield.

5.1.16 Theorem. *Let H be a Hilbert space and Y a subspace of H. Then:*

(a) *Y is complete if and only if Y is closed in H.*

(b) *$\overline{Y}$ is a closed subspace of H.*

5.1.17 Corollary. *A closed subspace of a Hilbert space is a Hilbert space.*

Problems

1. Let X be an inner product space over the field $\mathbb{K}$. If $\{x_n\}$ and $\{y_n\}$ be Cauchy sequences in X, prove that $\{<x_n, y_n>\}$ is a Cauchy sequence in $\mathbb{K}$.

2. In an inner product space X, if $\{x_n\}$ is a Cauchy sequence, prove that the sequence $\{\| x_n \|\}$ is convergent.

3. Let $(H_1, < \cdot , \cdot >_1)$ and $(H_2, < \cdot , \cdot >_2)$ be Hilbert spaces. Prove that the set of all pairs (x, y) with $x \in H_1$ and $y \in H_2$ is a Hilbert space with inner product given by

$$< (x_1, y_1), (x_2, y_2) > \, = \, < x_1, y_2 >_1 + < y_1, y_2 >_2.$$

4. Let $\{H_n\}$ be a sequence of Hilbert spaces. Let H denote the set of all sequences $\{x_n\}$ with $x_n \in H_n$ satisfying $\sum_{n=1}^{\infty} \| x_n \|_n^2 < \infty$. Prove that $H = \bigoplus_{n=1}^{\infty} H_n$ is a Hilbert space under the natural inner product.

5. In an inner product space X, for any elements $x, y, z \in X$, prove that

$$\| z - x \|^2 + \| z - y \|^2 = \frac{1}{2} \| x - y \|^2 + 2 \| z - \frac{1}{2} (x + y) \|^2.$$

[*Hint.* Use parallelogram equality.]

6. Consider the norm $\| \cdot \|$ on $\mathbb{C}^2$ given by

$$\| x \| = |x_1| + |x_2|, \quad (x_1, x_2) \in \mathbb{C}^2.$$

Prove that the space $\mathbb{C}^2$ is not an inner product space.

7. Let X be an inner product space and $\{x_n\} \subset X$. Prove that $x_n \to x$ if the following conditions hold:

    (i) $\| x_n \| \to \| x \|$, and

    (ii) $< x_n, x > \, \to \, < x, x >$.

8. Prove the result in Example 5.1.10 (9).

9. Prove the result in Example 5.1.10 (10).

5.2 Completion of Inner Product Spaces

By an application of Schwartz Inequality, it follows easily that the inner product is jointly continuous.

5.2.1 Lemma. *Let X be an inner product space. If $x_n \to x$ and $y_n \to y$ in X, then $< x_n, y_n > \, \to \, < x, y >$ in $\mathbb{K}$.*

Proof. We have

$$| < x_n, y_n > - < x, y > | = | < x_n, y_n > - < x_n, y > + < x_n, y > - < x, y > |$$

$$= |<x_n, y_n - y> + <x_n - x, y>|$$

$$\leq |<x_n, y_n - y>| + |<x_n - x, y>|$$

$$\leq \|x_n\| \|y_n - y\| + \|x_n - x\| \|y\| \quad \text{(Schwartz Inequality)}$$

$$\rightarrow 0, \text{ as } n \rightarrow \infty. \ \blacksquare$$

5.2.2 Corollary. *If* $x = \displaystyle\sum_{i=1}^{\infty} x_i$ *is an inner product space X, then*

$$<x, y> = \sum_{i=1}^{\infty} <x_i, y>, \ \forall \ y \in X.$$

As an application of Lemma 5.2.1, we prove that an inner product space can be completed so as to be a Hilbert space and this completion is unique except for isomorphisms. We first define isomorphism for inner product spaces.

5.2.3 Definition. Let X and Y be inner product spaces over the field $\mathbb{K}$. An operator $T: X \rightarrow Y$ is said to be an *isomorphism* of X onto Y if T is bijective and preserves the inner product, *i.e.*,

$$<Tx, Ty> = <x, y>, \quad \forall \ x, y \in X.$$

Such spaces X and Y are called *isomorphic* inner product spaces.

Note. The bijectivity and linearity of T guarantee that T is a linear space isomorphism of X onto Y so that T preserves the algebraic structure of the space X. T is also an isometry of X onto Y because the distances in X and Y are determined by the norms induced by the respective inner products. More precisely, we have

$$\|Tx\|^2 = <Tx, Tx>$$

$$= <x, x>$$

$$= \|x\|^2$$

$$\Rightarrow \qquad \|Tx\| = \|x\|, \quad \forall \ x \in X.$$

Thus, inner product isomorphism is an isometric isomorphism and something more. In fact, under such an isomorphism, the metric structure of X as an inner product space is preserved.

5.2.4 Theorem. *Let X be an inner product space. Then, $\exists$ a Hilbert space H and an isomorphism $T: X \rightarrow W$, where W is a dense subspace of H. Furthermore, the space H is unique except for isomorphisms.*

Proof. By Theorem 2.5.2, $\exists$ a Banach space H and an isometry $T : X \to W$, $W \subset H$ with $\overline{W} = H$. Under such an isometry sums and scalar multiples of elements of X and W correspond to each other, so that T is an isomorphism of X onto W, both regarded as normed spaces. In view of Lemma 5.2.1, we can define an inner product on H by setting

$$< \hat{x}, \hat{y} > = \lim_{n \to \infty} < x_n, y_n >,$$

that is, $\{x_n\}$ and $\{y_n\}$ are representatives of $\hat{x} \in H$ and $\hat{y} \in H$, respectively. By polarization identity, it follows that T is an isomorphism of X onto W, both regarded as inner product spaces. Theorem 2.5.2 also guarantees that H is unique except for isometries, *i.e.*, two completions H and $\hat{H}$ of X are related by an isometry $\hat{T} : H \to \hat{H}$. Under the same reasons as in the case of T, we conclude that $\hat{T}$ must be an isomorphism of H onto $\hat{H}$.

5.2.5 Examples

1. The completion of the inner product space Φ (Example 5.1.10 (3)) is the space l^2 (Example 5.1.10 (4)).

2. The completion of the inner product space $C\,[a,\, b]$ (Example 5.1.10 (7)) is the space $L^2\,[a,b]$ (Example 5.1.10 (8)).

5.3 Orthogonality of Vectors

Recall that if the dot product of two vectors in the space $\mathbb{R}^3$ is zero, the vectors are orthogonal or at least one of the vectors is the zero vector. We generalize this concept in an inner product space.

5.3.1 Definition. Let X be an inner product space. A vector $x \in X$ is said to be *orthogonal* to a vector $y \in X$ if

$$< x, y > = 0.$$

Such vectors x and y are called *orthogonal vectors*, written $x \perp y$ (the symbol $\perp$ is pronounced as "per"). Similarly, for subsets $A, B \subset X$, we write $x \perp A$ if $x \perp a,\ \forall\, a \in A$ and $A \perp B$ if $a \perp b,\ \forall\, a \in A$ and $b \in B$.

5.3.2 Observations

(i) $x \perp y \Leftrightarrow y \perp x.$

(ii) $x \perp 0,\ \forall\, x \in X.$

(iii) 0 is the only vector in X orthogonal to itself.

(iv) The parallelogram equality can be considerably simplified, for orthogonal pair of vectors, giving:

Pythagorean Theorem. In an inner product space X, if $x \perp y$, then

(10) $$\| x + y \|^2 = \| x \|^2 + \| y \|^2.$$

More generally, if $x_1, x_2, ..., x_n$ are pairwise orthogonal vectors in X then

$$\left\| \sum_{i=1}^{n} x_i \right\|^2 = \sum_{i=1}^{n} x_i \, \| x_i \|^2.$$

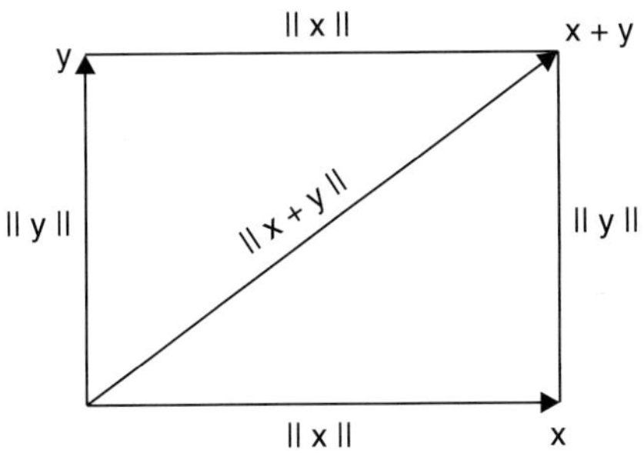

Fig. 5.2 Illustrating (10) in $\mathbb{R}^2$

For a subset A of an inner product space X, define the set

$$A^{\perp} = \{x \in X : x \perp A\}.$$

We write $(A^{\perp})^{\perp} = A^{\perp\perp}$, $(A^{\perp\perp})^{\perp} = A^{\perp\perp\perp}$ and so on.

5.3.3 Observations

(i) *$\{0\}^{\perp} = X$ and $X^{\perp} = \{0\}$, i.e., 0 is the only vector orthogonal to every vector.*

Note that

$$\{0\}^{\perp} = \{x \in X : <x, 0> = 0\} = X$$

since $<x, 0> = 0$, $\forall x \in X$. Also, if $x \neq 0$, then $<x, x> \neq 0$. In other words, a non-zero vector x cannot be orthogonal to the entire space X. Hence, $X^{\perp} = \{0\}$.

(ii) *If $A \neq \phi$ is a subset of X, then the set $A^{\perp}$ is a closed subspace of X. Furthermore, $A \cap A^{\perp}$ is either $\{0\}$ or empty (when $0 \notin A$).*

Let $x, y \in A^{\perp}$ and let $\alpha, \beta \in \mathbb{K}$. Then

$$<\alpha x + \beta y, z> = \alpha <x, z> + \beta <y, z> = 0, \quad \forall z \in A.$$

This implies that $\alpha x + \beta y \in A^{\perp}$. Hence, $A^{\perp}$ is a (linear) subspace of X.

Since $A^{\perp} \subset \overline{A^{\perp}}$, it remains to prove that $\overline{A^{\perp}} \subset A^{\perp}$.

Let $x, \in \overline{A^{\perp}}$. Then, $\exists$ a sequence $\{ x_n \}$ in $A^{\perp}$ such that $\lim_{n \to \infty} x_n = x$. Thus

$$< x, y > = \lim_{n \to \infty} < x_n, z > = 0, \forall z \in A.$$

Hence, $x \in A^{\perp}$ and this completes the proof of the first part. For the second part, suppose $A \cap A^{\perp} \neq \phi$ and let $x \in A \cap A^{\perp}$. Then, we must have $x \perp x$. Hence, $x = 0$.

(iii) *If A is a subset of X, then $A \subset A^{\perp\perp}$.*

Let $x \in A$. Then, $x \perp A^{\perp}$ which means that $x \in (A^{\perp})^{\perp}$.

(iv) *If A and B are subsets of X such that $A \subset B$, then $A^{\perp} \supset B^{\perp}$.*

Let $x \in B^{\perp}$. Then, $< x, y > = 0$, $\forall y \in B$ and, in particular, $\forall x \in A$ since $A \subset B$. This verifies that $x \in A^{\perp}$. Hence, $B^{\perp} \subset A^{\perp}$.

(v) *If $A \neq \phi$ is a subset of X, then $A^{\perp} = A^{\perp\perp\perp}$.*

Changing A by $A^{\perp}$ in (*iii*), we get

$$A^{\perp} \subset (A^{\perp})^{\perp\perp} = A^{\perp\perp\perp}.$$

The reverse inclusion follows by an application of (iv) in (iii) ∎

Note. There are two main ways of getting closed subspace of an inner product space X:

(i) Start with an arbitary subset $A \subset X$ and find span A. Then, $\overline{\text{span } A}$ is a closed subspace of X.

(ii) For any given subset $A \subset X$, find $A^{\perp}$. Then, $A^{\perp}$ is a closed subspace of X.

5.3.4 Definition. Let X be an inner product space. If M and N are subspaces of X with $M \perp N$, then the vector sum $M + N = \{x + y : x \in M, y \in N\}$ is called the *orthogonal sum* of M and N.

5.3.5 Lemma. *Let H be a Hilbert space. If M and N are orthogonal closed subspaces of H, then $M + N$ is a closed subspace of H.*

Proof. Clearly, $M + N$ is a subspace of H, regarded as an inner product space. It remains to prove that $M + N$ is closed in H. In other words, we need to show that $\overline{M + N} \subset M + N$.

Let $z \in \overline{M + N}$. Then $\exists$ a sequence $\{z_n\}$ in $M + N$ such that $\lim\limits_{n \to \infty} z_n = z$.

Since $M \perp N$, we have $M \cap N = \{0\}$. Therefore, each $z_n \in M + N$ can be uniquely expressed in the form

$$z_n = x_n + y_n, \quad x_n \in M, y_n \in N.$$

Since $x_n \perp y_n$, $n \in \mathbb{N}$, in view of Pythagorean Theorem, we have

$$\| z_m - z_n \|^2 = \| (x_m - x_n) + (y_m - y_n) \|^2.$$
$$= \| x_m - x_n \|^2 + \| y_m - y_n \|^2.$$

But $\{z_n\}$ being convergent is a Cauchy sequence and, therefore, $\{x_n\}$ and $\{y_n\}$ are Cauchy sequences in M and N, respectively. Since M and N are closed subspaces of H, they are complete. Hence, $\exists$ vectors $x \in M$ and $y \in N$ such that

$$\lim_{n \to \infty} x_n = x \text{ and } \lim_{n \to \infty} y_n = y.$$

Thus

$$z = \lim_{n \to \infty} z_n = \lim_{n \to \infty} (x_n + y_n) = x + y.$$

Hence, $z \in M + N$ and the proof is complete. ∎

Remark. An analogous statement of the preceding lemma applies for a finitely many mutually orthogonal subspaces. The proof of the lemma can also be extended to apply in case of countably many summands.

Problems

10. In an inner product space, prove that x and y are othogonal vectors if and only if

$$\| x + \alpha y \| = \| x - \alpha y \|, \quad \forall \alpha \in \mathbb{K}.$$

11. Let X be an inner product space and $x, y \in X$, Prove that

$$x \perp y \Rightarrow \| x + \lambda y \| \geq \| x \|, \quad \forall \lambda \in \mathbb{C}.$$

12. In a real inner product space X if

$$\| x + y \|^2 = \| x \|^2 + \| y \|^2$$

(Pythagorean identity) holds for $x, y \in X$, prove that x and y are orthogonal vectors.

13. Let x, y and $\{x_n\}$ be in an inner product space X. Prove that $x \perp y$ if

(i) $y \perp x_n$, $\forall\, n \in \mathbb{N}$, and

(ii) $x_n \to x$.

5.4 Orthogonal Complements and Projection Theorem

In three-dimensional Euclidean space $\mathbb{R}^3$ if M is a plane through the origin, then $M^\perp = \{\, x \in \mathbb{R}^3 : <x, y> = 0,\ \forall\, y \in M \,\}$ is the line through the origin perpendicular to M. The space $\mathbb{R}^3$ has two well known typical properties; namely:

(i) There exists a unique vector in M at minimum distance from a given vector.

(ii) Any vector x may be written uniquely in the form $x = y + z$ with $y \in M$, $z \in M^\perp$. In other words, $\mathbb{R}^3$ may be decomposed into the direct sum of orthogonal subspaces.

The purpose of this section is to generalize the geometric properties described above to general Hilbert spaces. We shall establish a few results which exhibit that the special setting of a Hilbert space is indispensable in contrast to the general Banach spaces where such results are not valid. We are mainly concerned with the representation of a Hilbert space as a direct sum by making use of orthogonality. However, we first need to establish a few results for our main theorem. The first result is that the minimum of the distances from any fixed vector to the vectors of any fixed closed subspace is always attained. We prove this approximation theorem with somethat more generality than we actually need in the main theorem. Recall that a subset A of a vector space X is said to be a convex set if for any $x, y \in A$ and $0 \le \alpha \le 1$, it follows that $\alpha\, x + (1 - \alpha)\, y \in A$. Any subspace of an inner product space is obviously convex and the intersection of convex sets is a convex set.

5.4.1 Theorem. *Let X be an inner product space and $C \neq \phi$ a convex subset of X which is complete in the metric induced by the inner product on X. Then, for every given $x \in X$, $\exists$ a unique $y_0 \in C$ such that*

$$\inf_{y \in C} \| x - y \| = \| x - y_0 \|.$$

Proof. Write

$$\inf_{y \in C} \| x - y \| = \lambda.$$

Existence. By the definition of λ, there is a sequence $\{\,y_n\,\}$ in C such that

$$\lim_{n \to \infty} \| x - y_n \| = \lambda.$$

We first show that $\{\,y_n\,\}$ is a Cauchy sequence in C. Writing $\| x - y_n \| = \lambda_n$ so that $\lambda_n \to \lambda$, as $n \to \infty$ and $y_n - x = v_n$, we have $\| v_n \| = \lambda_n$ and

$$\| v_n + v_m \| = \| y_n + y_m - 2x \| = 2 \left\| \frac{1}{2}\,(y_n + y_m) - x \right\|.$$

But C being a convex set, we have

$$y_n, y_m \in C \implies \frac{1}{2}\,(y_n + y_m) \in C.$$

Therefore

$$\left\| \frac{1}{2}\,(y_n + y_m) - x \right\| \geq \lambda$$

$$\implies \qquad \| v_n + v_m \| \geq 2\,\lambda.$$

Also, we note that $y_n - y_m = v_n - v_m$ so that

$$\| y_n - y_m \|^2 = \| v_n - v_m \|^2$$

$$= - \| v_n + v_m \|^2 + 2(\| v_n \|^2 + \| v_m \|^2)$$

$$\text{(Parallelogram equality)}$$

$$\leq - (2\lambda)^2 + 2(\lambda_n^2 + \lambda_m^2).$$

This shows that $\{\,y_n\}$ is a Cauchy sequence since $\lambda_n \to \lambda$, as $n \to \infty$. Since C is complete, $\{\,y_n\}$ converges in C. Suppose $y_n \to y_0$ in C. Since $y_0 \in C$, we have

$$\| x - y_0 \| \geq \lambda.$$

Also

$$\| x - y_0 \| \leq \| x - y_n \| + \| y_n - y_0 \|$$

$$= \lambda_n + \| y_n - y_0 \| \to \lambda, \text{ as } n \to \infty$$

$$\implies \qquad \| x - y_0 \| \leq \lambda.$$

Hence, $\qquad \| x - y_0 \| = \lambda.$

Uniqueness. Let there be y_0 and y' in C both satisfying

$$\| x - y_0 \| = \lambda = \| x - y' \|.$$

By the parallelogram equality, we have

$$\| y' - y_0 \|^2 = \| (y' - x) - (y_0 - x) \|^2$$
$$= 2 (\| y' - x \|^2 + \| y_0 - x \|^2) - \| (y' - x) + (y_0 - x) \|^2.$$

But

$$y', y_0 \in C \;\; \Rightarrow \;\; \frac{1}{2} (y' + y_0) \in C$$

so that

$$\| \frac{1}{2} (y' + y_0) - x \| \geq \lambda.$$

Consequently

$$\| y' - y_0 \|^2 \leq 2\lambda^2 + 2\lambda^2 - 4\lambda^2 = 0.$$

Hence, $y' = y_0$. ∎

Remark. The conclusion of Theorem 5.4.1 need not hold if X is taken to be a Banach space (Verify !).

Problem 14. Let A be a convex subset of a Hilbert space H and let $\{ x_n \}$ be a sequence in A such that

$$\lim_{n \to \infty} \| x_n \| = d = \inf_{x \in A} \| x \|.$$

Prove that $\{x_n\}$ converges in H.

Solution. For all $m, n \in \mathbb{N}$, we have

$$\| x_n + x_m \|^2 + \| x_n - x_m \|^2 = 2(\| x_n \|^2 + \| x_m \|^2)$$

$$\text{(Parallelogram equality)}$$

$$\Rightarrow \;\; \| x_n - x_m \|^2 = 2 (\| x_n \|^2 + \| x_m \|^2) - \| x_n + x_m \|^2.$$

Since A is convex, it follows that

$$\frac{1}{2} (x_n + x_m) \in A \text{ and } \| x_n + x_m \| \geq 2d.$$

Hence $\quad \| x_n - x_m \|^2 \leq 2 (\| x_n \|^2 + \| x_m \|^2) - 4d^2$

$$\to 2 (d^2 + d^2) - 4d^2 = 0, \text{ as } m, n \to \infty$$

$\Rightarrow \quad \{x_n\}$ is Cauchy in A and hence in H

$\Rightarrow \quad \{x_n\}$ converges in H.

5.4.2 Theorem. *Let X be an inner product space and $M \neq \phi$ be a complete proper subspace of X. Then $M^\perp \neq \phi$*

Proof. Let $0 \neq x \in X - M$. This is possible since M is proper subset of X. Note that M is a convex set. By Theorem 5.4.1, $\exists$ a unique $y_0 \in M$ such that

$$\| x - y_0 \| = \inf_{y \in M} \| x - y \| = \lambda \text{ (say).}$$

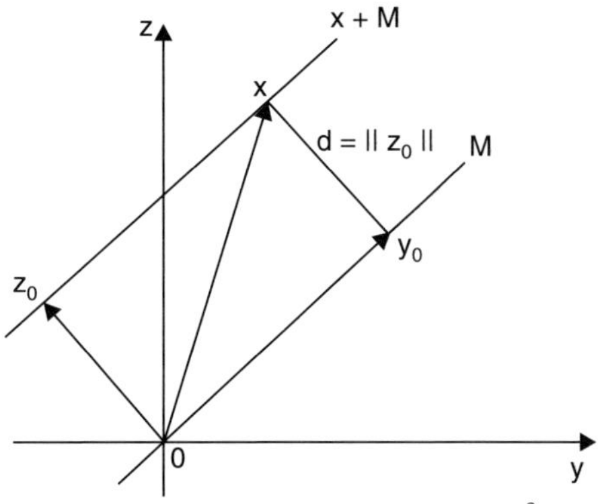

Fig. 5.3 Illustration of $z_0 = x - y_0$ in $\mathbb{R}^2$

Observe that $\lambda > 0$. Define $z_0 = x - y_0$. Clearly, $z_0 \neq 0$, $z_0 \in X$ and $\| z_0 \| = \lambda$. We shall prove that $z_0 \in M^\perp$. Let $y \in M$ and $\alpha \in \mathbb{K}$. Then, $y_0 + \alpha y \in M$ so that

$$\| z_0 - \alpha y \| = \| x - (y_0 + \alpha y) \| \geq \lambda = \| z_0 \|$$

$$\Rightarrow \quad \| z_0 - \alpha y \|^2 - \|z_0 \|^2 \geq 0$$

$$\Rightarrow \quad < z_0 - \alpha y, z_0 - \alpha y > - \| z_0 \|^2 \geq 0$$

$$\Rightarrow \quad - \bar{\alpha} < z_0, y > - \alpha < y, z_0 > + | \alpha |^2 \| y \|^2 \geq 0.$$

Putting $\alpha = \beta < z_0, y >$ for arbitrary real number β, we get

$$- 2\beta |< z_0, y >|^2 + \beta^2 |< z_0, y >|^2 \| y \|^2 \geq 0.$$

This can be expressed in the form

$$(11) \qquad \beta\, a\, (\beta\, b - 2) \geq 0$$

for all real β, where $a = |< z_0, y >|^2 \geq 0$, $b = \| y \|^2 \geq 0$. But $a \not> 0$ since otherwise (11) would be false for sufficiently small β. Thus, a must be zero. This implies that

$$| < z_0, y > | = 0$$

$$\Rightarrow \quad < z_0, y > = 0$$

$$\Rightarrow \quad z_0 \perp y.$$

But $y \in M$ is arbitrary. Hence, $z_0 \in M^\perp$. $\blacksquare$

Note. The preceding theorem generalizes the well-known fact of Euclidean geometry; namely, the unique point y_0 in a line M closest to a given point x in the plane is obtained by dropping a perpendicular from x to M (Fig. 5.3).

We recall that given a linear space X and its subspaces M and N, X is defined to be the direct sum of M and N, written $X = M \oplus N$, if every $x \in X$ can be expressed uniquely in the form

$$x = y + z, \quad y \in M \text{ and } z \in N.$$

Equivalently, it is easy to see that $X = M \oplus N$ if $X = M + N$ and $M \cap N = \{0\}$. We now prove the main theorem about decomposing a Hilbert space into a direct sum of mutually orthogonal closed subspaces.

5.4.3 Theorem (Projection). *Let M be a closed subspace of a Hilbert space H. Then, $H = M \oplus M^{\perp}$.*

Proof. Note that $M^{\perp}$ is a closed subspace of H (Observation 5.3.3 (*ii*)). Therefore, M and $M^{\perp}$ are orthogonal closed subspaces of H. Hence, it follows that $M + M^{\perp}$ is also a closed subspace of H (Lemma 5.3.5). Since H is complete and $M + M^{\perp}$ is closed in H, it follows that $M + M^{\perp}$ is complete (Theorem 5.1.16). We prove that $H = M + M^{\perp}$. Let, if possible, $M + M^{\perp}$ be a proper subspace of H. By Theorem 5.4.2, $(M + M^{\perp})^{\perp} \neq \{0\}$, *i.e.*, $\exists$ a vector $0 \neq z_0 \in H$ such that $z_0 \perp M + M^{\perp}$. This gives

$$< z_0, y + z > = 0, \quad \forall\, y \in M \text{ and } z \in M^{\perp}$$

$$\Rightarrow \quad < z_0, y > + < z_0, z > = 0, \quad \forall\, y \in M \text{ and } z \in M^{\perp}.$$

In particular, taking $z = 0$ and $y = 0$, respectively, we obtain

$$\begin{cases} < z_0, y > = 0, & \forall\, y \in M \\ < z_0, z > = 0, & \forall\, z \in M^{\perp} \end{cases}$$

Consequently

$$z_0 \in M^{\perp} \cap M^{\perp\perp}.$$

But $\quad M^{\perp} \cap M^{\perp\perp} = \{0\}$ $\hfill$ (Observation 5.3.3 (*ii*)).

Therefore, it follows that $z_0 = 0$, which contradicts the fact that $z_0 \neq 0$. Hence

$$H = M + M^{\perp}.$$

Since $M \cap M^{\perp} = \{0\}$, we have

$$H = M \oplus M^{\perp}$$

This completes the proof. ∎

Remark. The projection theorem shows that every closed linear subspace M of H has at least one complementary closed linear subspace and it gives an explicit description of one such subspace ; namely, $M^{\perp}$. One may note that in some Banach spaces a closed subspace may fail to have complementary closed linear subspace; for instance, the closed subspace c_0 of the Banach space l^{∞} is not complemented in l^{∞}.

The projection theorem, in fact, essentially means that if M is a closed subspace of H, then $M^{\perp}$ is just big enough to guarantee that $M \oplus M^{\perp} = H$. It leads to the concept of orthogonal projection.

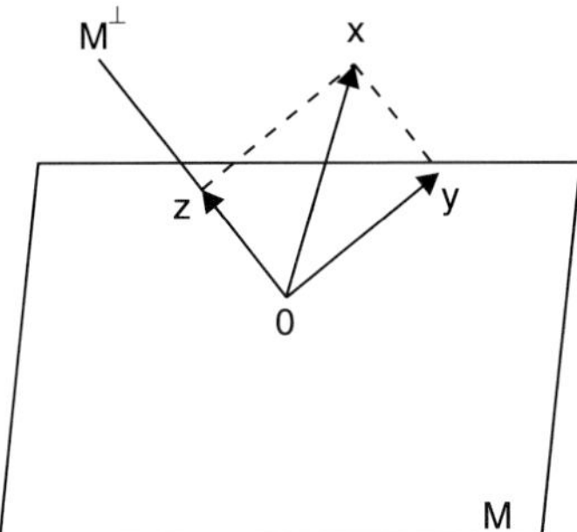

Fig. 5.4 Illustration of x = y + z in $\mathbb{R}^3$

5.4.4 Definition. Let M be a closed subspace of a Hilbert space H and let x be any given vector in H. Suppose $x = y + z$, with $y \in M$ and $z \in M^{\perp}$, is the unique decomposition of x.

(a) The vector y is called the *orthogonal projection* of x on M.

(b) The mapping $P_M : H \to M$ defined by $P_M x = y$ is called the *orthogonal projection operator* (briefly, *orthogonal projection*) of H onto M. Such an operator is also called *perpendicular projection operator.*

Note. The notation P_M is used to indicate that it relates to the closed subspace M. If it is clear to which closed subspace it relates to we simply write P in place of P_M.

5.4.5 Observations

The orthogonal projection operator $P : H \to H$ has the following properties:

(i) P is linear.

(ii) P is bounded and $\| P \| \leq 1$.

Suppose $x = y + z$, where $y \in M$ and $z \in M^{\perp}$, is the decomposition of x. Then

$$\| x \|^2 = \| y + z \|^2$$

$$= \| y \|^2 + \| z \|^2 \qquad \text{(Pythagorean Theorem)}$$

$$\geq \| y \|^2 = | Px \|^2$$

$$\Rightarrow \qquad \| Px \| \leq \| x \|, \quad \forall\, x \in H.$$

(iii) $\mathcal{R}(P) = M$ and $\mathcal{N}(P) = M^{\perp}$.

(iv) P maps H onto M, M onto itself and $M^{\perp}$ onto $\{0\}$.

(v) P is idempotent, *i.e.*, $P^2 = P$.

Since $Px \in M$ for all $x \in H$, it follows that

$$P^2 x = P(Px) = Px, \, \forall\, x \in H.$$

(vi) $\| P \| = 1$.

(vii) $P\big|_M$ is the identity operator on M.

(viii) $< P\, x_1, x_2 > \, = \, < x_1, Px_2 >, \quad \forall\, x_1, x_2 \in H.$

Let $x_1 = y_1 + z_1$ and $x_2 = y_2 + z_2$ be the orthogonal decompositions of x_1 and x_2, respectively where $y_1, y_2 \in M$ and $z_1, z_2 \in M^{\perp}$. Then

$$Px_1 = y_1, P\, x_2 = y_2.$$

Now $< P\, x_1, x_2 > \, = \, < y_1, y_2 + z_2 > \, = \, < y_1, y_2 > + < y_1, z_2 > \, = \, < y_1, y_2 >$ and similarly

$$< x_1, P\, x_2 > \, = \, < y_1, y_2 >. \quad \blacksquare$$

In the light of the preceding discussion, we observe that the projection theorem guarantees that a Hilbert space is always rich in projections. In fact, to each closed subspace M of H, there is associated a projection P_M of H with $\mathcal{R}(P_M) = M$ and $\mathcal{N}(P_M) = M^{\perp}$. This situation is more than satisfactory in contrast with in general Banach spaces. We shall discuss further details of such projections in Section 6.6.

The projection theorem provides a characterization of closed subspaces of a Hilbert space in terms of orthogonality.

5.4.6 Theorem. *A subspace M of a Hilbert space H is closed in H if and only if* $M = M^{\perp\perp}$.

Proof. If $M = M^{\perp\perp}$, then it is obviously true that M is a closed subspace of H, since $M^{\perp\perp}$ is always a closed subspace (Observation 5.3.3 (*ii*).

Conversely, we assume that M is a closed subspace of H. For any subset M of H, we have $M \subset M^{\perp\perp}$ (Observation 5.3.3 (*iii*). So it remains to prove that $M^{\perp\perp} \subset M$.

Let $x \in M^{\perp\perp}$. By projection theorem, $x = y + z$ with $y \in M$ and $z \in M^{\perp}$. Since $y \in M$, it follows that $y \in M^{\perp\perp}$. Now, since $x, y \in M^{\perp\perp}$ and $M^{\perp\perp}$ is a linear subspace of H, we have

$$z = x - y \in M^{\perp\perp}.$$

Consequently, $z \in M^{\perp} \cap M^{\perp\perp}$. As such $z \perp z$. This yields $z = x - y = 0$ and hence $x \in M$. ∎

The projection theorem provides a characterization of sets in Hilbert spaces whose span is dense in H.

5.4.7. Theorem. *Let A be a non-empty subset of a Hilbert space H. Then, span A is dense in H if and only if* $A^{\perp} = \{0\}$.

Proof. We assume that span A is dense in H. Write $M = $ span A so that $\overline{M} = H$. Clearly, $\{0\} \subset A^{\perp}$. To establish the reverse inclusion, let $x \in A^{\perp}$ and since $x \in \overline{M} = H$, $\exists$ a sequence $\{x_n\}$ in M such that

$$\lim_{n \to \infty} x_n = x.$$

Now, since $x \in A^{\perp}$ and $A^{\perp} \perp M$, we have

$$<x, x_n> = 0, \quad \forall n$$

Using the fact that the inner product is a continuous function, we have

$$\lim_{n \to \infty} <x, x_n> = <x, x>$$

$$\Rightarrow \qquad <x, x> = 0$$

$$\Rightarrow \qquad x = 0.$$

Hence, $A^{\perp} = \{0\}$.

Conversely, we assume that $A^\perp = \{0\}$. Let $x \in M^\perp$. Then, $x \perp M$ and, in particular, $x \perp A$. This verifies that $x \in A^\perp = \{0\}$, yielding $M^\perp = \{0\}$. But $\overline{M}$ is a closed subspace of H. Hence, by projection theorem, $H = \overline{M}$. ∎

Problems

15. If A is any non-empty subset of a Hilbert space H, prove that $A^{\perp\perp} = \overline{\text{span } A}$.

(That is, $A^{\perp\perp}$ is the smallest closed subspace containing A).

16. In the usual Hilbert space $\mathbb{R}^2$, find $M^\perp$ if

(a) $M = \{x\}$, where $x = (x_1, x_2) \neq 0$.

(b) $M = \{x, y\} \subset \mathbb{R}^2$ is a linearly independent set.

[Ans. : (a) $M^\perp = \{z : z = \alpha\, (x_2, -x_1), \alpha \in \mathbb{R}\}$

(b) $M^\perp = \{0\}$.]

5.5 Orthonormal Sets and Fourier Analysis

In this section, we study orthonormal sets and sequences which are singular both for their beauty and for the wealth of applications to science that they provide. We also develop a generalized approach to Fourier analysis on abstract Hilbert spaces.

The concept of orthogonality introduced in the previous section plays an important role in the development of the theory of inner product spaces and Hilbert spaces. We now investigate the special theory developed around the idea of mutually orthogonal sets of elements.

5.5.1. Definition. Let X be an inner product space and A be a subset of X. The set A is said to be an:

(a) *Orthogonal set* in X, if $x \perp y$ whenever, $x, y \in A$ with $x \neq y$.

(b) *Orthonormal set* in X, if A is orthogonal and $\| x \| = 1$, $\forall\, x \in A$.

If an orthogonal (orthonormal) set A is countable, we can arrange it in the form of a sequence $\{x_n\}$ and call it orthogonal (orthonormal) sequence in X. More generally, an indexed set (or family) $\{x_\alpha\}_{\alpha \in \Lambda}$ is orthogonal if and only if $x_\alpha \perp x_\beta$, $\forall\, \alpha, \beta \in \Lambda, \alpha \neq \beta$; and orthonormal if, in addition, $\| x_\alpha \| = 1$, $\forall\, \alpha \in \Lambda$.

5.5.2 Existence of orthonormal sets in X

If X contains only zero vector, then it has no orthonormal set. If it contains a non-zero vector x and if we normalize x by considering

$$e = \frac{x}{\|x\|},$$

then the single element set $\{e\}$ is clearly an orthonormal set in X. More generally, if $\{x_\alpha\}_{\alpha \in \Lambda}$ is a non-empty set of mutually orthogonal non-zero vectors in X, and if x_α's are normalized by replacing each of them by

$$e_\alpha = \frac{x_\alpha}{\|x_\alpha\|}, \quad \alpha \in \Lambda$$

then the resulting set $\{e_\alpha\}_{\alpha \in \Lambda}$ is an orthonormal set in X.

5.5.3 Examples

1. The unit vectors

$$e_1 = (1, 0, 0), \ e_2 = (0, 1, 0), \ e_3 = (0, 0, 1)$$

   in the direction of three axes of rectangular coordinate system form an orthonormal set in the Euclidean space $\mathbb{R}^3$.

2. In the inner product space Φ (Example 5.1.6 (1)), the sequence $\{e_n\}$, where $e_n = \{\delta_{nj}\}_j$, is an orthonormal sequence.

3. In the Hilbert space l^2 (Example 5.1.10 (4)), the sequence $\{e_n\}$, where $e_n = \{\delta_{nj}\}_j$, is an orthonormal sequence.

4. Consider the inner product space $C\,[0, 2\,\pi]$ of all real valued continuous functions defined on $[0, 2\,\pi]$ with the inner product defined by

$$<x, y> = \int_0^{2\pi} x\,(t)\,y(t)\,dt, \quad x, y \in C\,[0, 2\,\pi].$$

   An orthogonal sequence in $C\,[0, 2\pi]$ is $\{x_n\}$, where

$$x_n\,(t) = \cos nt, \quad n = 0, 1, 2, \dots$$

   Another orthogonal sequence in $C\,[0, 2\pi]$ is $\{y_n\}$, where

$$y_n(t) = \sin nt, \quad n = 1, 2, \dots$$

   In fact, we obtain

$$<x_m, x_n> = \int_0^{2\pi} \cos mt \cos nt \, dt$$

$$= \begin{cases} 0, & m \neq n \\ \pi, & m = n = 1, 2, \dots \\ 2\pi, & m = n = 0 \end{cases}$$

and so $\qquad \| x_n \| = \sqrt{\pi}.$

Similarly

$$<y_m, y_n> = \int_0^{2\pi} \sin mt \sin nt \, dt$$

$$= \begin{cases} 0, & m \neq n \\ \pi, & m = n = 1, 2, \dots \end{cases}$$

and so $\| y_n \| = \sqrt{\pi}.$

Hence, the sequences $\{e_n\}_0^\infty$ and $\{\tilde{e}_n\}_1^\infty$, where

$$e_0(t) = \frac{1}{\sqrt{2\pi}}$$

$$e_n(t) = \frac{x_n(t)}{\| x_n \|} = \frac{\cos nt}{\sqrt{\pi}}$$

and

$$\tilde{e}_n(t) = \frac{y_n(t)}{\| y_n \|} = \frac{\sin nt}{\sqrt{\pi}}$$

are orthonormal sequences in $C[0, 2\pi]$.

It may also be noted that $x_m \perp y_n$ for all m and n.

5. Consider the Hilbert space $L^2[-\pi, \pi]$ with the inner product defined by

$$<x, y> = \int_{-\pi}^{\pi} x(t) \, \overline{y(t)} \, dt.$$

Then $\{x_n\}$ is an orthonormal set in $L^2[-\pi, \pi]$, where

$$x_n(t) = \frac{1}{\sqrt{2\pi}} \, e^{int}, \quad n \in \mathbb{Z}.$$

5.5.4 Theorem. *An orthonormal set in an inner product space X is linearly independent.*

Proof. Let A be an orthonormal set in X. Assume that A be finite. Writing $A = \{e_1, e_2, \dots, e_n\}$, consider the equation

$$\sum_{i=1}^{n} \alpha_i\, e_i = 0,$$

where α_i's are scalars. Then

$$0 = \left< \sum_{i=1}^{n} \alpha_i\, e_i, e_j \right> = \sum_{i=1}^{n} \alpha_i <e_i, e_j>$$

$$= \alpha_j <e_j, e_j>$$

$$= \alpha_j.$$

Thus, $\alpha_j = 0$ ($j = 1, 2, \ldots n$). Hence, A is linearly independent.

Recall that an arbitrary set A (finite or infinite) is said to be linearly independent if every non-empty finite subset of A is linearly independent. Hence, it follows that our assertion is also valid for the case when A is infinite. ∎

5.5.5 Advantages of orthonormal sets over arbitrary linearly independent sets

(i) If we know that a given x can be represented as a linear combination of some elements of an orthonormal sequence, then the orthonormality makes the actual determination of the coefficients very easy. Indeed, if $\{e_i\}$ is an orthonormal sequence in an inner product space X and $x \in$ span $\{e_1, e_2, \ldots, e_n\}$, where n is fixed, then one may write

$$(12) \qquad\qquad x = \sum_{i=1}^{n} \alpha_i\, e_i,$$

where α_i's are scalars to be determined. Taking the inner product by a fixed e_j on both the sides of the above equation, we obtain

$$<x, e_j> = \alpha_j.$$

Therefore

$$(13) \qquad\qquad x = \sum_{i=1}^{n} <x, e_i> e_i.$$

This shows that the determination of the unknown coefficients in (12) is simple.

(ii) If we wish to add a term $\alpha_{n+1}\, e_{n+1}$ to (12) or (13) in order to obtain

$$\tilde{x} = x + \alpha_{n+1}\, e_{n+1} \in \text{span}\ \{e_1,\ ...,\ e_n,\ e_{n+1}\},$$

in that situation, we need to determine only one more coefficient since the other coefficients remain unchanged. In fact, we have

$$\alpha_j = <\tilde{x}, e_j>, \quad j = 1, 2, ..., n$$

$$= <x + \alpha_{n+1}\, e_{n+1}, e_j>$$

$$= <x, e_j> + \alpha_{n+1} <e_{n+1}, e_j>$$

$$= <x, e_j>, \quad j = 1, 2, ..., n.$$

(iii) If we consider any $x \in X$, not necessarily in $Y_n = \text{span}\ \{e_1, e_2, ..., e_n\}$, we can define $y \in Y_n$ by setting

$$y = \sum_{i=1}^{n} <x, e_i> e_i$$

where n is fixed. Then

$$\sum_{i=1}^{n} |<x, e_i>|^2 \leq \| x \|^2$$

and, furthermore, $x - y \perp y$ (Theorem 5.5.6).

Likewise, there are many more advantages.

5.5.6 Theorem. *Let $\{e_1, e_2, ..., e_n\}$ be a finite orthonormal set in an inner product space X. Then, for any x in X, we have*

(a) $\displaystyle\sum_{i=1}^{n} |<x, e_i>|^2 \leq \| x \|^2$ *(Bessel Inequality).*

(b) $\displaystyle x - \sum_{i=1}^{n} <x, e_i> e_i \perp e_j$ $(j = 1, 2, ..., n)$

Proof. (a) We have

$$0 \leq \| x - \sum_{i=1}^{n} <x, e_i> e_i \|^2$$

$$= <x - \sum_{i=1}^{n} <x, e_i> e_i, \, x - \sum_{j=1}^{n} <x, e_j> e_j>$$

$$= <x, x> - \sum_{i=1}^{n} <x, e_i> \overline{<x, e_i>} - \sum_{j=1}^{n} <x, e_j> \overline{<x, e_j>}$$

$$+ \sum_{i=1}^{n} \sum_{j=1}^{n} <x, e_i> \overline{<x, e_j>} <e_i, e_j>$$

$$= \| x \|^2 - \sum_{i=1}^{n} | <x, e_i> |^2$$

$$\Rightarrow \qquad \sum_{i=1}^{n} | <x, e_i> |^2 \leq \| x \|^2.$$

(b) Observe that

$$<x - \sum_{i=1}^{n} <x, e_i> e_i, \, e_j> = <x, e_j> - \sum_{i=1}^{n} <x, e_i> <e_i, e_j>$$

$$= <x, e_j> - <x, e_j>$$

$$= 0$$

$$\Rightarrow \quad x - \sum_{i=1}^{n} <x, e_i> e_i \perp e_j, \quad (j = 1, 2, ..., n). \quad \blacksquare$$

Remarks

1. The inequality (a) can be given the following look motivating geometric interpretation : "The sum of the squares of the components of a vector in various perpendicular directions does not exceed the square of the length of the vector itself".

2. The relation (b) would mean similarly "If we subtract from a vector its components in several perpendicular directions, then the resultant has no

component left in any of these directions". Hence, the resultant vector is perpendicular to each of these perpendicular directions and, in general,

$$x - \sum_{i=1}^{n} <x, e_i> e_i \perp \sum_{i=1}^{n} <x, e_i> e_i$$

i.e., $x - y \perp y.$

3. Bessel Inequality for $n = 1$ is essentially the Schwarz Inequality.

If X is finite dimensional, then every orthonormal set in X must be finite since it is linearly independent and as such Bessel Inequality has a finite sum only. Next, we shall be interested to generalize the results in Theorem 5.5.6 to the case of arbitrary orthonormal set. The main problem here is to show that the sums involved can be defined in a reasonable way when no restriction is placed on e_i's under consideration. We first prove the following result.

5.5.7 Lemma. *Let $\{e_\alpha\}_{\alpha \in \Lambda}$ be an orthonormal set in an inner product space X and $x \in X$. Then, the set*

$$S = \{e_\alpha : <x, e_\alpha> \neq 0, \alpha \in \Lambda\}$$

is either empty or countable.

Proof. For each positive integer n, consider the set

$$S_n = \{e_\alpha : |<x, e_\alpha>|^2 > \frac{\|x\|^2}{n}, \alpha \in \Lambda\}.$$

By Bessel Inequality (Theorem 5.5.6 (*a*)), S_n contains at most $n - 1$ vectors, for otherwise, if S_n contains $m \, (\geq n)$ vectors, then for these m vectors, we have

$$\sum_{\alpha \in \Lambda} |<x, e_\alpha>|^2 > \frac{m}{n} \|x\|^2 \geq \|x\|^2$$

which is untrue.

The conclusion now follows from the fact that

$$S = \bigcup_{n=1}^{\infty} S_n. \ \blacksquare$$

5.5.8 Theorem (Bessel Generalised Inequality). *Let $\{e_\alpha\}_{\alpha \in \Lambda}$ be an orthonormal set in an inner product space X and $x \in X$. Then*

$$\sum_{\alpha \in \Lambda} |<x, e_\alpha>|^2 \leq \|x\|^2,$$

for every vector $x \in X$.

Proof. Since, for any $x \in X$ and an orthonormal set $\{e_\lambda\}_{\lambda \in \Lambda}$ (countable or uncountable), the set

$$S = \{e_\lambda : <x, e_\lambda> \neq 0, \lambda \in \Lambda\}$$

is empty or countable, we have the following three possibilities:

(i) $S = \phi$.

(ii) S is finite.

(iii) S is countably infinite.

In each of the above three possibilities, our basic point is to explain what is meant by the sum on the left of the inequality in question. If S is empty, we define the sum $\sum_{\alpha \in \Lambda} |<x, e_\alpha>|^2$ to be number 0, and in this case the result is obviously true. In case S is non-empty but finite, it can be written as $S = \{e_1, e_2, ..., e_n\}$, for some positive integer n. We define $\sum_{\alpha \in \Lambda} |<x, e_\alpha>|^2$ in this case to be $\sum_{i=1}^{n} |<x, e_i>|^2$. This sum is clearly independent of the order in which the terms of S are arranged and as such it is a well-defined sum. Our desired result to be proved reduces to

$$\sum_{i=1}^{n} |<x, e_i>|^2 \leq \|x\|^2,$$

which is proved already in Theorem 5.5.6 (a).

Finally, it remains to consider the case when S is countably infinite. Let the vectors in S be arranged in a definite order:

$$S = \{ e_1, e_2, ..., e_n, ... \}.$$

Since, by the theory of absolutely convergent series, if $\sum_{i=1}^{\infty} |<x, e_i>|^2$ converges, then every series obtained from this by rearranging its terms also converge, and all such series have the same sum. We, therefore, can safely define

$$\sum_{\alpha \in \Lambda} |<x, e_\alpha>|^2 \text{ to be } \sum_{i=1}^{\infty} |<x, e_i>|^2$$

and the problem reduces to prove

$$\sum_{i=1}^{\infty} |<x, e_i>|^2 \leq \| x \|^2.$$

To do so consider the partial sums

$$s_n = \sum_{i=1}^{n} |<x, e_i>|^2.$$

These sums have non-negative terms and so form a monotonically increasing sequence. Moreover, by Theorem 5.5.6 (a), the sequence $\{s_n\}$ is bounded above by $\| x \|^2$. Hence the sequence $\{s_n\}$ is convergent and the result follows. ∎

5.5.9 Theorem. *Let* $\{e_\alpha\}_{\alpha \in \Lambda}$ *be an orthonormal set in a Hilbert space H and* $x \in H$. *Then*

$$x - \sum_{\alpha \in \Lambda} <x, e_\alpha> e_\alpha \perp e_\beta, \quad \forall \beta \in \Lambda.$$

Proof. First we define $\sum_{\alpha \in \Lambda} <x, e_\alpha> e_\alpha$ for each of the three possible cases, when the set $S = \{ e_\alpha : <x, e_\alpha> \neq 0, \alpha \in \Lambda\}$ is empty, finite or countably infinite.

Case (i). When $S = \phi$. We define $\sum_{\alpha \in \Lambda} <x, e_\alpha> e_\alpha$ to be the vector 0, and observe that the result to be proved reduces to the statement that

$$x - 0 \perp e_\beta, \quad \forall \beta \in \Lambda.$$

This is precisely what is meant by saying that S is empty.

Case (ii). When S *is finite.* S can be written in the form

$$S = \{e_1, e_2, ..., e_n\},$$

for some $n \in \mathbb{N}$. Then, we define

$$\sum_{\alpha \in \Lambda} < x, e_\alpha > e_\alpha \ \text{ to be } \ \sum_{i=1}^{n} < x, e_i > e_i \ .$$

This sum is clearly independent of the order in which the terms of S are arranged and as such it is a well defined sum. Our desired result to be proved reduces to

$$x - \sum_{i=1}^{n} < x, e_i > e_i \perp e_j \quad (j = 1, 2,..., n)$$

which, in fact, is already proved (Theorem 5.5.6 (b)).

Case (iii). When S is countably infinite. Let the vectors in S be listed in a definite order:

$$S = \{e_1, e_2,..., \}.$$

Write
$$s_n = \sum_{i=1}^{n} < x, e_i > e_i \ .$$

Note that

$$\| s_m - s_n \|^2 = \| \sum_{i=n+1}^{m} < x, e_i > e_i \ \|^2$$

$$= < \sum_{i=n+1}^{m} < x, e_i > e_i, \ \sum_{j=n+1}^{m} < x, e_j > e_j >$$

$$= \sum_{i=n+1}^{m} \sum_{j=n+1}^{m} < x, e_i > \overline{< x, e_j >} < e_i, e_j >$$

$$= \sum_{i=n+1}^{m} |< x, e_i >|^2, \quad (m > n).$$

In view of Bessel Inequality (Theorem 5.5.8), the series $\displaystyle\sum_{i=1}^{\infty} |< x, e_i >|^2$

converges and so the sequence $\{s_n\}$ is Cauchy in H. Since H is complete, the sequence $\{s_n\}$ converges to a vector in H. Let

$$s = \sum_{i=1}^{\infty} <x, e_i> e_i .$$

We now define $\displaystyle\sum_{\alpha \in \Lambda} <x, e_\alpha> e_\alpha$ to be $\displaystyle\sum_{i=1}^{\infty} <x, e_i> e_i$ and observe that

$$<x - \sum_{i=1}^{\infty} <x, e_i> e_i, e_j> = <x, e_j> - <\sum_{i=1}^{\infty} <x, e_i> e_i, e_j>$$

$$= <x, e_j> - < \lim_{n \to \infty} \sum_{i=1}^{n} <x, e_i> e_i, e_j>$$

$$= <x, e_j> - \lim_{n \to \infty} < \sum_{i=1}^{n} <x, e_i> e_i, e_j>$$

$$= <x, e_j> - \lim_{n \to \infty} \sum_{i=1}^{n} <x, e_i> <e_i, e_j>$$

$$= <x, e_j> - <x, e_j> = 0$$

$$\Rightarrow x - \sum_{i=1}^{\infty} <x, e_i> e_i \perp e_j, \quad j = 1, 2,...$$

It now remains to show that the definition of $\displaystyle\sum_{\alpha \in \Lambda} <x, e_\alpha> e_\alpha$ is valid in the sense that it does not depend on the arrangements of the vectors in S. Let us consider another arrangement of vectors in S as

$$S = \{f_1, f_2, f_3,...\}.$$

Write
$$s'_n = \sum_{i=1}^{n} <x, f_i> f_i.$$

As above, we can show that the sequence $\{s'_n\}$ converges to s' in H, so that

$$s' = \sum_{i=1}^{\infty} <x, f_i> f_i.$$

We shall now prove that $s = s'$.

Let $\varepsilon > 0$ be given and let N be a positive integer so large that if $n \geq N$, then

$$\begin{cases} \| s_n - s \| < \varepsilon \\ \| s_n' - s' \| < \varepsilon \\ \sum_{i=N+1}^{\infty} |<x, e_i>|^2 < \varepsilon^2. \end{cases}$$

For some positive integer $M > N$, all terms of s_n occur among those of s'_M. Consequently, $s'_M - s_N$ is a finite sum of terms of the form

$$<x, e_i> e_i, \quad i = N + 1, N + 2,...,$$

and so

$$\| s'_M - s_N \|^2 \leq \sum_{i=N+1}^{\infty} |<x, e_i>|^2 < \varepsilon^2$$

$$\Rightarrow \quad \| s'_M - s_N \| < \varepsilon.$$

Hence

$$\| s' - s \| \leq \| s' - s'_M \| + \| s'_M - s_N \| + \| s_N - s \| < 3\,\varepsilon.$$

Since $\varepsilon > 0$ is arbitrary, letting $\varepsilon \to 0$, we get $s' = s$. ∎

Let $\{ e_\alpha \}_{\alpha \in \Lambda}$ be an orthonormal set in an inner product space X and $x \in X$ be arbitrary. The numbers $<x, e_\alpha>$, $\alpha \in \Lambda$, are called the *Fourier Coefficients* of x with respect to the orthonormal set $\{e_\alpha\}_{\alpha \in \Lambda}$. In general, the sum of the squares of the absolute values of these coefficients does not exceed $\| x \|^2$ (Bessel Inequality). In case the equality

$$\sum_{\alpha \in \Lambda} |<x, e_\alpha>|^2 = \| x \|^2$$

occurs, then x is said to satisfy *Parseval Relation*. Furthermore, with an $x \in X$, we can associate a series of the form

$$\sum_{\alpha \in \Lambda} <x, e_\alpha> e_\alpha$$

which, of course, is convergent, not necessarily in X, called the *Fourier series expansion* of x with respect to $\{e_\alpha\}_{\alpha \in \Lambda}$ and is written as

$$x \sim \sum_{\alpha \in \Lambda} <x, e_\alpha> e_\alpha.$$

If $x = \sum_{\alpha \in \Lambda} <x, e_\alpha> e_\alpha$, then we say that x is represented by its Fourier series expansion with respect to $\{e_\alpha\}_{\alpha \in \Lambda}$.

We now show that an element in an inner product space is represented by its Fourier series if and only if it satisfies the Parseval Relation.

5.5.10 Theorem. *Let $\{e_\alpha\}_{\alpha \in \Lambda}$ be an orthonormal set in an inner product space X and let $x \in X$. Then*

$$x = \sum_{\alpha \in \Lambda} <x, e_\alpha> e_\alpha \quad \Leftrightarrow \quad \sum_{\alpha \in \Lambda} |<x, e_\alpha>|^2 = \| x \|^2.$$

Proof. By Lemma 5.5.7, the set

$$S = \{e_\alpha : <x, e_\alpha> \neq 0, \alpha \in \Lambda\}$$

is either empty or countable. If $S = \phi$ there is nothing to prove. Let S be countable and we assume it to be infinite since the case of finite will then follow automatically. Let $x = \sum_{\alpha \in \Lambda} <x, e_\alpha> e_\alpha$. Then, $\exists$ an arrangement of vectors e_α in S, $\{e_1, e_2, ..., e_n, ...\}$, such that

$$x = \sum_{i=1}^{\infty} <x, e_i> e_i.$$

We have

$$\| x \|^2 = < \sum_{i=1}^{\infty} <x, e_i> e_i, \sum_{j=1}^{\infty} <x, e_j> e_j >.$$

By the continuity of the inner product, we get

$$\| x \|^2 = \sum_{i=1}^{\infty} |<x, e_i>|^2.$$

Since the above series is absolutely convergent, the sum is independent of the ordering of the terms. Hence, we can write

$$(14) \qquad \| x \|^2 = \sum_{\alpha \in \wedge} |< x, e_\alpha >|^2 .$$

Conversely, we assume that (14) be true. Then, for any given arrangement of vectors in S, $\{e_1, e_2,...., e_n,.....\}$, for a given $\varepsilon > 0$, $\exists\, N \in \mathbb{N}$ such that

$$\left| \| x \|^2 - \sum_{i=1}^{n} |< x, e_i >|^2 \right| < \varepsilon, \quad \forall\, n \geq N.$$

As in the proof of Theorem 5.5.6, we obtain

$$0 \leq \| x - \sum_{i=1}^{n} < x, e_i > e_i \|^2 = \left| \| x \|^2 - \sum_{i=1}^{n} |< x, e_i >|^2 \right|$$

$$< \varepsilon, \qquad \forall \quad n \geq N$$

$$\Rightarrow \qquad x = \sum_{i=1}^{\infty} < x, e_i > e_i .$$

Since this is true for any arrangement of vectors in S, we can write

$$x = \sum_{\alpha \in \wedge} < x, e_\alpha > e_\alpha . \; \blacksquare$$

Keeping in view the usefulness and convenience of orthonormal sequences over the linearly independent sequences, one is interested to generate the former one if the later is given. This is done by a constructive procedure, known as the *Grahm-Schmidt Process,* given below.

Let $\{x_n\}$ be a (finite or countably infinite) linearly independent set of vectors in an inner product space X. The problem is to convert this set into an orthonormal set $\{e_n\}$ such that span $\{e_1, e_2,..., e_n\}$ = span $\{x_1, x_2,..., x_n\}$, for each n. The process is inductive. It will suffice to indicate a few steps.

Step 1. Normalise x_1, which is necessarily non-zero, so as to obtain e_1 as

$$e_1 = \frac{x_1}{\| x_1 \|}.$$

Step 2. Write

$$x_2 = <x_2, e_1> e_1 + v_2,$$

so that

$$v_2 = x_2 - <x_2, e_1> e_1.$$

Clearly $v_2 \neq 0$, since $\{x_i\}$ is linearly independent. Also, $v_2 \perp e_1$, since

$$<v_2, e_1> = <x_2, e_1> - <x_2, e_1> <e_1, e_1> = 0.$$

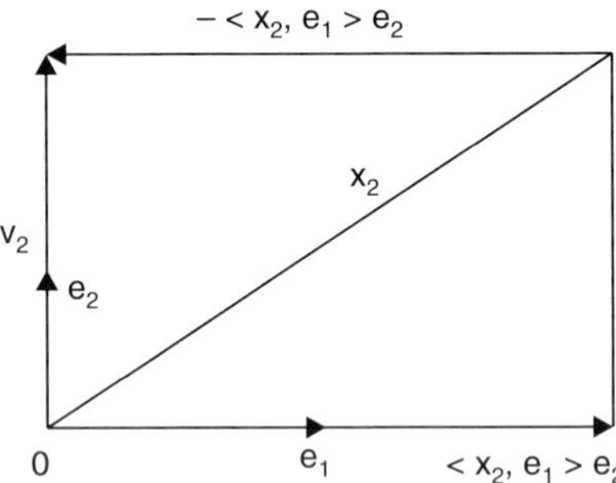

Fig. 5.5 Illustrating Step 2

So we can take e_2 by normalizing v_2; namely,

$$e_2 = \frac{v_2}{\| v_2 \|},$$

or, in other words,

$$e_2 = \frac{x_2 - <x_2, e_1> e_1}{\| x_2 - <x_2, e_1> e_1 \|}.$$

Now, since $\| e_2 \| = 1$ and $<e_1, e_2> = 0$, it is clear that $\{e_1, e_2\}$ is an orthonormal set in X. Further, e_2 being a linear combination of x_1 and x_2, and x_2 a linear combination of e_1 and e_2, we have span $\{x_1, x_2\}$ = span $\{e_1, e_2\}$.

Step 3. Similarly, the vector v_3 given by

$$v_3 = x_3 - <x_3, e_1> e_1 - <x_3, e_2> e_2$$

is non-zero, and $v_3 \perp e_1$, $v_3 \perp e_2$. Therefore, we may take

$$e_3 = \frac{v_3}{\| v_3 \|} = \frac{x_3 - <x_3, e_1> e_1 - <x_3, e_2> e_2}{\| x_3 - <x_3, e_1> e_1 - <x_3, e_2> e_2 \|}.$$

Obviously, $\{e_1, e_2, e_3\}$ is an orthonormal set in X and it can be easily verified that

$$\text{span } \{e_1, e_2, e_3\} = \text{span } \{x_1, x_2, x_3\}.$$

Step 4. Proceeding in the same way, construct e_4, and so on.

Step n. The vector

$$v_n = x_n - \,< x_n, e_1 > e_1 \ldots - \,< x_n, e_{n-1} > e_{n-1}$$

$$= x_n - \sum_{i=1}^{n-1} < x_n, e_i > e_i$$

is non-zero, and $v_n \perp e_i$ ($i = 1, 2,\ldots, n-1$). Thus, we take

$$e_n = \frac{v_n}{\|v_n\|} = \frac{x_n - \displaystyle\sum_{i=1}^{n-1} < x_n, e_i > e_i}{\|x_n - \displaystyle\sum_{i=1}^{n-1} < x_n, e_i > e_i \|}.$$

Hence, this process leads to an orthonormal set $\{e_1, e_2,\ldots,e_n,\ldots\}$ with the required property. ∎

Note. This process was designed by E. Schmidt (1907) and also J.P. Grahm (1883).

Remark. The sum $\displaystyle\sum_{i=1}^{n-1} < x_n, e_i > e_i$ which is subtracted from x_n in order to obtain

e_n is, in fact, the projection of x_n on span $\{e_1, e_2,\ldots,e_{n-1}\}$. In other words, in each step we subtract from x_n its components in directions of previously

orthonormalized vectors giving v_n which is then multiplied by $\dfrac{1}{\|v_n\|}$, so that

we get a vector of norm 1. The vectors v_n cannot be zero for any n. If n were the smallest subscript for which $v_n = 0$, then the relation

$$v_n = x_n - \sum_{i=1}^{n-1} < x_n, e_i > e_i$$

show that x_n would be a linear combination of $x_1, x_2,\ldots, x_{n-1}$, contradicting the assumption that $\{ x_1, x_2,\ldots, x_n\}$ is linearly independent.

Problems

17. Let $\{e_n\}$ be an orthonormal sequence in a Hilbert space H and $\{\alpha_n\}$ be a sequence of scalars. Prove that

(a) The series $\sum\limits_{n=1}^{\infty} \alpha_n e_n$ converges if and only if $\sum\limits_{n=1}^{\infty} |\alpha_n|^2 < \infty$, equivalently, if and only if $\{\alpha_n\} \subset l^2$.

(b) The series $\sum\limits_{n=1}^{\infty} <x, e_n> e_n$ converges for all $x \in H$.

18. Let $\{e_n\}$ be an orthonormal sequence in an inner product space X. Prove that

$$\sum_{n=1}^{\infty} |<x, e_n><y, e_n>| \le \|x\| \, \|y\|, \quad x, y \in X$$

[*Hint:* Apply Cauchy Schwartz Inequality

$$\sum_{n=1}^{\infty} |<x, e_n><y, e_n>| \le [\sum_{n=1}^{\infty} |<x, e_n>|^2]^{\frac{1}{2}} [\sum_{n=1}^{\infty} |<y, e_n>|^2]^{\frac{1}{2}}.$$

and then use Bessel Inequality (Theorem 5.5.8) for each term on the right.]

19. Consider the inner product space $C[-1, 1]$ (Example 5.1.6 (3) with $a = -1, b = 1$).

(a) Let $x_n(t) = t^n$, $n = 0, 1, 2, \ldots$ Then, orthonormalize the first three terms of the sequence $\{x_0, x_1, x_2, \ldots\}$.

[*Hint:* Use the method of Example 5.5.3 (4).

$$\text{Ans.: } \frac{1}{\sqrt{2}}, \sqrt{\frac{3}{2}}\, t, \sqrt{\frac{5}{8}}\, (3t^2 - 1)].$$

(b) Let $x_1(t) = t^2$, $x_2(t) = t$ and $x_3(t) = 1$. Orthonormalize x_1, x_2 and x_3.

20. Work out the details for Example 5.5.3 (5).

21. Show that the Legendre polynomials defined by

$$P_0(x) = 0$$

$$P_n(x) = \frac{1}{2^n \, n!} \frac{d^n (x^n - 1)^n}{dx^n}, \quad n \in \mathbb{N}$$

form an orthonormal system in $L^2[-1, 1]$.

22. Consider the Hermite polynomials of degree n, *i.e.*,

$$H_n(x) = (-1)^n \, e^{x^2} \, \frac{d^n}{dx^n} (e^{-x^2}), \quad n \in \mathbb{N}.$$

Prove that the sequence $\{G_n\}$, where

$$G_n(x) = e^{-\frac{1}{2}x^2} \, H_n(x), \quad n \in \mathbb{N}$$

forms an orthogonal sequence, in $L^2(\mathbb{R})$.

5.6 Complete Orthonormal Sets

We continue to draw inspiration from our experience of the familiar Euclidean space $\mathbb{R}^3$. Recall that the set of three unit vectors $\{e_1, e_2, e_3\}$ in the positive direction of the axes of a rectangular coordinate system form a basis for $\mathbb{R}^3$. The set $\{e_1, e_2, e_3\}$ is such that the vectors in it are pairwise orthogonal and each vector has the unit norm. A new concept will be introduced in this section somewhat similar to the notion of a basis in an Euclidean space; namely, the notion of a complete orthonormal set in an inner product space. We now investigate into the existence of complete orthonormal sets in an inner product space.

Let $X \neq \{0\}$ be an inner product space. Then, the collection $\mathscr{C}$ of all orthonormal subsets of X is clearly non-empty. It can be easily seen that the class $\mathscr{C}$ can be partially ordered under 'set inclusion' relation.

5.6.1 Definition. An orthonormal set $\{e_\alpha\}_{\alpha \in \Lambda}$ in an inner product space X is said to be a *complete* (or *maximal*) *orthonormal set* in X if it is maximal as a set in $\mathscr{C}$ the family of all orthonormal subsets of X, partially ordered by set inclusion.

In other words, an orthonormal set $\{e_\alpha\}_{\alpha \in \Lambda}$ is a complete orthonormal set in X if and only if it is impossible to adjoin an additional element $e \in X$, $e \neq 0$, to $\{e_\alpha\}_{\alpha \in \Lambda}$ such that $\{e_\alpha, e\}_{\alpha \in \Lambda}$ is an orthonormal set in X. Such sets are also called *total sets*.

5.6.2 Examples. The orthonormal sets described in Examples 5.5.3 (1), 5.5.3 (2), 5.5.3 (3) and 5.5.3 (5) are complete while that in Example 5.5.3 (4) is not complete.

5.6.3 Theorem. *Every inner product space $X \neq \{0\}$ contains a complete orthonormal set.*

Proof. Consider the family $\mathscr{C}$ of orthonormal sets in X partially ordered by set inclusion. For any $x \in X$, $x \neq 0$, the set $\left\{ \dfrac{x}{\|x\|} \right\}$ is an orthonormal set. Therefore, $\mathscr{C} \neq \phi$. Now consider any totally ordered sub-family in $\mathscr{C}$. The union of sets in this sub-family is clearly an orthonormal set and is an upper bound for the totally ordered sub-family. By Zorn's Lemma 1.3.5, it follows that C has a maximal element which, in fact, is a complete orthonormal set in X. ∎

The next theorem provides a nice characterisation of complete orthonormal sets in a Hilbert space.

5.6.4 Theorem. *Let $\{e_\alpha\}_{\alpha \in \Lambda}$ be an orthonormal set in a Hilbert space H and let $x \in H$ be arbitrary. Then, $\{e_\alpha\}_{\alpha \in \Lambda}$ is a complete orthonormal set if and only if $x \perp \{e_\alpha\}_{\alpha \in \Lambda} \Rightarrow x = 0$.*

Proof. Suppose $\{e_\alpha\}_{\alpha \in \Lambda}$ is a complete orthonormal set. Let, if possible, the conclusion of the theorem do not hold. Then, $\exists$ a vector $x \neq 0$ such that $x \perp \{e_\alpha\}_{\alpha \in \Lambda}$. Define

$$e = \frac{x}{\|x\|}.$$

Clearly, $\{e_\alpha, e\}_{\alpha \in \Lambda}$ is an orthonormal set and $\{e_\alpha, e\}_{\alpha \in \Lambda} \supsetneq \{e_\alpha\}_{\alpha \in \Lambda}$.

This contradicts the assumption that $\{e_\alpha\}_{\alpha \in \Lambda}$ is complete.

Conversely, we assume that $x \perp \{e_\alpha\}_{\alpha \in \Lambda} \Rightarrow x = 0$. If $\{e_\alpha\}_{\alpha \in \Lambda}$ is not complete, then there must exist an $e \in H$ such that $\{e_\alpha, e\}_{\alpha \in \Lambda}$ is an orthonormal set and $\{e_\alpha, e\}_{\alpha \in \Lambda} \supsetneq \{e_\alpha\}_{\alpha \in \Lambda}$. Since $e \perp e_\alpha$, $\forall \alpha \in \Lambda$, it follows, by the hypothesis, that $e = 0$ which contradicts the fact that e is a unit vector. This complete the proof. ∎

We are now in a position to offer an important Hilbert space property that every element in the Hilbert space can be represented by its Fourier series expansion with respect to a complete orthonormal set.

5.6.5 Theorem. *Let $\{e_\alpha\}_{\alpha \in \Lambda}$ be an orthonormal set in a Hilbert space H. Then, $\{e_\alpha\}_{\alpha \in \Lambda}$ is complete if and only if*

$$(15) \qquad x = \sum_{\alpha \in \Lambda} <x, e_\alpha > e_\alpha , \quad \forall \, x \in H.$$

Proof. Suppose $\{e_\alpha\}_{\alpha \in \Lambda}$ is complete. Since, for any orthonormal set $\{e_\alpha\}_{\alpha \in \Lambda}$, the vector $x - \sum_{\alpha \in \Lambda} <x, e_\alpha > e_\alpha$ is orthogonal to e_β, for all β (Theorem 5.5.9), it follows that

$$x - \sum_{\alpha \in \Lambda} <x, e_\alpha > e_\alpha = 0, \quad \forall \, x \in H \text{ (Theorem 5.6.4)}$$

$$\Rightarrow \qquad x = \sum_{\alpha \in \Lambda} <x, e_\alpha > e_\alpha , \quad \forall \, x \in H.$$

Conversely, assume that (15) holds. Let, if possible, $\exists$ an $e \in H$ such that $\{e_\alpha, e\}_{\alpha \in \Lambda}$ is an orthonormal set and $\{e_\alpha, e\}_{\alpha \in \Lambda} \supsetneq \{e_\alpha\}_{\alpha \in \Lambda}$. But

$$e = \sum_{\alpha \in \Lambda} <x, e_\alpha > e_\alpha.$$

Since $<e, e_\alpha> = 0$, $\forall \, \alpha \in \Lambda$, it follows that $e = 0$. This contradicts the fact that e is a unit vector. Hence, the result follows. ∎

5.6.6 Definition. An orthonormal set $\{e_\alpha\}_{\alpha \in \Lambda}$ in an inner product space X is said to be an *orthonormal basis* for X if for every $x \in X$

$$x = \sum_{\alpha \in \Lambda} <x, e_\alpha > e_\alpha ,$$

that is, every $x \in X$ can be represented by its Fourier series expansion with respect to $\{e_\alpha\}_{\alpha \in \Lambda}$.

Remark. The two notions of complete orthonormal set and orthonormal basis in a Hilbert space are equivalent (Theorem 5.6.5).

Our next result summarizes the various characterisations of a complete orthonormal set (Theorem 5.5.10, 5.6.4 and 5.6.5).

5.6.7 Theorem. *Let $\{e_\alpha\}_{\alpha \in \Lambda}$ be an orthonormal set in a Hilbert space H. Then, the following statements are equivalent:*

(a) $\{e_\alpha\}_{\alpha \in \Lambda}$ *is complete.*

(b) $x \perp \{e_\alpha\}_{\alpha \in \Lambda} \implies x = 0.$

(c) $x = \displaystyle\sum_{\alpha \in \Lambda} <x, e_\alpha> e_\alpha, \quad \forall x \in H.$

   (or equivalently, $\{e_\alpha\}_{\alpha \in \Lambda}$ is an orthonormal basis for H)

(d) $\| x \|^2 = \displaystyle\sum_{\alpha \in \Lambda} |<x, e_\alpha>|^2, \quad \forall x \in H.$

Finally, we prove a significant cardinality result on complete orthonormal sets for a Hilbert space.

5.6.8 Theorem. *Any two complete orthonormal sets in a Hilbert space H have the same cardinal number.*

Proof. Suppose $\exists$ a finite complete orthonormal set $\{e_1, e_2, ..., e_n\}$. Then, for any $x \in X$, write

$$\eta = x - \sum_{i=1}^{n} <x, e_i> e_i.$$

Clearly

$$<\eta, e_i> = 0, \qquad \text{(Theorem 5.5.6)}$$

$$\implies \qquad \eta \perp e_i, \quad (i = 1, 2, ..., n).$$

But $\{e_1, e_2, ..., e_n\}$ is complete. Therefore, $\eta = 0$ (Theorem 5.6.4) and

$$x = \sum_{i=1}^{n} <x, e_i> e_i.$$

It follows that $\{e_1, e_2, ..., e_n\}$ spans H. Also, $\{e_1, e_2, ..., e_n\}$ is linearly independent set (Theorem 5.5.4). Hence, $\{e_1, e_2, ..., e_n\}$ is a (Hamel) basis for H. Thus, in view of Theorem 1.6.10, every complete orthonormal set in H has n elements. This completes the proof for the case when H has a finite complete orthonormal set.

Furthermore, suppose that $\exists$ two infinite complete orthonormal sets, $B_1 = \{e_\alpha\}_{\alpha \in \Lambda_1}$ and $B_2 = \{f_\beta\}_{\beta \in \Lambda_2}$ in H. For each $e_\alpha \in B_1$, write

$$B_2(e_\alpha) = \{f_\beta \in B_2 : <f_\beta, e_a> \neq 0\} .$$

Then $B_2(e_\alpha)$ is countable (Lemma 5.5.7), for each $\alpha \in \Lambda_1$. But every member of B_2 is also a member of at least one of the sets $B_2(e_\alpha)$; for otherwise, if $\exists$ an $f_{\beta_0} \in B_2$ such that

$$< f_{\beta_0}, e_\alpha > = 0, \quad \forall\, \alpha \in \Lambda_1$$

then B_1 is not maximal. Therefore

$$B_2 \subset \bigcup_{\alpha \in \Lambda_1} B_2(e_\alpha).$$

So $\# B_2 \leq \aleph_0 . (\# B_1)$. By symmetry of arguments, we have

$$\# B_1 \leq \# B_2.$$

By Schröeder-Bernstein Theorem, it follows that $\# B_1 = \# B_2$. This completes the proof. ∎

Theorem 5.6.8 enables us to make the following definition.

5.6.9 Definitions. The cardinal number of a complete orthonormal set (orthonormal basis) in a Hilbert space H is called the *orthogonal* (or *Hilbert*) *dimension of H.*

For a finite dimensional Hilbert space H, a complete orthonormal set is also a (Hamel) basis for the space, and thus, in this case the orthogonal dimension and the dimension are the same. If H has no complete orthonormal set, its orthogonal dimension is defined to be 0 (by convention).

5.6.10 Theorem. A Hilbert space H is separable if and only if it has a countable orthonormal basis.

Proof. Suppose that H has an uncountable complete orthonormal set $\{e_\alpha\}_{\alpha \in \Lambda}$. Then

$$\| e_\alpha - e_\beta \| > 1, \quad \forall\, \alpha, \beta \in \Lambda, \alpha \neq \beta.$$

Therefore

$$S\left(e_\alpha; \frac{1}{2}\right) \cap S\left(e_\beta; \frac{1}{2}\right) = \phi, \quad \forall\, \alpha, \beta \in \Lambda,\ \alpha \neq \beta$$

and hence $\exists$ an uncountable family of disjoint open spheres in H each with radius $\frac{1}{2}$. This verifies that H is not separable.

Conversely, suppose that H has a countable complete orthonormal set $\{e_n\}$. Then, by Theorem 5.6.5, it follows that

$$x = \sum_{n=1}^{\infty} <x, e_n> e_n, \quad \forall\, x \in H.$$

It implies that $x \in H$ is a cluster point of the set of linear combinations of elements of $\{e_n\}$. But the set of linear combinations of elements of $\{e_n\}$ contains the countable dense set of linear combinations of elements of $\{e_n\}$ with rational coefficients. Hence, H is separable. ∎

5.6.11 Observations

Note that since each of the following Hilbert spaces possesses a countable orthonormal basis is separable:

  (i) l^2 (Example 5.5.3 (3))

  (ii) $l^2 [-\pi, \pi]$ (Example 5.5.3 (5))

  (iii) $L^2 [-1, 1]$ (Problem 21)

  (iv) $L^2 (\mathbb{R})$ (Problem 22).

And likewise many others.

We give below an example of a non-separable Hilbert space.

5.6.12 Example. Let H be the space of all complex valued functions defined on $\mathbb{R}$ which vanish everywhere except on a countable number of point of $\mathbb{R}$ and such that the values of these point form a square summable sequence. Consider the inner product on H given by

$$<f, g> = \sum_{x \in \mathbb{R}} f(x)\, \overline{g(x)}.$$

Note that:

(i) The inner product is well defined since the summation is actually over a countable set of points of $\mathbb{R}$.

(ii) The space H is not separable since for any sequence $\{f_n\}$ of functions is H, there are non-zero functions $f \in H$ such that

$$<f,f_n> = 0, \quad \forall\, n \in \mathbb{N}.$$

Remark. An orthonormal basis is a separable Hilbert space is a special example of a Schander basis for H (considered as a Banach space).

5.6.13 Example. Schander basis which is not an orthonormal basis for H.

Consider an orthonormal basis $\{e_n\}$ in a separable Hilbert space H and a sequence $\{x_k\}$ of element in H, where

$$x_k = \sum_{i=1}^{k} \alpha_i \, e_i \, , \ (k \in \mathbb{N}).$$

One can verify that $\{x_k\}$ forms a (Schander) basis for H if α_i satisfy the following conditions:

(i) $|\alpha_i| > 0, i \in \mathbb{N}$

$$\text{(ii)} \ \ \frac{\displaystyle\sum_{i=1}^{n} \alpha_i^2}{\alpha_{n+1}^2} \leq M, \ (n \in \mathbb{N})$$

for some constant M independent of n.

Also, the sequence $\{f_n\} \subset H^*$ forming the biorthogonal system with $\{x_n\}$ is given by

$$f_n = \frac{1}{\alpha_n} \, e_n - \frac{1}{\alpha_{n+1}} \, e_{n+1}.$$

Problem 23. Let H be a separable Hilbert space. Prove that:

(a) Every orthonormal basis in H is unconditional

(b) An arbitrary unconditional basis in H can be represented in the form of $\{Ue_k\}$, where $\{e_k\}$ is an orthonormal basis in H and $U : H \to H$ is an unitary operator.

[*cf.* §6.5 for unitary operator]

5.6.14 Theorem. *Let H be an infinite dimensional separable Hilbert space over the field $\mathbb{K}$. Then H is isometrically isomorphic to l^2.*

Proof. Since H is separable, by Theorem 5.6.10, it has a countable orthonormal basis. Let the orthonormal basis be $\{e_n\}$. Let $x \in H$. Then

$$x = \sum_{n=1}^{\infty} <x, e_n> e_n \quad \text{(Theorem 5.6.7)}$$

Write $\lambda_n = <x, e_n>$. By Problem 17, it follows that $\{\lambda_n\} \in l^2$.

Define $T : H \to l^2$ by $T x = \{\lambda_n\}$.

Note that:

(i) T is linear.

(ii) T is one-one. Indeed, if $T x = 0$, then $\{\lambda_n\} = 0$ and so $\lambda_n = <x, e_n> = 0$ for all $n \in \mathbb{N}$. But $\{e_n\}$ being an orthonormal basis, by Theorem 5.6.7 $x = 0$.

(iii) T is onto. Indeed, if $\mu = \{\mu_n\} \in l^2$, then there is an element $y \in H$ with

$$y = \sum_{n=1}^{\infty} \mu_n e_n \text{ and } \mu_n = <y, e_n>.$$

Thus $T y = \mu$.

(iv) T is an isometry. We have

$$\| T x \|^2 = \sum_{n=1}^{\infty} |\lambda_n|^2 = \sum_{n=1}^{\infty} |<x, e_n>|^2 = \| x \|^2, \quad \forall \, x \in H.$$

This complete the proof.

Remark: Since isomorphism of Hilbert space is an equivalence relation, any two infinite dimensional separable Hilbert spaces over the same field $\mathbb{K}$ ($\mathbb{R}$ or $\mathbb{C}$) are isomorphic, and each one is isomorphic to the space of absolute square summable scalar sequences in $\mathbb{K}$. Thus, in some sense, there is only one real and one complex infinite dimensional separable Hilbert space.

6
CHAPTER

Functionals and Operators on Hilbert Spaces

In the preceding chapter, we initiated the study of special type of Banach spaces; namely, Hilbert spaces which possess the additional structure of an inner product enabling us to say when two vectors are perpendicular. More precisely, we discussed there the geometric implications of the inner product structure on Hilbert spaces. In this chapter, we study the bounded linear operators on Hilbert spaces. We shall demonstrate here that there is a natural correspondence between H and its dual H^*. This leads to a representation theorem, called the Riesz-Fréchet Theorem, for bounded linear functionals. The Riesz-Fréchet Theorem enables us to introduce the Hilbert-adjoint operator of a bounded linear operator on a Hilbert space. Finally, the adjoint operators motivate us to design and study several other operators; namely, self-adjoint, positive, normal, unitary and projection operators. The study of such operators is vast and is beyond the scope of the present book. Here, we only give the motivation and the introductory results about them.

6.1 Bounded Linear Functionals

In Section 4.2, we investigated the general form of the bounded linear functionals on various Banach spaces. But the derivation of such representations of bounded linear functionals becomes quite difficult in several other Banach spaces. However, in the special setting of Hilbert spaces, we get a representation theorem in terms of a fixed vector and the inner product, for any bounded linear functional on the space. Thus, our aim in this section is to explore the relationship between the vectors in the Hilbert space H and the bounded linear functionals on H.

Since a Hilbert space is a special case of Banach space, most of the concepts developed in Chapters 3 and 4 for Banach spaces are also valid for Hilbert spaces. For instance, if f is in $B(H, \mathbb{K})$, we refer to f as a bounded linear functional on the Hilbert space H over the scalar field $\mathbb{K}$. The space $B(H, \mathbb{K})$ is called the dual (or conjugate) space of H and we denote it by H^* (Definition 4.1.13).

287

Let z be any fixed vector in an inner product space X. Define

(1) $$f_z(x) = <x, z>, \quad \forall\, x \in X.$$

Then:

(a) f_z is a functional on X since $f_z : X \to \mathbb{K}$.

(b) f_z is linear since

$$f_z(\alpha x + \beta y) = <\alpha x + \beta y, z>$$

$$= \alpha <x, z> + \beta <y, z>$$

$$= \alpha f_z(x) + \beta f_z(y), \quad x, y \in X \text{ and } \alpha, \beta \in \mathbb{K}.$$

(c) f_z is bounded and $\|f_z\| = \|z\|$.

In fact, we have

$$\left| f_z(x) \right| = \left| <x, z> \right| \leq \|x\|\,\|z\|, \quad \forall\, x \in X.$$

Therefore, f_z is bounded. Moreover

$$\|f_z\| \leq \|z\|.$$

On the other hand, if $z \neq 0$, then

$$\|f_z\| = \sup\{ \left| <x, z> \right| : x \in X, \|x\| = 1 \}$$

$$\geq < \frac{z}{\|z\|}, z >$$

$$= \|z\|.$$

In case $z = 0$, $\|f_z\| \geq \|z\|$ is obvious. Hence

$$\|f_z\| = \|z\|.$$

Clearly such a functional f_z defined by (1) is unique. Hence, for each $z \in X$, there is a (unique) bounded linear functional f_z on X with $\|f_z\| = \|z\|$.

Below, we prove the converse.

6.1.1 Theorem (Riesz-Fréchet). *Let H be a Hilbert space and H^* be its dual space. If $f \in H^*$ is an arbitrary but fixed functional, then $\exists$ a unique vector $z \in H$ such that*

(2) $$f(x) = <x, z>, \quad \forall\, x \in H$$

where z depends on f and has the norm

(3) $$\|z\| = \|f\|.$$

Proof. If $f = 0$, then (2) and (3) hold if we take $z = 0$. Let $f \neq 0$. We prove the result in three steps.

Step (*i*) (Existence of z). Since f is linear and continuous, the null space $\mathcal{N}(f)$ is a closed subspace of H (Theorem 4.1.8). Because $f \neq 0$, it follows that $\mathcal{N}(f) \neq H$. But, by the projection theorem (Theorem 5.4.3) $H = \mathcal{N} \oplus \mathcal{N}^{\perp}$ with $\mathcal{N} \equiv \mathcal{N}(f)$ and, therefore, $\mathcal{N}^{\perp} \neq \{0\}$. This implies that $\exists$ a vector $z_0 \in \mathcal{N}^{\perp}$ such that $z_0 \neq 0$. Consider the set

$$S = \{ \, v = z_0 f(x) - x f(z_0) : x \in H \, \}.$$

Then, $S \subset \mathcal{N}$ since

$$\begin{aligned}
f(v) &= f[\, z_0 f(x) - x f(z_0) \,] \\
&= f(z_0) f(x) - f(x) f) f(z_0) \\
&= 0, \quad \forall \, x \in H.
\end{aligned}$$

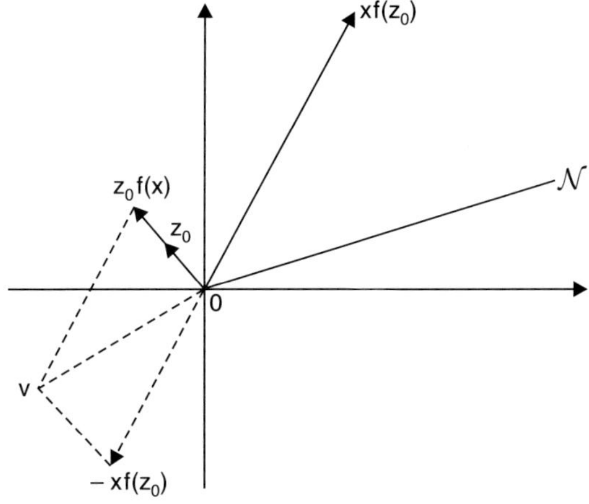

Fig. 6.1. Illustration of S in $\mathbb{R}^2$

Therefore, $z_0 \perp S$. This gives

$$< z_0 f(x) - x f(z_0), z_0 > \; = 0$$

$$\Rightarrow \qquad f(x) \, \| z_0 \|^2 = f(z_0) < x, z_0 >$$

$$\Rightarrow \qquad f(x) = \frac{f(z_0)}{\| z_0 \|^2} < x, z_0 >$$

$$= < x, z >, \quad \forall \, x \in H$$

where

$$z = \frac{\overline{f(z_0)}\, z_0}{\| z_0 \|^2} \in H.$$

This proves (2).

   Step (*ii*) (Uniqueness). Let there be $z_1, z_2 \in H$ such that

$$f(x) = \,<x, z_1> \,= \,<x, z_2>, \quad \forall\, x \in H.$$

Then

$$<x, z_1 - z_2> \,= 0, \quad \forall\, x \in H.$$

In particular, for $x = z_1 - z_2$, we have

$$<z_1 - z_2, z_1 - z_2> \,= 0$$

$$\Rightarrow \qquad \| z_1 - z_2 \|^2 = 0$$

$$\Rightarrow \qquad z_1 - z_2.$$

This proves the uniqueness of z in (2).

   Step (*iii*) (Norm is preserved). With $x = z$ in (2), we have

$$f(z) = \,<z, z> \,= \| z \|^2.$$

But, f being bounded, we get

$$\| z \|^2 = \left| f(z) \right| \le \| f \| \, \| z \|,$$

so that

$$\| z \| \le \| f \|.$$

   On the other hand, using Schwartz Inequality, we obtain

$$\left| f(x) \right| = \left| <x, z> \right| \le \| x \| \, \| z \|,$$

$$\Rightarrow \qquad \| f \| = \sup_{\substack{x \in H \\ \| x \| \le 1}} \left| <x, z> \right| \le \| z \|.$$

Hence $\| f \| = \| z \|.$

   This completes the proof of the theorem. ∎

Remark. The bounded linear functional on H are precisely the inner products on it.

Note. Theorem 6.1.1 is oftenly called Riesz-Fréchet Representation Theorem or simply Riesz Representation Theorem.

6.1.2 Examples

1. Let $\{e_1, e_2, ...\}$ be an orthonormal basis for H. Then, the vector z corresponding to the functional f in the Riesz Representation Theorem is given by

$$z = \sum_{i=1}^{\infty} \overline{f(e_i)}\, e_i \, .$$

Indeed, since $f(e_i) = <e_i, z>$, we have

$$z = \sum_{i=1}^{\infty} <z, e_i>e_i = \sum_{i=1}^{\infty} \overline{<e_i, z>}\, e_i = \sum_{i=1}^{\infty} \overline{f(e_i)}\, e_i .$$

2. A functional f on l^2 is bounded and linear if and only if $\exists$ a vector $z = \{\beta_i\} \in l^2$ such that

$$f(x) = <x, z> = \sum_{i=1}^{\infty} \alpha_i \,\overline{\beta}_i \, , \quad \forall\, x = \{\alpha_i\} \in l^2.$$

3. Let $\{\mu_i\}$ be a bounded sequence of positive numbers. For $x = \{\alpha_i\}$ and $z = \{\beta_i\}$ in l^2, define

$$<x, z>_{\mu} = \sum_{i=1}^{\infty} \mu_i\, \alpha_i \,\overline{\beta}_i \, , \quad \mu \equiv \{\mu_i\}.$$

It is easy to verify that $< \cdot , \cdot >_{\mu}$ is an inner product on l^2 (This inner product is called the *weighted inner product* on l^2). Suppose $\inf\{\mu_i\} > 0$. Then, $(l^2, < \cdot , \cdot >_{\mu})$ is complete which can be proved by similar arguments as in Example 2.2.5. Hence, $(l^2, < \cdot , \cdot >_{\mu})$ is a Hilbert space. By Riesz Representation Theorem, f is bounded and linear on $(l^2, < \cdot , \cdot >_{\mu})$ if and only if $\exists$ a vector $z = \{\beta_i\} \in l^2$ such that

$$f(x) = <x, z>_{\mu} = \sum_{i=1}^{\infty} \mu_i\, \alpha_i \,\overline{\beta}_i \, , \quad \forall\, x = \{\alpha_i\} \in l^2.$$

Also

$$\|f\| = \sum_{i=1}^{\infty} \mu_i\, |\beta_i|^2 \, .$$

Remark. The dual space H^* can in natural way be identified with the Hilbert space H (Riesz Representation Theorem), *i.e.*, a Hilbert space is self conjugate; it is this "self-duality" of Hilbert spaces which distinguishes the theory of bounded linear functionals on Hilbert spaces from that of such functionals on Banach spaces.

6.1.3 Example. The dual of the real space l^2 is l^2.

Recall that l^2 is a Hilbert space (but all other spaces l^p, $p \neq 2$, are not Hilbert spaces (Example 5.1.10 (5)). By Riesz Representation Theorem, for each fixed $f \in (l^2)^*$, $\exists$ a unique vector $z = \{\beta_i\} \in l^2$ such that

$$f(x) = <x, z>, \quad \forall\, x = \{\alpha_i\} \in l^2.$$

Further, the mapping $f \to z$ is an isometric isomorphism of $(l^2)^*$ onto l^2 (Verify !). Hence, we conclude that the dual space of l^2 is l^2.

Note. For the complex space l^2, the mapping $f \to z$ of $(l^2)^*$ onto l^2 is conjugate linear since $\lambda f \to \bar{\lambda} z$.

   More precisely, we have

6.1.4 Theorem. *Let H be a Hilbert space. Then, the mapping $z \to f_z$ of H to H^* is bijective and conjugate linear which preserves the norm.*

   Observe that $H^* = B(H, \mathbb{K})$ is a Banach space with the norm defined by

$$\| f_z \| = \sup \{ |f_z(x)| : x \in H, \| x \| = 1 \}.$$

$$= \sup \{ | <x, z> | : x \in H, \| x \| = 1 \}.$$

One can easily establish that this norm on H^* is induced by the inner product on H given by

$$<f, g>_{H^*} = <v, z>_H,$$

where z and v are the uniquely determined vectors in H corresponding to f and g, respectively (Theorem 6.1.1). Consequently, H^* is a Hilbert space. Further, as a Hilbert space, H^* has its own dual space $(H^*)^*$, denoted by H^{**}. As above one can see that H^{**} is also a Hilbert space with the inner product given by

$$< \varphi, \psi >_{H^{**}} = < g, f >_{H^*},$$

where f and g are in H^* corresponding to φ and ψ, respectively. If z and v in H correspond to f and g, respectively, then

$$< \varphi, \psi >_{H^{**}} = < z, v >_H .$$

This in view of Theorem 4.4.1, leads to

6.1.5 Theorem. *Any Hilbert space H is isometrically isomorphic to its second dual H^{**}, the isomorphism $\pi : H \to H^{**}$ being given by*

$$\pi(x) = \varphi f_x$$

where $f_x \in H^*$ corresponds to x in H, which also preserves the inner products.

Proof. One can easily verify that the mapping $\pi : H \to H^{**}$ is canonical. ∎

6.1.6 Corollary. *Every Hilbert space is reflexive.*

In view of the general definition of weak convergence (Definition 4.6.1) and invoking to Riesz-Fréchet Theorem (Theorem 6.1.1), a sequence $\{x_n\}$ in H is said to be weakly convergent to x_0 if

$$< x_n, y > \to < x_0, y >, \quad \forall \, y \in H.$$

Using reflexivity of H, one can easily verify the following:

6.1.7 Theorem. *Let H be a Hilbert space and let $\{x_n\}$, x_0 be in H. Then:*

(a) If $x_n \xrightarrow{w} x_0$ and $\| x_n \| \to \| x_0 \|$, then $\| x_n - x_0 \| \to 0$, *i.e.*, $x_n \to x_0$ strongly.

(b) H is weakly complete.

(c) Bounded sets in H are weakly compact, *i.e.*, from an arbitrary infinite set of elements in H which is bounded in the norm, a weakly convergent sequence can be chosen.

Finally, in this section we present a general representation of sesquilinear functional on Hilbert spaces. Before doing it we introduce the concept of bounded sesquilinear functional.

6.1.8 Definition. A sesquilinear functional h on $X \times Y$, where X and Y are normed spaces over the field $\mathbb{K}$, is said to be *bounded* if $\exists$ a constant $K > 0$ such that

$$| h (x, y) | \le K \| x \|_X \| y \|_Y, \quad \forall \, x \in X, y \in Y.$$

The number

$$\| h \| = \sup_{\substack{0 \ne x \in X \\ 0 \ne y \in X}} \frac{| h (x, y) |}{\| x \|_X \| y \|_Y}$$

is called the *norm* of h.

Note that

$$| h (x, y) | \leq \| h \| \| x \|_X \| y \|_Y, \quad \forall \ x \in X \text{ and } y \in Y.$$

6.1.9 Theorem. *Let H_1 and H_2 be Hilbert spaces over the field $\mathbb{K}$ and*

$$h : H_1 \times H_2 \to \mathbb{K}$$

a bounded sesquilinear functional. Then, h has a representation

(4) $$h (x, y) = < S x, y >$$

where $S : H_1 \to H_2$ is a uniquely determined bounded linear operator and has the norm $\| S \| = \| h \|$.

Proof. Consider $\overline{h (x, y)}$ which is linear in y, because of the bar. Let x be fixed. Then

$$\overline{h (x, y)} = < y, z >, \qquad \text{(Theorem 6.1.1).}$$

Therefore

(5) $$h (x, y) = < z, y >.$$

Here $z \in H_2$ is unique but depends on fixed $x \in H_1$. Clearly, equation (5) defines an operator $S : H_1 \to H_2$ given by

$$z = S x.$$

Substituting $z = S x$ in (5), we obtain (4). Also, note that:

(i) *S is linear.*

    For $x_1, x_2 \in H_1$ and $\alpha, \beta, \in \mathbb{K}$, we have

$$< S (\alpha x_1 + \beta x_2), y > = h (\alpha x_1 + \beta x_2, y)$$
$$= \alpha \, h (x_1, y) + \beta \, h (x_2, y)$$
$$= \alpha < S x_1, y > + \beta < S x_2, y >$$
$$= <\alpha S x_1 + \beta S x_2, y >, \forall \ y \in H_2$$
$$\Rightarrow \qquad S (\alpha x_1 + \beta x_2) = \alpha S x_1 + \beta S x_2.$$

$$\text{(by Lemma 6.2.1 to be proved in § 6.2)}$$

(ii) *S is bounded and $\| S \| = \| h \|$.*

    We have

$$\| h \| = \sup_{\substack{0 \neq x \in H_1 \\ 0 \neq y \in H_2}} \frac{|< S x, y >|}{\| x \| \, \| y \|} \geq \sup_{\substack{0 \neq x \in H_1 \\ 0 \neq S x \in H_2}} \frac{|< Sx, Sx >|}{\| x \| \, \| Sx \|}$$

$$= \sup_{\substack{x \in H_1 \\ x \neq 0}} \frac{\| Sx \|}{\| x \|} = \| S \|.$$

This proves that S is bounded. Moreover, $\| S \| \leq \| h \|$.

Further, by applying Schwartz Inequality, we have

$$\| h \| = \sup_{\substack{0 \neq x \in H_1 \\ 0 \neq y \in H_2}} \frac{|< Sx, y >|}{\| x \| \, \| y \|} \leq \sup_{\substack{0 \neq x \in H_1 \\ 0 \neq x \in H_2}} \frac{\| Sx \| \, \| y \|}{\| x \| \, \| y \|} = \| S \|$$

Hence, $\qquad \| S \| = \| h \|.$

(iii) *S is unique.*

Let there be a linear operator $T : H_1 \to H_2$ such that

$$h\,(x, y) \;=\; < S\,x, y > \;=\; < T\,x, y >, \quad \forall\, x \in H_1, y \in H_2.$$

Then $\qquad\qquad Sx = Tx, \quad \forall\, x \in H_1$

$\Rightarrow \qquad\qquad S = T.$

Problems

1. Prove Theorem 6.1.4.

2. Verify the result in Corollary 6.1.6.

3. Let X and Y be normed spaces. Prove that a bounded sesquilinear functional h on $X \times Y$ is jointly continuous in both variables.

4. Obtain a condition for a Hermitian sesquilinear functional (Definition 1.6.33) to be an inner product on X.

5. Let $(X, < . , . >)$ be an inner product space. Prove that the inner product $< . , . >$ is a bounded sesquilinear functional. Determine the bound of it.

6.2 Hilbert-Adjoint Operators

In Section 4.5, we demonstrated that with each bounded linear operator T from a Banach space X into a Banach space Y is associated a bounded linear operator $T^{\times}$, called its adjoint operator, from the dual space Y^* into the dual space X^*. However, it was observed in the previous section that in case of a Hilbert space

H, there is a natural correspondence between H and its dual space H^*. So if $T : H_1 \to H_2$ is a bounded linear operator, where H_1 and H_2 are Hilbert spaces over the field $\mathbb{K}$, we can consider the adjoint $T^\times$ as acting from H_2 into H_1 (instead of from H_2^* and H_1^*), where it can be compared with T. These ideas lead to the concept of the Hilbert-adjoint of bounded linear operators on Hilbert spaces. These operators were suggested by problems in matrices, linear differential equations and integral equations.

The Riesz Representation Theorem will now help us to prove the following theorem which, in turn, enables us to introduce the concept of Hilbert-adjoint of bounded linear operators of Hilbert spaces. First, we prove an elementary result.

6.2.1 Lemma. *Let $(X, <., .>)$ be an inner product space. If $<y, x> = <z, x>$, $\forall \, x \in X$, then $y = z$.*

Proof. We have

$$<y - z, x> = 0, \quad \forall \, x \in X.$$

Choosing $x = y - z$, we get

$$\| y - z \|^2 = 0$$

$$\Rightarrow \qquad y = z. \ \blacksquare$$

6.2.2 Corollary. *If $<z, x> = 0$, $\forall \, x \in X$, then $z = 0$.*

6.2.3 Theorem. *Let H_1 and H_2 be Hilbert spaces and $T : H_1 \to H_2$ be a bounded linear operator. Then, $\exists$ a unique bounded linear operator $T^* : H_2 \to H_1$ such that*[i]

$$(6) \qquad <T x, y> = <x, T^* y>, \quad \forall \, x \in H_1 \text{ and } y \in H_2.$$

Moreover, $\| T^* \| \leq \| T \|$.

Proof. For each fixed $y \in H_2$, let f_y be the functional on H_1 defined by

$$f_y(x) = <T x, y>, \quad \forall \, x \in H_1.$$

Clearly, f_y is linear. Also, by Schwartz Inequality

$$\left| f_y(x) \right| = \left| <T x, y> \right|$$

---

[i] The inner product on left side is on H_2 and that on the right side is on H_1. Thus, if there is no confusion, we may denote the inner products on two Hilbert spaces by the same symbol.

$$\leq \| T x \| \| y \|$$

$$\leq \| T \| \| x \| \| y \|, \quad \forall\, x \in H_1$$

$\Rightarrow$ f_y is bounded and

$$(7) \qquad \| f_y \| \leq \| T \| \| y \|.$$

Thus, f_y is a bounded linear functional on H_1. By Riesz Representation Theorem, $\exists$ a unique vector $y^* \in H_1$ such that

$$< T x, y > = f_y(x) = < x, y^* >, \quad \forall\, x \in H_1$$

and

$$(8) \qquad \| f_y \| = \| y^* \|.$$

This gives rise to a mapping $T^* : H_2 \to H_1$ defined by $T^* y = y^*$. Thus

$$< T x, y > = < x, y^* > = < x, T^* y >, \quad \forall\, x \in H_1$$

which establishes the existence of a mapping $T^* : H_2 \to H_1$ satisfying (6).

Further, we show that T^* is linear and bounded. Let $y, z \in H_2$ and $\alpha, \beta \in \mathbb{K}$. Then

$$< x, T^*(\alpha y + \beta z) > = < T x, \alpha y + \beta z >$$

$$= \bar{\alpha} < T x, y > + \bar{\beta} < T x, z >$$

$$= \bar{\alpha} < x, T^* y > + \bar{\beta} < x, T^* z >$$

$$= < x, \alpha T^* y > + < x, \beta T^* z >$$

$$= < x, \alpha T^* y + \beta T^* z >, \quad \forall\, x \in H_1$$

$\Rightarrow$ $T^*(\alpha y + \beta z) = \alpha T^* y + \beta T^* z$ (Lemma 6.2.1)

$\Rightarrow$ T^* is a linear operator.

Also, by (7) and (8), we have

$$\| T \| \| y \| \geq \| f_y \| = \| y^* \| = \| T^* y \|.$$

But $y \in H_2$ being arbitrary, T^* is bounded and $\| T^* \| \leq \| T \|$.

Finally, T^* is unique because if $S : H_2 \to H_1$ is a mapping such that

$$< T x, y > = < x, S y >, \quad \forall\, x \in H_1 \text{ and } y \in H_2$$

then

$$< x, T^* y > = < x, S y >$$

$$\Rightarrow \quad < x, T^* y - S y > = 0, \quad \forall \, x \in H_1 \text{ and } y \in H_2$$

$$\Rightarrow \quad T^* y = S y, \quad \forall \, y \in H_2 \text{ (Corollary 6.2.2)}$$

$$\Rightarrow \quad S = T^*.$$

This completes the proof of the theorem. ∎

Although every bounded linear operator $T : H_1 \to H_2$ need not possess an inverse, but by Theorem 6.2.3, T always has a twin brother $T^* : H_2 \to H_1$ (uniquely) which is bounded and linear, concerned with T by the equation (6). Such an operator is called the Hilbert-adjoint operator of T. More precisely

6.2.4 Definition. Let H_1 and H_2 be Hilbert spaces and $T : H_1 \to H_2$ a bounded linear operator. The operator $T^* : H_2 \to H_1$ satisfying

$$< T x, y > = < x, T^* y >, \quad \forall \, x \in H_1 \text{ and } y \in H_2$$

is called the *Hilbert-adjoint operator* of T.

6.2.5 Examples

1. Let O and I be zero and identity operators on a Hilbert space H. Then, $O^* = O$ because

$$< x, O^* y > = < O x, y > = < 0, y > = 0 = < x, O y >,$$

   and since the adjoint is unique. Similarly, we can show that $I^* = I$.

2. Let $S_r : l^2 \to l^2$ be defined by

$$S_r (\alpha_1, \alpha_2,....) = (0, \alpha_1, \alpha_2,......), \quad \{\alpha_i\} \in l^2.$$

   Obviously, S_r is a bounded linear operator with $\| S_r \| = 1$ and is called the *right shift operator.* Similarly, we define $S_l : l^2 \to l^2$ by

$$S_l (\alpha_1, \alpha_2,....) = (\alpha_2, \alpha_3,....), \quad \{\alpha_i\} \in l^2.$$

   It may be noted again that S_l is a bounded linear operator with $\| S_l \| = 1$ and is called the *left shift operator.*

   Let $x = \{ \alpha_i \}$ and $y = \{ \beta_i \}$ be in l^2. Then

$$S_r (x) = (0, \alpha_1, \alpha_2, ...)$$

and therefore

$$< S_r(x), y > = \alpha_1 \bar{\beta}_2 + \alpha_2 \bar{\beta}_3 + \ldots\ldots$$

$$= < x, S_l(y) >.$$

Thus $\qquad\qquad S_r^* = S_l.$

Similarly, note that

$$S_l(x) = (\alpha_2, \alpha_3, \ldots\ldots)$$

and

$$< S_l(x), y > = \alpha_2 \bar{\beta}_1 + \alpha_3 \bar{\beta}_2 + \ldots\ldots$$

$$= < x, S_r(y) >$$

which gives $S_l^* = S_r.$

Thus, S_l is the Hilbert-adjoint operator of S_r and S_r is that of S_l.

3. Let $\eta(t)$ be a complex valued bounded Lebesgue measurable function defined on $[a, b]$, and let $T : L^2[a, b] \to L^2[a, b]$ be the bounded linear operator defined by

$$(Tf)(t) = \eta(t) f(t).$$

Note that

$$< Tf, g > = \int_a^b \eta(t) f(t) \bar{g}(t) \, dt$$

$$= <f, \bar{\eta} \, g>, \qquad \forall f, g \in L^2[a, b].$$

Then, the Hilbert-adjoint operator T^* of T is given by

$$(T^* g)(t) = \bar{\eta}(t) g(t).$$

4. Consider the Hilbert space H of all differentiable functions on $\mathbb{R}$ vanishing at infinity with the inner product given by

$$< x, y > = \int_{-\infty}^{\infty} x(t) \, \overline{y(t)} \, dt$$

Then:

(i) The adjoint of the differential operator D is $-D$.

(ii) The adjoint of the operator $T = i D$ is T.

5. Let T be a linear operator on the Hilbert space $\mathbb{K}^n$ ($\mathbb{K}^n = \mathbb{R}^n$, Example 5.1.3; and $\mathbb{K}^n = \mathbb{C}^n$, Example 5.1.5) with inner product defined by

$$<x, y> = \sum_{i=1}^{n} \xi_i \, \overline{\eta}_i \, ,$$

for $x = (\xi_1, \xi_2, ..., \xi_n)$ and $y = (\eta_1, \eta_2, ..., \eta_n)$ in $\mathbb{K}^n$. Let $\{e_1, e_2, ..., e_n\}$ be an orthonormal basis for $\mathbb{K}^n$. Then with respect to the basis, the linear operator T has matrix representation

$$T = [\alpha_{ij}], \quad 1 \le i, j \le n.$$

For $x = \sum_{i=1}^{n} \lambda_i \, e_i$ and $y = \sum_{i=1}^{n} \mu_i \, e_i$, we have

$$Tx = \sum_{i=1}^{n} \alpha_{ij} \, \lambda_i$$

and $\qquad <Tx, y> = \sum_{j=1}^{n} \sum_{i=1}^{n} \alpha_{ij} \, \lambda_i \overline{\mu}_j.$

Similarly, if $T^* = [\beta_{ij}]$, $1 \le i, j \le n$, then

$$T^* y = \sum_{j=1}^{n} \beta_{ij} \, \mu_j$$

and $\qquad <x, T^* y> = \sum_{j=1}^{n} \sum_{i=1}^{n} \lambda_i \, \overline{\beta}_{ij} \, \overline{\mu}_j.$

Since, $<Tx, y> = <x, T^* y>$, we have

$$\beta_{ij} = \lambda_{ji}, \quad 1 \le i, j \le n.$$

Hence, $T^* = [\overline{\alpha}_{ji}]$, $1 \le i, j \le n$.

This shows that the matrix that represents Hilbert-adjoint T^* of T is the Hermitian transpose of the matrix representation of T.

On the other hand, the (Banach) adjoint operator $T^\times$ is a bounded linear operator on the dual space $(\mathbb{K}^n)^*$. Let $\{g_1, g_2, ..., g_n\}$ be the dual basis for $(\mathbb{K}^n)^*$. Then, for $g = \sum_{i=1}^{n} \mu_i \, g_i$, we have

$$g\,(Tx) = g\left(\sum_{i=1}^{n} \alpha_{ij}\, \lambda_i\right) = \sum_{j=1}^{n} \sum_{i=1}^{n} \alpha_{ij}\, \lambda_i\, \mu_j.$$

But, by the definition of $T^\times$, $g\,(Tx) = T^\times\,(g\,(x))$. Therefore

$$T^\times\,(g\,(x)) = \sum_{j=1}^{n} \sum_{i=1}^{n} \alpha_{ij}\, \lambda_i\, \mu_j$$

$$\Rightarrow \qquad T^\times g = \sum_{j=1}^{n} \alpha_{ij}\, \mu_i$$

$$\Rightarrow \qquad T^\times = [\alpha_{ji}], \quad 1 \le i, j \le n.$$

Hence, the matrix that represents the (Banach) adjoint $T^\times$ is the transpose of the matrix that represents T. ∎

We now give below the relationship between the (Banach) adjoint operator $T^\times$ (Definition 4.5.1) and the Hilbert-adjoint operator T^* (Definition 6.2.4).

Consider a bounded linear operator $T : H_1 \rightarrow H_2$, where H_1 and H_2 are Hilbert spaces. The adjoint operator $T^\times : H_2^* \rightarrow H_1^*$ (in the sense of Definition 4.5.1) is given by

$$T^\times g = f$$

$$(T^\times g)\,(x) = f\,(x) = g\,(Tx),$$

where $f \in H_1^*$ and $g \in H_2^*$. Since f and g are bounded linear operator on H_1 and H_2, respectively, by Riesz Representation Theorem, $\exists\ x_0 \in H_1$ and $y_0 \in H_2$ (uniquely) such that

$$\begin{cases} f\,(x) = \ <x, x_0> \\ g\,(y) = \ <y, y_0> \end{cases}.$$

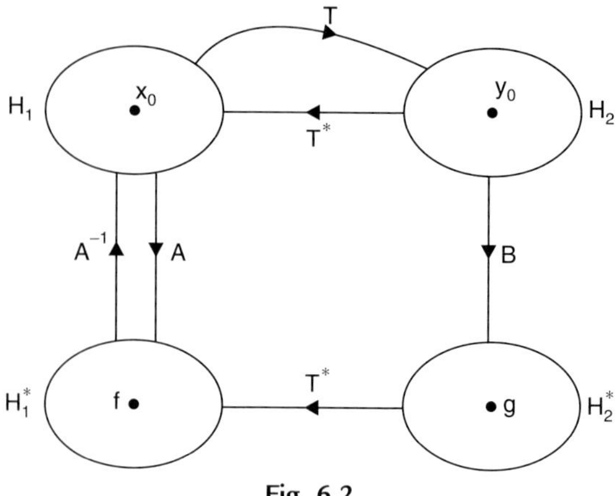

Fig. 6.2

These relations give rise to the operators

$$\begin{cases} A : H_1 \to H_1^*, & A\, x_0 = f \\ B : H_2 \to H_2^*, & B\, y_0 = g \end{cases}$$

One may note, in view of Theorem 6.1.4, that A and B are bijective, conjugate linear and isometries. Define T^* $(= A^{-1}\, T^{\times}\, B) : H_2 \to H_1$ by $T^*\, y_0 = x_0$. Note that:

(a) T^* is linear since it is the composition of two conjugate linear mappings together with the linear operator $T^{\times}$.

(b) T^* is the Hilbert-adjoint operator of T. In fact, we have

$$< T\, x,\, y_0 > \, = g\, (T\, x) = f(x) = \, < x,\, x_0 >.$$

But

$$x_0 = A^{-1} f = A^{-1}\, T^{\times} g = (A^{-1}\, T^{\times}\, B)\, (y_0) = T^*\, y_0.$$

Consequently

$$< T\, x,\, y_0 > \, = \, <x,\, T^*\, y_0 >.$$

(c) The assertion $\| T^* \| = \| T \|$ follows immediately from Theorem 4.5.2 and the isometries A and B.

We have, thus, established that

6.2.6 Theorem. *Let H_1 and H_2 be Hilbert spaces and $T : H_1 \to H_2$ a bounded linear operator. Then, the Hilbert-adjoint operator T^* of T is given in terms of* *the adjoint operator $T^{\times}$ of T by the formula $T^{*} = A^{-1} T^{\times} B$, where A and B are bijective, conjugate linear and isometric operators given above. Furthermore,*

$\| T^{*} \| = \| T^{\times} \| = \| T \|.$

A Special Case For Hilbert-Adjoint Operators

Recall that $B(H)$ is a Banach algebra of bounded linear operators on a Hilbert space H (Example 3.2.7 (3)). The class $B(H)$ is of particular importance. As noticed above, rather than the adjoint operator $T^{\times} : H^{*} \to H^{*}$, it is convenient to exploit the identification of H^{*} with H and then consider the Hilbert-adjoint operator T^{*} on H itself (Definition 6.2.4 with $H_1 = H_2 = H$). Furthermore, since the Hilbert-adjoint of an operator T in $B(H)$ itself lies in $B(H)$, T^{*} is a more convenient object to deal with than the adjoint operator $T^{\times}$. We shall provide below the motivation for considering the Hilbert-adjoint operators by taking any given bounded linear operator T on H.

Let $y \in H$ and f_y its corresponding functional in H^{*}. Operate with the adjoint operator $T^{\times}$ on f_y to obtain a functional f_z $(= T^{\times} f_y)$ in H^{*} and return to its corresponding vector z in H. In the process, we have considered three mappings:

$$y \xrightarrow{\ A\ } f_y \xrightarrow{\ T^{\times}\ } T^{\times}f_y \ (=f_z) \xrightarrow{\ A^{-1}\ } z.$$

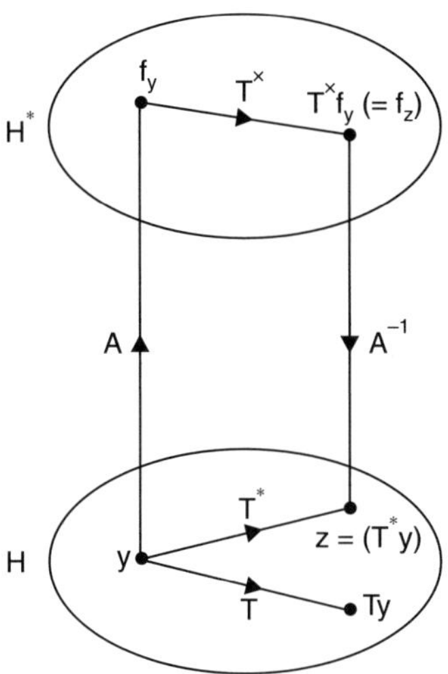

Fig. 6.3

Note that the mapping A is bijective, conjugate linear and norm preserving. As such A^{-1} exists and is also so. Now, forming the product of these three mappings, we write $T^* = A^{-1} T^{\times} A$ so that

$$z = T^* y.$$

The mapping T^* is the Hilbert-adjoint operator of T. Indeed, we have

$$\begin{cases} (T^{\times} f_y)(x) = f_y(Tx) = \,<Tx, y> \\ (T^{\times} f_y)(x) = f_z(x) = \,<x, T^* y> \end{cases}$$

$$\Rightarrow \qquad <Tx, y> = \,<x, T^* y>, \quad \forall\, x, y \in H.$$

We now give some general properties of Hilbert-adjoint operators which are used frequently in applying these operators. We first state and prove a lemma.

6.2.7 Lemma. *Let H_1 and H_2 be Hilbert spaces and $Q \in B\,(H_1, H_2)$. Then*

$$<Qx, y> = 0 \iff Q = 0, \quad \forall x \in H_1 \text{ and } y \in H_2.$$

Proof. It follows obviously in view of Corollary 6.2.2. ∎

6.2.8 Theorem. *Let H_1 and H_2 be Hilbert spaces, $T, S \in B\,(H_1, H_2)$ and $\alpha \in \mathbb{K}$. Then:*

(a) $<T^* y, x> = \,<y, Tx>, \quad \forall\, x \in H_1 \text{ and } y \in H_2.$

(b) $(T + S)^* = T^* + S^*.$

(c) $(\alpha T)^* = \bar{\alpha}\, T^*.$

(d) $(TS)^* = S^* T^*, \; (H_1 = H_2).$

(e) $T^{**} = T.$ (Notation : $T^{**} = (T^*)^*$)

(f) $\| T^* \| = \| T \|.$

(g) $\| T^* T \| = \| T \|^2 = \| T T^* \|.$

Proof. (a) From the definition of T^*, we have

$$<T^* y, x> = \,\overline{<x, T^* y>} = \,\overline{<Tx, y>} = \,<y, Tx>.$$

(b) Note that

$$<x, (T + S)^* (y)> = \,<(T + S)(x), y>$$

$$= \,<Tx, y> + \,<Sx, y>$$

$$= \,<x, T^* y> + \,<x, S^* y>

$$= <x, (T^* + S^*)(y)>, \quad \forall\, x \in H_1 \text{ and } y \in H_2$$

$$\Rightarrow \quad (T + S)^*(y) = (T^* + S^*)(y), \quad \forall\, y \in H_2 \text{ (Lemma 6.2.1).}$$

This verifies (b).

(c) We have

$$< (\alpha\, T)^*(y), x > = < y, (\alpha\, T)(x) >$$

$$= < y, \alpha\, Tx >$$

$$= \bar{\alpha} < y, Tx >$$

$$= \bar{\alpha} < T^*y, x >$$

$$= < (\bar{\alpha}\, T^*)(y), x >, \quad \forall\, x \in H_1 \text{ and } y \in H_2$$

The result follows by using Lemma 6.2.7 with

$$Q = (\alpha\, T)^* - \bar{\alpha}\, T^*.$$

(d) We have

$$< (T S)^*(y), x > = < y, (T S)(x) >$$

$$= < y, T(S x) >$$

$$= < T^*y, S x >$$

$$= < (S^*\, T^*)(y), x >, \quad \forall\, x \in H_1 \text{ and } y \in H_2.$$

The result follows by using Lemma 6.2.7 with

$$Q = (TS)^* - S^*\, T^*.$$

(e) We have

$$< T^{**}(x), y > = < x, T^*y >$$

$$= < Tx, y >, \quad \forall\, x \in H_1 \text{ and } y \in H_2.$$

The result follows by using Lemma 6.2.7 with

$$Q = T^{**} - T.$$

(f) We already have

$$\| T^* \| \le \| T \| \qquad\qquad \text{(Theorem 6.2.3).}$$

Using T^* instead of T and applying to (e), we have

$$\| T \| = \| T^{**} \| = \| (T^*)^* \| \le \| T^* \|.$$

Hence $\qquad\qquad \| T^* \| = \| T \|.$

(g) Observe that $T^* T : H_1 \to H_1$ and $T T^* : H_2 \to H_2$. Then, we have

$$\| T \|^2 = \sup_{\substack{x \in H_1 \\ \| x \| \le 1}} \| Tx \|^2$$

$$= \sup_{\substack{x \in H_1 \\ \| x \| \le 1}} < Tx, Tx >$$

$$= \sup_{\substack{x \in H_1 \\ \| x \| \le 1}} < (T^* T) (x), x >$$

$$\le \sup_{\substack{x \in H_1 \\ \| x \| \le 1}} \| T^* T \| \| x \|^2 \quad \text{(Schwartz Inequality)}$$

$$\le \| T^* T \|$$

$$\le \| T^* \| \| T \| \quad \text{(Theorem 3.2.3)}$$

$$= \| T \|^2$$

$$\Rightarrow \quad \| T^* T \| = \| T \|^2.$$

Further applying this equality to the operator T^*, we get

$$\| T T^* \| = \| (T^*)^* T^* \| = \| T^* \|^2 = \| T \|^2. \quad \blacksquare$$

6.2.9 Corollary. *The mapping $T \to T^*$ of $B\,(H_1, H_2)$ into $B\,(H_2, H_1)$ has the additional property*

$$T^* T = 0 \quad \Leftrightarrow \quad T = 0.$$

Proof. It follows from (g). $\blacksquare$

Now we give below certain main differences between the adjoint operator $T^\times$ and the Hilbert-adjoint operator T^* of a bounded linear operator T:

1. $T^\times$ is defined for any $T \in B\,(X, Y)$, where X and Y are, in general, normed spaces whereas T^* is defined only for $T \in B\,(H_1, H_2)$, where H_1 and H_2 are Hilbert spaces.

2. $T^\times$ is defined on the dual of a space which contains the range of T whereas T^* is directly defined on the space which contains the range of T.

3. For $T^\times$, we have

$$(\alpha T)^\times = \alpha T^\times,$$

but for T^*, we have

$$(\alpha T)^* = \overline{\alpha}\ T^*.$$

4. In finite dimensional case $T^{\times}$ is represented by the transpose of the matrix representing T whereas T^* is represented by the conjugate (Hermitian) transpose of that matrix (Example 6.2.5 (4)). ∎

At this stage, motivated by the properties of the operator $T \to T^*$ on $B\,(H)$ (with $H_1 = H_2 = H$ in Theorem 6.2.8), we are tempted to introduce a few concepts and finally we show that the Banach algebra $B\,(H)$ is a B^*-algebra.

6.2.10 Definition. A mapping $x \to x^*$ of an algebra A into itself is said to be an *involution* if the following conditions are satisfied:

(i) $(x + y)^* = x^* + y^*$

(ii) $(\alpha x)^* = \bar{\alpha}\, x^*$

(iii) $(x y)^* = y^*\, x^*$

(iv) $x^{**} = x.$

6.2.11 Definition. A Banach algebra A is called a *Banach**-*algebra* if it has an involution. A Banach*-algebra is called a B^*-*algebra* if the involution and the norm are related by

$$\| x^*\, x \| = \| x \|^2.$$

6.2.12 Example. For $H \neq \{0\}$, $B\,(H)$ is a B^*-algebra with adjoint operation $T \to T^*$ taken as the involution. Indeed, $B\,(H)$ is a Banach*-algebra, the adjoint operation $T \to T^*$ on $B\,(H)$ satisfies all the properties of being an involution on $B\,(H)$ (Theorem 6.2.8 with $H_1 = H^2 = H$) and we have

$$\| T^*\, T \| = \| T^* \| \, \| T \| = \| T \|^2.$$

Problems

6. Prove that the adjoint operation $T \to T^*$ in $B\,(H)$ is bijective.

7. Let $T \in B\,(H)$ be bijective and its inverse be bounded. Prove that $(T^*)^{-1}$ exists and

$$(T^*)^{-1} = (T^{-1})^*.$$

[*Hint:* Take the adjoint of $T\,T^{-1} = I = T^{-1}\,T$ and use $I^* = I$.]

8. Let $\{T_n\} \subset B\,(H)$ be such that $T_n \to T$. Prove that $T_n^* \to T^*$.

[*Hint:* $\| T_n^* - T^* \| = \| T_n - T \|$.]

9. Let $T \in B(H_1, H_2)$. Prove that:

   (a) $\ker T = [T^*(H_2)]^\perp$.

   (b) $\ker T^* = [T(H_1)]^\perp$.

   (c) $\overline{T(H_1)} = [\ker T^*]^\perp$.

   (d) $\overline{T^*(H_2)} = [\ker T]^\perp$.

10. Let $T \in B(H_1, H_2)$. If $A \subset H_1$, $B \subset H_2$ and $T(A) \subset B$, prove that

$$T^*(B^\perp) \subset A^\perp.$$

11. Let $M \subset H_1$, $N \subset H_2$ be closed subspaces and $T \in B(H_1, H_2)$. Prove that

$$T(M) \subset N \quad \Leftrightarrow \quad T^*(N^\perp) \subset M^\perp.$$

6.3 Self-Adjoint Operators

We have observed that the set $B(H)$ of all bounded linear operators on a non-trivial Hilbert space H forms a Banach algebra together with the mapping $T \to T^*$ of $B(H)$ onto itself. Also, it is well known that the set $\mathbb{C}$ of all complex numbers is a Banach algebra with the mapping $z \to \bar{z}$ of $\mathbb{C}$ onto itself. The mappings $T \to T^*$ and $z \to \bar{z}$ have some analogous properties which we list in the following table. (The properties of the mapping $T \to T^*$ we have proved in Theorem 6.2.8 with $H_1 = H_2 = H$).

Mapping $z \to \bar{z}$ of $\mathbb{C}$ onto $\mathbb{C}$	*Mapping $T \to T^*$ of $B(H)$ onto $B(H)$*
(i) $\overline{z_1 + z_2} = \bar{z}_1 + \bar{z}_2$	$(T_1 + T_2)^* = T_1^* + T_2^*$
(ii) $\overline{\alpha z} = \bar{\alpha}\,\bar{z}$	$(\alpha T)^* = \bar{\alpha}\,T^*$
(iii) $\overline{z_1 z_2} = \bar{z}_1\,\bar{z}_2 \; (= \bar{z}_2\,\bar{z}_1)$	$(T_1 T_2)^* = T_2^*\,T_1^* \; (\neq T_1^*\,T_2^*)$
(iv) $\overline{(\bar{z})} = z$	$(T^*)^* = T$
(v) $|\bar{z}| = |z|$	$\|T^*\| = \|T\|$
(vi) $|\bar{z}\,z| = |z|^2 = |z\bar{z}|$	$\|T^*T\| = \|T\|^2 = \|TT^*\|$

    The main difference between these two systems is that multiplication in Banach algebra $B(H)$, in general, is non-commutative. The analogy of the properties of $B(H)$ with those of the complex plane motivates the study of certain classes of bounded linear operators which are of great practical importance. These are described as follows:

(I) The real line $\mathbb{R}$ is one of the most significant subset of the complex plane $\mathbb{C}$ and is given by $\mathbb{R} = \{z \in \mathbb{C} : z = \bar{z}\}$. By analogy, we consider the subfamily

$$S(H) = \{T \in B(H) : T = T^*\}.$$

Such operators are called *self-adjoint operators.*

(II) Non-commutativity of operators in $B(H)$ leads to a special class of operators; namely,

$$\mathcal{N}(H) = \{T \in B(H) : TT^* = T^*T\}.$$

Such operators are called *normal operators.*

(III) Another important subsystem of complex plane is the unit circle, *i.e.,* the set $\{z \in \mathbb{C} : z\bar{z} = \bar{z}z = 1\}$. By analogy, we consider the class

$$\mathcal{U}(H) = \{T \in B(H) : T^*T = TT^* = I\}.$$

These operators are called *unitary operators.*

We shall deal with these special classes of operators in $B(H)$ one by one. This section concerns with the self-adjoint operators. We observe that there is a natural definition of "Self-adjointness" in Hilbert spaces whereas it is difficult to find a meaningful formulation of such a concept in Banach spaces.

6.3.1 Definition. A bounded linear operator $T : H \to H$ on a Hilbert space H is said to be *self-adjoint* (or *Hermitian*) if $T^* = T$.

Clearly, bounded linear operator T in $B(H)$ is self-adjoint if and only if it satisfies the equation

$$<Tx, y> = <x, Ty>, \quad \forall\, x, y \in H.$$

6.3.2 Examples

1. The zero operator and the identity operator in $B(H)$ are self-adjoint since $O^* = O$ and $I^* = I$.

2. If $T \in B(H)$, then T^*T is self-adjoint since $(T^*T)^* = T^*T^{**} = T^*T$, and similarly, TT^* is self-adjoint.

3. The operator T defined in Example 6.2.5 (3) is self-adjoint if and only if

$$\eta(t) = \overline{\eta(t)}, \text{ a.e.}$$

4. Define $L : l^2 \to l^2$ by

$$(Lf)_i = \sum_{j=1}^{\infty} \alpha_{ij} f_j, \quad (i = 1, 2, ...)$$

where $f = \{ f_j \} \in l^2$ and α_{ij}'s are scalars in $\mathbb{K}$. If $\displaystyle\sum_{i=1}^{\infty} \sum_{i=1}^{\infty} |\alpha_{ij}|^2 < \infty$, then

the operator L is bounded. Note that the inner product in l^2 is given by

$$<f, g> = \sum_{i=1}^{\infty} f_i \overline{g_i}$$

and therefore, we have

$$<Lf, g> = \sum_{i=1}^{\infty} (Lf)_i \, \overline{g_i}$$

$$= \sum_{i=1}^{\infty} \sum_{i=1}^{\infty} \alpha_{ij} f_i \, \overline{g_i}$$

$$= \sum_{j=1}^{\infty} f_j \, \overline{\left(\sum_{i=1}^{\infty} \overline{\alpha_{ij}} \, g_i \right)}$$

$$= <f, L^* g>, \qquad (L^* g)_j = \sum_{i=1}^{\infty} \overline{\alpha_{ji}} \, g_i.$$

Thus L^* is represented by the conjugate transpose of the infinite matrix $A = [\alpha_{ij}]$. Hence, L is self-adjoint if and only if $\alpha_{ij} = \overline{\alpha_{ji}}$ ($i = 1, 2, ...$). We conclude that L is self-adjoint if and only if the infinite matrix A is Hermitian.

5. Consider the operator T defined on $L^2(\mathbb{R})$ given by

$$(Tx)(t) = \overline{e}^{|t|} x(t).$$

Clearly T is linear and it is bounded since

$$\| Tx \|^2 = \int_{-\infty}^{\infty} |x(t)|^2 \, \overline{e}^{\,2|t|} \, dt$$

$$= 2 \int_{0}^{\infty} |x(t)|^2 \, \overline{e}^{\,2|t|} \, dt$$

$$\leq 2 \max_{t \in [0, \infty)} \overline{e}^{\,2|t|} \int_{0}^{\infty} |x(t)|^2 \, dt$$

$$\Rightarrow \qquad \| Tx \|^2 \leq M \| x \|^2, \qquad M = \max_{t \in [0, \infty)} \overline{e}^{\,2|t|}.$$

Also

$$< Tx, y > = \int_{-\infty}^{\infty} \overline{e}^{\,|t|} \, x(t) \, \overline{y(t)} \, dt$$

$$= \int_{-\infty}^{\infty} x(t) \left(\overline{\overline{e}^{\,|t|} \, y(t)} \right) dt$$

$$= < x, Ty >.$$

Thus T is a self-adjoint operator.

6. Consider the operator T on $L^2(\mathbb{R})$ given by

$$(Tx)(t) = t \, x(t)$$

Then T is self-adjoint.

7. In Example 6.2.5 (4), the operator D is not self-adjoint while the operator $i\,D$ is self-adjoint.

6.3.3 Lemma. *Let H be a complex Hilbert space and $Q \in B(H)$. Then*

$$Q = 0 \quad \Leftrightarrow \quad < Q x, x > = 0, \quad \forall \, x \in H.$$

Proof. Assume first that $< Q x, x > = 0, \forall \, x \in H$. Then, for $x, y \in H$ and $\alpha, \beta \in \mathbb{C}$, we have

$$< Q\,(\alpha x + \beta y),\ \alpha x + \beta y > = 0$$

$$\Rightarrow \quad |\alpha|^2 < Q\,x, x > + \alpha\,\overline{\beta} < Q\,x, y > + \beta\overline{\alpha} < Q\,y, x >$$

$$+ |\beta|^2 < Q\,y, y > = 0$$

$$\Rightarrow \quad \alpha\overline{\beta} < Q\,x, y > + \beta\overline{\alpha} < Q\,y, x > = 0.$$

Taking $\alpha = \beta = 1$, we get

$$(9) \qquad < Q\,x, y > + < Q\,y, x > = 0;$$

and $\alpha = i$ and $\beta = 1$, we get

$$(10) \qquad < Q\,x, y > - < Q\,y, x > = 0.$$

By adding (9) and (10), we obtain

$$< Q\,x, y > = 0, \quad \forall\,x, y \in H$$

$$\Rightarrow \qquad Q = 0. \qquad\qquad \text{(Lemma 6.2.7)}$$

The proof of the converse part is straightforward again in view of Lemma 6.2.7 with $y = x$. ∎

We now give simple criterion for self-adjointness.

6.3.4 Theorem. *Let H be a Hilbert space and $T \in B\,(H)$. Then:*

(a) *If T is self-adjoint, then $< Tx, x >$ is real $\forall\,x \in H$.*

(b) *If H is a complex Hilbert space and $< Tx, x >$ is real $\forall\,x \in H$, then the operator T is self-adjoint.*

Proof. (a) Let T be self-adjoint. Then

$$\overline{< T\,x, x >} = < x, T\,x > = < T\,x, x >, \quad \forall\,x \in H$$

$$\Rightarrow \qquad < T\,x, x > \text{ is real } \forall\,x \in H.$$

(b) Let $< T\,x, x >$ be real $\forall\,x \in H$. Then

$$< T\,x, x > = \overline{< T\,x, x >}$$

$$= \overline{< x, T^*\,x >}$$

$$= < T^*\,x, x >$$

$$\Rightarrow \qquad\qquad <Tx - T^*x, x> = 0$$

$$\Rightarrow \qquad\qquad <(T - T^*)(x), x> = 0, \quad \forall\, x \in H.$$

Hence, by Lemma 6.3.3, $T - T^* = 0$ since H is a complex Hilbert space. This verifies that T is self-adjoint. ∎

Note. In part (a) H may be real or complex. But in part (b) it is essential that H be a complex Hilbert space since Lemma 6.3.3 holds for complex H. Moreover, when H is real, the inner product is real valued which makes $< T x, x >$ real automatically for every $T \in B(H)$ irrespective of any further assumption on T.

Remark. The assertion in Theorem 6.3.4 further strengthens the idea of considering the class of self-adjoint operators in analogy with $\mathbb{R}$.

6.3.5 Theorem. *Let H be a Hilbert space. Then:*

(a) $S, T \in S(H) \qquad \Rightarrow \quad S + T \in S(H).$

(b) $\alpha \in \mathbb{R}, T \in S(H) \quad \Rightarrow \quad \alpha\, T \in S(H).$

(c) *If $\{T_n\} \subset S(H)$ is a sequence such that $T_n \to T$ in the norm topology of $B(H)$, then $T \in S(H)$.*

Proof. The assertions (a) and (b) follow in view of the fact that

$$(S + T)^* = S^* + T^* = S + T,$$

and

$$(\alpha\, T)^* = \overline{\alpha}\, T^* = \alpha\, T.$$

    (c) Note that

$$\| T_n^* - T^* \| = \| (T_n - T)^* \| = \| T_n - T \|,$$

and so

$$\| T - T^* \| \leq \| T - T_n \| + \| T_n - T_n^* \| + \| T_n^* - T^* \|$$

$$= 2 \| T_n - T \|$$

$$\to 0, \text{ as } n \to \infty. \ \blacksquare$$

6.3.6 Corollary. *The class $S(H)$ of all self-adjoint operators in $B(H)$ forms a closed (real), subspace of $B(H)$ and hence $S(H)$ is a real Banach space which contains the identity operator.*

6.3.7 Theorem. *Let H be a complex Hilbert space. Then, every $T \in B(H)$ can be expressed uniquely as*

$$(11) \qquad\qquad T = A + i\, B,$$

where $A, B \in \mathcal{S}(H)$.

Proof. Write

$$A = \frac{1}{2}\,(T + T^*), \quad B = \frac{1}{2i}\,(T - T^*).$$

Clearly　　　　$T = A + i\, B$ and $A^* = A$, $B^* = B$.

To establish the uniqueness of this representation, let $T = C + i\, D$ be another such representation, where $C, D \in \mathcal{S}(H)$. Then

$$T^* = C^* - i\, D^* = C - i\, D.$$

Therefore

$$C = \frac{1}{2}\,(T + T^*) = A$$

$$D = \frac{1}{2i}\,(T - T^*) = B. \;\blacksquare$$

It is quite interesting to compare the above decomposition of an arbitrary bounded linear operator in $B\,(H)$ with that of any complex number of the form $\lambda = \alpha + i\,\beta$, where α and β are real. The self-adjoint operators A and B in (11) are called, respectively, the real part and the imaginary part of T.

The product of two self-adjoint operators may not be a self-adjoint operator. However, we have the following.

6.3.8 Theorem. *Let $T, S \in \mathcal{S}(H)$. Then*

$$TS \in \mathcal{S}(H) \quad \Leftrightarrow \quad T\,S = ST.$$

Proof. Straight forward. $\blacksquare$

The following theorem gives a useful alternative expression for the norm of a self-adjoint operator.

6.3.9 Theorem. *If $T \in \mathcal{S}(H)$, then*

$$\| T \| = \sup \{ \, | <T x, x> | : x \in H, \| x \| = 1 \}.$$

Proof. By Schwartz Inequality (Lemma 5.1.7), we have

$$| <T x, x> | \leq \| Tx \| \, \| x \| \leq \| T \| \, \| x \|^2, \quad \forall \, x \in H.$$

Write $\qquad \alpha = \sup \{ | <T x, x> | : x \in H, \| x \| = 1 \}.$

Then, $\alpha \leq \| T \|$. To prove the converse inequality, first note that every $x \in H$ can be represented in the form $x = \| x \| \, x'$, where $\| x' \| = 1$. Hence

$$(12) \qquad | <T x, x> | = \| x \|^2 \, | <T x', x'> | \leq \alpha \| x \|^2, \quad \forall \, x \in H.$$

Note that

$$<T x, y> = <x, T y>, \quad \forall \, x, y \in H$$

and therefore

$$\begin{cases} <T(x+y), x+y> = <T x, x> + 2 \, \mathrm{Re} <T x, y> + <T y, y> \\ <T(x-y), x-y> = <T x, x> - 2 \, \mathrm{Re} <T x, y> + <T y, y> \end{cases}$$

$$\Rightarrow \quad <T(x+y), x+y> - <T(x-y), x-y> = 4 \, \mathrm{Re} <T x, y>, \quad \forall \, x, y \in H.$$

This, by using (12) and parallelogram equality, gives

$$| \, \mathrm{Re} <T x, y> | \leq \frac{1}{4} \, \alpha \, (\| x+y \|^2 + \| x-y \|^2)$$

$$(13) \qquad = \frac{1}{2} \, \alpha \, (\| x \|^2 + \| y \|^2), \quad \forall \, x, y \in H.$$

Let $x \in H$ with $\| x \| \leq 1$ and $T x \neq 0$. Setting $y = \dfrac{T x}{\| T x \|}$ in (13), we get

$$\| T x \| = | \, \mathrm{Re} <T x, \frac{T x}{\| T x \|}> | \leq \frac{1}{2} \, \alpha \, (\| x \|^2 + 1) \leq \alpha.$$

This inequality is trivially valid for $Tx = 0$. Consequently, $\|T\| \leq \alpha$. Hence, the equality $\|T\| = \alpha$ is proved. ∎

By the analogy of $B(H)$ into $\mathbb{C}$, we can think of $S(H)$ as a generalization of $\mathbb{R}$. This fact coupled with Theorem 6.3.4, we may define order relation on the set $S(H)$ of all self-adjoint operators on H.

6.3.10 Definition. Let H be a complex Hilbert space and let $T, S \in S(H)$. We say that $T \leq S$ if

$$< Tx, x > \; \leq \; < Sx, x >, \quad \forall x \in H.$$

6.3.11 Theorem (*Order properties of self-adjoint operators*)

(a) $(S(H), \leq)$ *is a partially ordered set.*

(b) $A \leq B$ *and* $\alpha \geq 0 \quad \Rightarrow \quad \alpha A \leq \alpha B, A, B \in S(H).$

The fact that the order structure being defined on $S(H)$ suggests another concept for self-adjoint operators similar to that for real numbers; namely:

6.3.12 Definition. An operator $T \in S(H)$ is said to be *positive* if

$$< Tx, x > \; \geq 0, \quad \forall x \in H.$$

We may denote the set of all positive operators on H by $S^{+}(H)$. If T is a positive operator, we write $T \geq O$.

6.3.13 Examples

1. The self-adjoint operators O and I are positive.
2. $T^{*}T$ and TT^{*} are positive operators for each $T \in B(H)$ since

$$< (T^{*}T)(x), x > \; = \; < Tx, Tx > \; = \; \|Tx\|^{2} \geq 0, \quad \forall x \in H,$$

and similarly, for TT^{*}.

6.3.14 Theorem. *If* $T \in S^{+}(H)$, *then* $I + T$ *is a topological isomorphism of* H *onto* H.

Proof. Note that Hilbert space is a special Banach space. In view of a consequence of the Open Mapping Theorem (Corollary 3.5.6), we need only to show that $I + T$ is bijective.

Let $(I + T)(x) = 0$. Then, $Tx = -x$ and so

$$< Tx, x > \; = \; < -x, x > \; = \; -\|x\|^{2} \leq 0.$$

But T being positive, it follows that $x = 0$. Thus, we have established that

$$x \in \ker (I + T) \implies x = 0.$$

This proves that $I + T$ is injective. We need to prove that $I + T$ is onto. Now

$$\| (I + T)(x) \|^2 = \, <x + Tx, x + Tx>$$

$$= \| x \|^2 + \| Tx \|^2 + 2 <Tx, x>$$

$$\geq \| x \|^2 + \| Tx \|^2, \qquad\qquad (\because \quad T \geq O)$$

$$\geq \| x \|^2$$

$$\implies \quad \| (I + T)(x) \| \geq \| x \|, \quad \forall\, x \in H$$

i.e., $\qquad\qquad \| x \| \leq \| I + T \| \, \| x \|, \quad \forall\, x \in H$

But H is complete. Therefore, the range space $M = (I + T)(H)$ is complete. Hence, M is a closed subspace of H. We claim that $M = H$. If $M \neq H$, then $\exists\, x_0 \in H$, $x_0 \neq 0$, $x_0 \perp M$ (Theorem 5.4.2) so that

$$< (I + T)(x_0), x_0 > \, = 0$$

$$\implies \qquad \| x_0 \|^2 + <T x_0, x_0> \, = 0$$

$$\implies \qquad\qquad \| x_0 \|^2 = - <T x_0, x_0>$$

which contradicts the fact that T is a positive operator. Thus, $M = H$ and hence $I + T$ is onto. This completes the proof of the theorem. ∎

6.3.15 Corollary. *If $T \in \mathcal{S}^+(H)$, then $I + T$ is non-singular. In particular, $I + T^* T$ and $I + T T^*$ are non-singular operators for each $T \in B(H)$.*

Problems

12. Let $T \in \mathcal{S}(H)$. Then, the numbers M_T and m_T, given by

$$M_T = \sup \, \{<Tx, x> : x \in H, \| x \| = 1\}$$

$$m_T = \inf \, \{<Tx, x> : x \in H, \| x \| = 1\}$$

are called, respectively, the *upper bound* and the *lower bound* of the self-adjoint operator T. Prove that:

(a) $\| T \| = \max \, \{| M_T |, | m_T |\}$.

(b) $m_T \| x \|^2 \leq <Tx, x> \, \leq M_T \| x \|^2, \quad \forall\, x \in H.$

13. If $T \in \mathcal{S}(H)$, prove that $T^n \in \mathcal{S}(H)$, where n is a positive integer.

[*Hint:* With $S = T$ in Theorem 6.3.8, $T^2 \in \mathcal{S}(H)$. Define $T^n = T^{n-1} T$ and use induction.]

14. Let $T : \mathbb{C}^2 \to \mathbb{C}^2$ be defined by

$$Tx = (\xi_1 + i\,\xi_2,\ \xi_1 - i\,\xi_2),$$

where $x = (\xi_1, \xi_2)$. Find T^* and prove that

$$T^* T = T T^* = 2\,I.$$

Also find T_1 and T_2 such that $T = T_1 + i\,T_2$.

15. An operator $T \in B(H)$ is said to be *skew-adjoint if* $T^* = -T$. Prove that:

(a) T is skew-adjoint $\Leftrightarrow$ $i\,T$ is self-adjoint.

(b) For each $T \in B(H)$, $T - T^*$ is skew-adjoint.

16. An operator $T \in B(H)$ is said to be a *contraction* if

$$\| T x \| \le \| x \|, \quad \forall\, x \in H.$$

Prove, for an operator $T \in B(H)$, that the following conditions are equivalent:

(a) $\| T \| \le 1$.

(b) $T^* T \le I$.

(c) $T T^* \le I$.

(d) T^* is a contraction.

(e) $T^* T$ is a contraction.

17. If $T \in \mathcal{S}(H)$ and $S \in B(H)$, prove that $S^* T S \in \mathcal{S}(H)$.

18. If $T \in \mathcal{S}(H)$, prove that

$$T0 = 0 \quad \Leftrightarrow \quad (T T)\,(0) = 0.$$

Give an example of an operator T on H such that $T \ne O$ but $T T = O$.

19. For $T, S \in \mathcal{S}(H)$, define

$$T o S = \frac{1}{2}\,(T S + S T).$$

Note that $T o S$ is always self-adjoint, and also $T o S = T S$ whenever $T S = S T$. Prove that the operation 'o' satisfies the following properties:

    (a) $ToS = SoT$.

    (b) $To(S + U) = ToS + ToU$.

    (c) $\alpha\,(ToS) = (\alpha\,T\,)oS = To(\alpha\,S)$.

    (d) $ToI = T = IoT$.

    (e) $To(SoU) = (ToS)oU$.

20. Let $S, T \in \mathcal{S}^+(H)$ and let $\alpha \geq 0$ be in $\mathbb{R}$. Prove that

    (a) $S + T \in \mathcal{S}^+(H)$.

    (b) $\alpha\,T \in \mathcal{S}^+(H)$.

    (c) $T^n \in \mathcal{S}^+(H)$, n is any positive integer.

21. Prove that the relation $\leq$ is a partially ordered relation on $\mathcal{S}^+(H)$. Further, prove that:

    (a) $T_1 \leq S_1$ and $T_2 \leq S_2 \;\Rightarrow\; T_1 + T_2 \leq S_1 + S_2$.

    (b) $0 \leq T \leq S$ and $0 \leq \alpha \leq \beta \;\Rightarrow\; 0 \leq \alpha\,T \leq \beta\,S$.

    (c) $T \leq S \;\Leftrightarrow\; -S \leq -T$.

22. Prove that:

    (a) $T \in \mathcal{S}(H) \;\Rightarrow\; T^2 \in \mathcal{S}^+(H)$.

    (b) $S, T \in \mathcal{S}^+(H)$ and $S + T = 0 \;\Rightarrow\; S = T = 0$.

    (c) If $T \in \mathcal{S}^+(H)$, then $\exists$ a unique operator $S \in \mathcal{S}^+(H)$ such that $T = SS$ (Such an operator is called the *positive square root* of T and is denoted by $S = \sqrt{T}$).

    (d) If $S, T \in \mathcal{S}^+(H)$, then $S\,T \in \mathcal{S}^+(H)$ if and only if $S\,T = T\,S$. In this case

$$\sqrt{ST} = \sqrt{S}\,\sqrt{T}\,.$$

    (e) If $T \in \mathcal{S}(H)$, then $\exists$ operators $A, B \in \mathcal{S}^+(H)$ such that

$$T = A - B \text{ and } AB = 0.$$

23. The operators T and S are paid to be *commutating operators* if $T\,S = S\,T$. Show that:

    (a) Consider the space of differentiable functions on $\mathbb{R}$ and the operators T and D given by

$$(Tf)(x) = x f(x), \quad D = \frac{d}{dx}.$$

Then $TD \neq DT$, *i.e.*, T and D are non-commutating operators.

(b) The positive square root of a self adjoint operator T commutes with every operator that commutes with T.

6.4 Normal Operators

The normal operators are the most general operators on Hilbert spaces for which an interesting structure theory is possible. In this section, we present a few elementary properties of normal operators. Recall that an operator $T \in B(H)$ is said to be normal if $T^* T = TT^*$, *i.e.*, if T commutes with its Hilbert-adjoint operator T^*.

6.4.1 Theorem. *Let $\mathcal{N}(H)$ be the set of all normal operators on H. Then:*

(a) $\mathcal{N}(H) \supset \mathcal{S}(H)$.

(b) $\mathcal{N}(H)$ *is closed under scalar multiplication.*

(c) $\mathcal{N}(H)$ *is a closed subset of $B(H)$.*

Proof. (a) Let $T \in \mathcal{S}(H)$. Then

$$T^* T = T^2 = T T^*$$

and so $T \in \mathcal{N}(H)$.

   (b) Let $\alpha \in \mathbb{K}$ and $T \in \mathcal{N}(H)$. Then

$$(\alpha T)(\alpha T)^* = (\alpha T)(\bar{\alpha} T^*)$$

$$= \alpha \bar{\alpha} T T^*$$

$$= \alpha \bar{\alpha} T^* T$$

$$= (\alpha T)^* (\alpha T)$$

$$\Rightarrow \qquad \alpha T \in \mathcal{N}(H).$$

   (c) Let $\{T_n\}$ be a convergent sequence in $\mathcal{N}(H)$, and let $T_n \to T$. Then $T_n^* \to T^*$. Therefore

$$\| T^* T - T T^* \| \leq \| T^* T - T_n^* T_n \| + \| T_n^* T_n - T_n T_n^* \|$$

$$+ \| T_n T_n^* - T T^* \|$$

$$= \| T^* T - T_n^* T_n \| + \| T_n T_n^* - T T^* \|$$

$$\to 0, \text{ as } n \to \infty$$

$$\Rightarrow \qquad T^* T - T T^* = 0$$

$\Rightarrow$ T is normal.

Hence, $\mathcal{N}(H)$ is a closed subset of $B(H)$. ∎

Keeping in view the results of the preceding theorem, it is natural to ask whether the set $\mathcal{N}(H)$ forms a closed subspace of the Banach space $B(H)$. Unfortunately, $\mathcal{N}(H)$ is not closed under usual addition of operators and hence $\mathcal{N}(H)$ is not a subspace of $B(H)$. Also, the product of two normal operators need not be a normal operator. However, we have the following:

6.4.2 Theorem. *Let $S, T \in \mathcal{N}(H)$ and suppose that $S T^* = T^* S$. Then, $S + T$ and ST are in $\mathcal{N}(H)$.*

Proof. First note that

$$S T^* = T^* S \quad \Leftrightarrow \quad S^* T = T S^*,$$

which may be seen by taking adjoints.

Now, we have

$$(S + T)(S + T)^* = (S + T)(S^* + T^*)$$
$$= S S^* + S T^* + T S^* + T T^*$$
$$= S^* S + T^* S + S^* T + T^* T \quad \text{(by assumption)}$$
$$= (S^* + T^*)(S + T)$$
$$= (S + T)^* (S + T)$$

$$\Rightarrow \qquad S + T \in \mathcal{N}(H).$$

Again

$$(S T)(S T)^* = (ST)(T^* S^*)$$
$$= S (T T^*) S^*$$
$$= S (T^* T) S^*$$
$$= T^* S S^* T \qquad \text{(by assumption)}$$
$$= T^* S^* S T$$
$$= (S T)^* (S T)$$

$$\Rightarrow \qquad S T \in \mathcal{N}(H). \text{ ∎}$$

We now give a couple of characterizations of normal operators.

6.4.3 Theorem. *Let H be a complex Hilbert space and $T \in B(H)$. Then, the following statements are equivalent:*

(a) *T is normal*

(b) *T^* is normal.*

(c) $\| T^* x \| = \| T x \|, \quad \forall x \in H.$

Proof. (a) $\Leftrightarrow$ (b). Use the fact that $T^{**} = T$.

(c) $\Leftrightarrow$ (a). Note that

$$\| T^* x \| = \| T x \| \quad \Leftrightarrow \quad <T^* x, T^* x> = <T x, T x>$$

$$\Leftrightarrow \quad <(T T^*)(x), x> = <(T^* T)(x), x>$$

$$\Leftrightarrow \quad <(T T^* - T^* T)(x), x> = 0, \quad \forall x \in H$$

$$\Leftrightarrow \qquad\qquad T^* T = T T^* \qquad\qquad \text{(Lemma 6.3.3)}$$

$$\Leftrightarrow \quad T \text{ is normal.}$$

6.4.4 Corollary. *If $T \in \mathcal{N}(H)$, then*

$$\| T^2 \| = \| T \|^2.$$

Proof. By Theorem 6.4.3, we have

$$\| T^2 x \| = \| T(T x) \| = \| T^*(T x) \| = \| (T^* T)(x) \|$$

$$\Rightarrow \quad \sup_{\substack{x \in H \\ \|x\|=1}} \| T^2 x \| = \sup_{\substack{x \in H \\ \|x\|=1}} \| (T^* T)(x) \|$$

$$\Rightarrow \qquad \| T^2 \| = \| T^* T \|.$$

But $\| T^* T \| = \| T \|^2$ (Theorem 6.2.8 (g)). Hence

$$\| T^2 \| = \| T \|^2. \quad \blacksquare$$

Recall that if $T \in B(H)$, then we can uniquely express T in the form $T = A + i B$, where A and B are in $S(H)$ (Theorem 6.3.7). This gives rise to another characterization of normal operators.

6.4.5 Theorem. *Let $T \in B(H)$ and suppose that $T = A + i B$, where $A, B \in S(H)$. Then, T is normal if and only if $A B = B A$.*

Proof. Note that $T^* = A - i B$. Then

$$\begin{cases} T^* T = (A - i B)(A + i B) = A^2 + B^2 + i(A B - B A) \\ T T^* = (A + i B)(A - i B) = A^2 + B^2 + i(B A - A B). \end{cases}$$

Thus, T is normal if and only if $T^* T = T T^*$

$$\Leftrightarrow \qquad AB - BA = BA - AB$$

$$\Leftrightarrow \qquad AB = BA. \; \blacksquare$$

Problems

24. If $T \in \mathcal{N}(H)$, prove that

$$\ker T = \ker T^* = (T(H))^{\perp}.$$

[*Hint:* Use Problem 9 (a) and (b) with $H_1 = H_2 = H$.]

25. If $T \in \mathcal{N}(H)$ and $T x = \alpha x$ for some $x \in H$ and $\alpha \in \mathbb{K}$, prove that

$$T^* x = \overline{\alpha} \; x.$$

[*Hint:* Apply the result of Problem 23 to $T - \alpha I$.]

26. Let $T \in B(H)$. If α and β are scalars such that $|\alpha| = |\beta|$, prove that

$$\alpha T + \beta T^* \in \mathcal{N}(H).$$

27. If $T \in \mathcal{N}(H)$ and f is a polynomial with complex coefficients, prove that

$$f(T) \in \mathcal{N}(H).$$

6.5 Unitary Operators

The class of unitary operators in $B(H)$ is the analogue of the unit circle in $\mathbb{C}$ characterized by the equation $z \overline{z} = \overline{z} z = 1$ (or $|z| = 1$). Recall that an operator $T \in B(H)$ is said to be unitary if $T^* T = T T^* = I$, where I is the identity operator.

6.5.1 Theorem. *Let S and T be unitary operators on a Hilbert space H. Then:*

(a) *T is isometric.*

(b) $\| T \| = 1$, *provided* $H \neq \{0\}$.

(c) *T is normal.*

(d) *S T is unitary.*

(e) $T^{-1} (= T^*)$ *is unitary.*

Proof. (a) We have

$$\| T x \|^2 = \; < T x, T x >$$

$$= <x, (T^* T)(x)>$$
$$= <x, I x>$$
$$= \| x \|^2, \quad \forall x \in H.$$

Hence

(14) $$\| T x \| = \| x \|, \quad \forall x \in H.$$

(b) It follows by taking supremum on both sides of (14) for $\| x \| = 1$.

(c) Obvious.

(d) It follows by observing that

$$(S T)^* (S T) = (T^* S^*)(ST)$$
$$= T^* (S^* S) T$$
$$= T^* T$$
$$= I,$$

and

$$(S T)(S T)^* = (S T)(T^* S^*)$$
$$= S (T T^*) S^*$$
$$= S S^*$$
$$= I.$$

(e) Since

$$T^* T = T T^* = I,$$

it immediately follows that $T^{-1} = T^*$.

Further, note that

$$T^* (T^*)^* = T^* T = I,$$

and

$$(T^*)^* T^* = T T^* = I.$$

Hence, T^* and consequently T^{-1} is unitary. ∎

6.5.2 Theorem. *Let $T \in B(H)$. Then, the following statements are equivalent:*

(a) $T^* T = I.$

(b) $<T x, T y> = <x, y>, \quad \forall x, y \in H.$

(c) $\| T x \| = \| x \|, \quad \forall x \in H.$

Proof. (a) $\Rightarrow$ (b). We have

$$(T^* T)(x) = x, \quad \forall x \in H$$

$$\Rightarrow \qquad <(T^*\,T)\,(x),\,y> = <x,\,y>, \quad \forall,\,x,\,y \in H$$

$$\Rightarrow \qquad <T\,x,\,T\,y> = <x,\,y>, \quad \forall\,x,\,y \in H.$$

(b) $\Rightarrow$ (c). Take $y = x$ in (b).

(c) $\Rightarrow$ (a). We have

$$\|\,T\,x\,\| = \|\,x\,\|, \quad \forall\,x \in H$$

$$\Rightarrow \qquad <T\,x,\,T\,x> = <x,\,x>$$

$$\Rightarrow \qquad <(T^*\,T)\,(x),\,x> = <x,\,x>$$

$$\Rightarrow \qquad <(T^*\,T - I)\,(x),\,x> = 0, \quad \forall\,x \in H$$

$$\Rightarrow \qquad T^*\,T = I \qquad\qquad \text{(Lemma 6.3.3).}$$

This completes the proof of the theorem. ∎

Remark. An operator $T \in B\,(H)$ satisfying any one of the equivalent conditions in Theorem 6.5.2 may not be unitary operator.

6.5.3 Example. Consider the right shift operator S_r on l^2 as in Example 6.2.5 (2). Clearly, for each $x = \{\alpha_i\} \in l^2$, we have

$$\|\,S_r\,(x)\,\| = \|\,x\,\|,$$

and so $S_r^*\,S_r = I$ (Theorem 6.5.2). But S_r is not surjective. Therefore, S_r has no inverse. We conclude that S_r is not an unitary operator.

However, we improve Theorem 6.5.2 as

6.5.4 Theorem. *Let $T \in B\,(H)$. Then, the following statements are equivalent:*

(a) *T is unitary.*

(b) *T is surjective and $<T\,x,\,T\,y> = <x,\,y>, \quad \forall\,x,\,y \in H$.*

(c) *T is surjective and*

$$\|\,T\,x\,\| = \|\,x\,\|, \qquad \forall\,x \in H.$$

Proof. The implications (a) $\Rightarrow$ (b) $\Rightarrow$ (c) are obvious. Assume that (c) be true. Then $T^*\,T = I$ (Theorem 6.5.2) and since T is clearly bijective, T^{-1} exists. Therefore

$$(T^*\,T)\,T^{-1} = T^{-1}$$

$$\Rightarrow \qquad T^* = T^{-1}$$

$$\Rightarrow \qquad T\,T^* = T\,T^{-1} = I.$$

Thus, $T^*\,T = T\,T^* = I$. Hence, T is unitary which is (a). This completes the proof of the theorem. ∎

6.5.5 Corollary. *An operator $T \in B(H)$ is unitary if and only if T is an isometric isomorphism of H onto itself.*

Remark. The unitary operators are precisely linear isomorphisms of H that also preserve the inner products.

6.5.6 Definition. An operator $T \in B(H)$ satisfying one of the equivalent conditions in Theorem 6.5.2 is called an *isometric operator.*

Note. An isometric operator is a generalization of an unitary operator.

6.5.7 Theorem. An isometric linear operator $T : H \to H$ which is not unitary maps the Hilbert space H onto a proper closed subspace of H.

Proof. Let $M = T(H)$ be the range of T. Then M is a subspace of H. Let $y \in \overline{M}$. Then, $\exists$ a sequence $\{y_n\}$ in M such that $y_n \to y$. Set $y_n = Tx_n$. Since T is isometric, we have

$$\| x_m - x_n \| = \| T(x_m - x_n) \|$$

$$= \| Tx_m - Tx_n \|$$

$$= \| y_m - y_n \|$$

$$\to 0, \text{ as } m, n \to \infty.$$

This shows that $\{x_n\}$ is a Cauchy sequence in H. But H is complete. Therefore, $y = Tx \in M$. Hence, M is closed. Further, if T is not unitary, then $M \neq H$ and the result follows. ∎

Remark. In general, an isometric operator maps a subspace H_1 of H into a subspace H_2 of H with preservation of the inner product (and hence the norm). In case $H_1 = H = H_2$, the isometric operator becomes unitary operator.

Problems

28. Let $\eta(t)$ be a Lebesgue measurable function on $[a, b]$ such that $|\eta(t)| = 1$ *a.e.* Prove that the operator T defined on $L^2[a, b]$ by $(Tf)(t) = \eta(t)f(t)$ is unitary.

29. Given a real number r and let $T : L^2[a, b] \to L^2[a, b]$ be a bounded linear operator given by

$$Tf = f_r,$$

where $f_r(t) = f(t + r)$. Prove that T is unitary.

30. Consider $T = 2iI$, where I is the identity operator on a Hilbert space H. Prove that the operator T is normal but it is neither self-adjoint nor unitary.

31. Let $\mathcal{U}(H)$ be the class of all unitary operators on a Hilbert space H. Prove that:

(a) $T, S \in \mathcal{U}(H) \implies TS \in \mathcal{U}(H)$.

(b) $T(SU) = (TS)U, \quad \forall\, T, S, U \in \mathcal{U}(H)$.

(c) $TI = T = IT, \qquad \forall\, T \in \mathcal{U}(H)$.

(d) For each $T \in \mathcal{U}(H), \exists\, T^{-1} \in \mathcal{U}(H)$ such that $TT^{-1} = I = T^{-1}T$.

(e) $T, S \in \mathcal{U}(H) \implies TS = ST$.

[In fact, $\mathcal{U}(H)$ forms a commutative group of unitary operators on H.]

32. An operator $S \in B(H)$ is said to be *unitarily equivalent* to the operator $T \in B(H)$ if $\exists$ a unitary operator U such that $T = U^* S U$. We write $S \cong T$ to say that S is unitarily equivalent to T. Prove that:

(a) $\cong$ is an equivalence relation in $B(H)$.

(b) $S \cong T \implies S^* \cong T^*$.

33. Prove that an operator $T \in B(H)$ is unitary if and only if $\{Te_\lambda\}_{\lambda \in \Lambda}$ is a complete orthonormal set whenever $\{e_\lambda\}_{\lambda \in \Lambda}$ is a complete orthonormal set in H.

6.6 Orthogonal Projection Operators

In Section 3.6, we have examined the properties of the class of projection operators on normed spaces (and Banach spaces). We recall that a linear operator P on a linear space is called a projection operator (algebraic projection) if it is idempotent, *i.e.,* $P^2 = P$. In a normed space X, the natural projection operators to be considered are the bounded projection operators. We note that for a bounded projection operator P on a normed space X, the range M and the null space N of P are both closed subspaces of X such that $X = M \oplus N$. However, if X is a Banach space, and M, N are closed subspaces of X such that $X = M \oplus N$, then there is a unique bounded projection operator P on X whose range and null spaces are M and N, respectively (Theorem 3.6.6).

In Hilbert spaces, we have in addition an orthogonality relation; namely, the projection theorem (Theorem 5.4.3) which gives us a natural preferred way of forming direct sum decompositions of the space. The projection theorem leads to the concept of an orthogonal projection operator P on a Hilbert space H (Definition 5.4.4).

According to Observations 5.4.5, an orthogonal projection operator P on H is a bounded linear operator which is idempotent and is such that $H = M \oplus M^{\perp}$, where M and $M^{\perp}$ are the range and the null space of P. In Hilbert spaces, we are thus naturally directed to examine those bounded linear projections whose range and null spaces are mutually orthogonal subspaces. We have the following important property.

6.6.1 Theorem. *Let P be a bounded linear projection operator on a complex Hilbert space H with range M and the null space N. Then, $M \perp N$ if and only if P is self-adjoint, and in this case, $N = M^{\perp}$.*

Proof. By Theorem 3.6.6, each $z \in H$ can be uniquely expressed as

$$z = x + y, \quad x \in M \text{ and } y \in N.$$

Assume first that $M \perp N$. Then

$$<x, y> = 0, \quad \forall\, x \in M, y \in N.$$

Now, we have

$$<P^{*}z, z> = <z, Pz>$$

$$= <x + y, x>$$

$$= <x, x>$$

$$= <x, x + y>$$

$$= <Pz, z>$$

$$\Rightarrow \qquad P^{*} = P. \quad \text{(Lemma 6.3.3)}$$

Conversely, let P be self-adjoint. Then

$$<x, y> = <Px, y>$$

$$= <x, P^{*}y>$$

$$= <x, Py>$$

$$= 0, \quad \forall\, x \in M, y \in N$$

$$\Rightarrow \qquad M \perp N. \ \blacksquare$$

6.6.2 Corollary. *A bounded linear operator $P : H \to H$ is an orthogonal projection on H if and only if P is self- adjoint and idempotent.*

6.6.3 Definition. A projection operator on a Hilbert space H whose range and null spaces are orthogonal is called *orthogonal* (or *perpendicular*) *projection operator.*

Note. In fact, it is the same concept as given in Definition 5.4.4 but in a different context.

6.6.4 Theorem. *Let P be an orthogonal projection operator on a Hilbert space H. Then:*

(a) *P is positive.*

(b) $< P x, x > = \| P x \|^2, \quad \forall\, x \in H.$

(c) $\| P \| = 1$ *unless $P = O$.*

Proof. We have

$$< P x, x > = < P^2 x, x >$$
$$= < P x, P x >$$
$$= \| P x \|^2 \geq 0. \quad \forall\, x \in H.$$

This proves (a) and (b).

To prove (c), note that, $I - P$ is also an orthogonal projection operator. Therefore

$$< (I - P)\,(x), x > \geq 0$$

$$\Rightarrow \qquad < P x, x > \leq \| x \|^2$$

$$\Rightarrow \qquad \| P x \|^2 \leq \| x \|^2, \quad \forall\, x \in H$$

$$\Rightarrow \qquad \| P \| \leq 1.$$

On the other hand, for $x \in H$ such that $P x \neq 0$, we have

$$\| P x \| = \| P\,(P x) \| \leq \| P \| \, \| P x \|$$

$$\Rightarrow \qquad \| P \| \geq 1.$$

Hence, we conclude that $\| P \| = 1$ unless $P = O$ which is (c). ∎

Note. The only projection operators considered in the theory of Hilbert spaces are those which are orthogonal, we may call orthogonal projection operators on a Hilbert space H, simply as projections on H. If the range of such a projection is $M \subset H$, we refer it to as a projection on M.

6.6.5 Theorem. *Let P and Q be projections on closed subspaces M and N of a Hilbert space H, respectively. Then, the following statements are equivalent:*

(a) $P Q = Q P$.

(b) $P Q$ *is a projection.*

(c) $Q P$ *is a projection.*

Proof. (a) $\Leftrightarrow$ (b). Note that

$$(P Q)^* = Q^* P^* = Q P = P Q$$

and

$$(P Q)^2 = P Q P Q = P P Q Q = P Q.$$

Hence, $P Q$ is a projection (Corollary 6.6.2)

Let (b) be true. Then

$$Q P = Q^* P^* = (P Q)^* = P Q$$

which is (a).

(a) $\Leftrightarrow$ (c). Similarly. ∎

6.6.6 Corollary. *If $P Q$ is a projection, then its range is $M \cap N$.*

Proof. Note that

$$\begin{cases} (P Q)(H) = P(Q(H)) \subset P(H) = M \\ (P Q)(H) = (Q P)(H) \subset Q(H) = N \end{cases}$$

$$\text{(Theorem 6.6.5 with (b) } \Rightarrow \text{ (a))}$$

$$\Rightarrow \qquad (P Q)(H) \subset M \cap N.$$

Moreover, if $x \in M \cap N$, then

$$x = Q x = (P Q)(x)$$

$$\Rightarrow \qquad x \in (P Q)(H).$$

Consequently, $M \cap N \subset (P Q)(H)$. Hence, the result follows. ∎

6.6.7 Definition. Two projections P and Q on a Hilbert space H are said to be *orthogonal if $P\,Q = O$.*

Note. $P\,Q = O \iff Q\,P = O$.

6.6.8. Theorem. *Let P and Q be projections on closed subspaces M and N of a Hilbert space H, respectively. Then, P and Q are orthogonal if and only if $M \perp N$.*

Proof. Let $P\,Q = O$. Then

$$< x, y > = < P\,x,\, Q\,y >$$

$$= < (Q\,P)\,(x),\, y >$$

$$= 0, \quad \forall\, x \in M,\, y \in N$$

$$\Rightarrow \qquad M \perp N.$$

Conversely, assume that $M \perp N$. Clearly, $N \subset M^{\perp}$. Then

$$Q\,x \in M^{\perp}$$

$$\Rightarrow \qquad (P\,Q)\,(x) = 0, \quad \forall\, x \in H.$$

Hence $\qquad P\,Q = O.\ \blacksquare$

Recall that the set $B\,(H)$ of all bounded linear operators on a Hilbert space H is a Banach algebra. The subset of all projections does not even form a linear space. For example, if $P \neq O$ is a projection, then clearly $\lambda\,P$ is a projection if and only if $\lambda = 1$ or $\lambda = 0$. Thus, the set of all projections does not possess nice algebraic or topological structure. However, we have observed in Theorem 6.6.5 that the product $P\,Q$ of two projections P and Q is a projection if and only if $P\,Q = Q\,P$. We shall now discuss the question under what circumstances the sum or difference of two projections is again a projection.

6.6.9 Theorem. *The sum of two projections P_M and P_N is a projection if and only if $P_M\,P_N = 0$. In that case, $P_M + P_N = P_{M \oplus N}$.*

Proof. Let $P = P_M + P_N$ be a projection on H. Then, $P^2 = P$. It gives

$$(P_M + P_N)\,(P_M + P_N) = P_M + P_N$$

$$\Rightarrow \qquad P_M\,P_N + P_N\,P_M = O$$

$$\Rightarrow \qquad P_M\,(P_M\,P_N + P_N\,P_M) = P_M\,O = O$$

$$\Rightarrow \qquad P_M P_N + P_M P_N P_M = O$$

$$\Rightarrow \qquad (P_M P_N + P_M P_N P_M)\, P_M = O\, P_M = O$$

$$\Rightarrow \qquad 2\, P_M P_N P_M = O$$

Hence $\qquad P_M P_N = O.$

Conversely, suppose $P_M P_N = O$. Then, also $P_N P_M = O$. Now

$$P^2 = (P_M + P_N)^2 = P_M + P_N = P$$

and $\qquad P^* = (P_M + P_N)^* = P_M + P_N = P.$

Hence, P is a projection on H.

Further, since for all $x \in H$, we have

$$Px = P_M x + P_N x$$

with $P_M x \in M$ and $P_N x \in N$, it follows that

$$P = P_{M \oplus N}. \ \blacksquare$$

6.6.10 Definition. Let H be a Hilbert space and let $T \in B(H)$. A closed subspace M of H is said to be *invariant* under T if $T(M) \subset M$.

Remark. If a closed subspace M of H is invariant under T, then $T|_M$ can be regarded as an operator on M alone and the action of T on $H - M$ can be ignored.

6.6.11 Theorem. *Let H be a Hilbert space and let $T \in B(H)$. Then, a closed subspace M of H is invariant under T if and only if $M^\perp$ is invariant under T^*.*

Proof. M is invariant under T

$$\Rightarrow \qquad T(M) \subset M$$

$$\Rightarrow \qquad <Tx, y> = 0, \quad \forall\, x \in M, y \in M^\perp$$

$$\Rightarrow \qquad <x, T^* y> = 0, \quad \forall\, x \in M, y \in M^\perp$$

$$\Rightarrow \qquad T^*(M^\perp) \subset M^\perp$$

$\Rightarrow$ $M^\perp$ is invariant under T^*.

The converse follows by using the fact that $M^{\perp\perp} = M$ and $T^{**} = T$. $\blacksquare$

6.6.12 Theorem. *Let H be a Hilbert space, $T \in B(H)$ and let P be the projection on a closed subspace M of H. Then, M is invariant under the operator T if and only if $TP = PTP$.*

Proof. Let M be invariant under T. Then

$$(T P)(x) \in M, \qquad \forall\, x \in H$$

$$\Rightarrow \qquad (P T P)(x) = (T P)(x), \quad \forall\, x \in H$$

$$\Rightarrow \qquad P T P = T P.$$

Conversely, let $T P = P T P$. Then, for $x \in M$

$$T x = (T P)(x) = (P T P)(x) \in M.$$

Hence, $T(M) \subset M$ and therefore M is invariant under T. ∎

6.6.13 Definition. Let H be a Hilbert space and let $T \in B(H)$. A closed subspace M of H is said to reduce T (or T is reduced by M) if both M and $M^{\perp}$ are invariant under T.

In view of Definition 6.6.13 and Theorem 6.6.12, the following result follows immediately.

6.6.14 Theorem. *Let P be the projection on a closed subspace M of H and $T \in B(H)$. Then, M reduces T if and only if $T P = P T$.*

Problems

34. Let $P_{M_1}, P_{M_2}, \ldots, P_{M_n}$ be the projections on a Hilbert space H. Prove that

$$P = \sum_{i=1}^{n} P_{M_i} \text{ is a projection if and only if } P_{M_i} P_{M_j} = O, \forall\, i, j = 1, 2, \ldots,$$

$n, i \neq j$. In that case $P = P_M$, where $M = M_1 \oplus M_2 \oplus \ldots \oplus M_n$.

35. A projection P_M on a Hilbert space H is said to be a *part of a projection P_N* on H if $M \subset N$. If P_M and P_N be the projections on H, prove that the following statements are equivalent:

(a) P_M is a part of P_N.

(b) $P_N P_M = P_M$.

(c) $P_M P_N = P_M$.

(d) $P_N - P_M$ is a projection.

(e) $< (P_N - P_M)(x), x > \geq 0, \qquad \forall\, x \in H.$

(f) $\| P_M x \| \leq \| P_N x \|, \qquad \forall\, x \in H.$

7

Introduction to Spectral Theory

<u>CHAPTER</u>

While solving systems of linear algebraic equations, differential equations or integral equations, we come across the problems related to inverse operations. In particular, if $T \in B(X)$, where X is a normed space and λ a scalar, we are generally concerned with the inverse of the linear operator $T - \lambda I$. The spectral theory mainly deals with a systematic study of the inverse operators, their general properties and their relation to the original operators. Theory of spectral analysis is one of the most powerful branches of modern functional analysis and its applications. The spectral theory in finite dimensional spaces is much more simpler than in infinite dimensional spaces. Also, the finite dimensional spectral theory is of great practical importance, and furthermore, this motivates the part of the general setting of spectral theory in infinite dimensional spaces.

In this chapter we shall discuss the eigenvalues and spectrums of linear operators and some basic concepts of spectral theory. For further study, an interested reader may refer to advanced texts on the subject.

7.1 Eigenvalues of a Linear Operator

The spectral theory includes the study of eigenvalue problems. In this section, we discuss the eigenvalues, eigenvectors and eigenspaces of linear operators on a linear space.

Let X be an arbitrary linear space over the field $\mathbb{K}$ ($\mathbb{R}$ or $\mathbb{C}$). Note that the trivial subspace $\{0\}$ of X is always invariant under a linear operator T on X and so it is hardly interesting. Then the smallest non-trivial invariant subspace of X may be a one dimensional one. In this case, $\exists$ a non-zero vector $x \in X$ which generates the one dimensional invariant subspace under T and has the property that $T x = \lambda x$, for some $\lambda \in \mathbb{K}$.

334

7.1.1 Definition. Let X be a linear space over the field $\mathbb{K}$ and $T \in L(X)$. A scalar $\lambda \in \mathbb{K}$ is said to be an *eigenvalue* (*characteristic value* or *proper value*) of T if $\exists$ a non-zero vector $x \in X$ such that $Tx = \lambda x$.

Clearly, λ is an eigenvalue of T if and only if the null space of $T - \lambda I$ is non-trivial.

7.1.2 Definition. Let X be a linear space over the field $\mathbb{K}$ and $T \in L(X)$. A vector $x \in X$ is said to be an *eigenvector* (*characteristic vector* or *proper vector*) of T if

(i) $x \neq 0$, and

(ii) $Tx = \lambda x$, for some $\lambda \in \mathbb{K}$.

Remarks:

1. Given $T \in L(X)$, an eigenvector x is uniquely associated with an eigenvalue λ because if $Tx = \mu x$, $\mu \in \mathbb{K}$, then $(\lambda - \mu) x = 0$, and hence $\lambda = \mu$. Thus to each eigenvector there corresponds precisely one eigenvalue.

2. Each eigenvalue λ of $T \in L(X)$ may have one or more eigenvectors associated with λ. For example, if $T = cI$, then every non-zero vector of X is an eigenvector of T but the only eigenvalue of T is c.

3. If $X = \{0\}$, then any $T \in L(X)$ has no eigenvectors. Therefore, we assume that $X \neq \{0\}$.

7.1.3 Example. Let $T : \mathbb{R}^2 \to \mathbb{R}^2$ be the linear operator defined by

$$T\left(\begin{bmatrix} a_1 \\ a_2 \end{bmatrix}\right) = \begin{bmatrix} a_2 \\ a_1 \end{bmatrix}.$$

Then

$$T\left(\begin{bmatrix} a \\ a \end{bmatrix}\right) = \begin{bmatrix} a \\ a \end{bmatrix} \text{ and } T\left(\begin{bmatrix} a \\ -a \end{bmatrix}\right) = -\begin{bmatrix} a \\ -a \end{bmatrix}.$$

Thus a vector of the form $\begin{bmatrix} a \\ a \end{bmatrix}$, where a is any non-zero real number, is an eigenvector of T associated with the eigenvalue $\lambda = 1$ and a vector of the form $\begin{bmatrix} a \\ -a \end{bmatrix}$ is an eigenvector of T associated with the eigenvalue $\lambda = -1$.

The following example shows that a linear operator T on a linear space may not have any eigenvalue.

7.1.4 Example. Let $T : \mathbb{R}^2 \to \mathbb{R}^2$ be the linear operator performing a counter clockwise rotation by $90°$. Note that, the only vector x mapped by T into a multiple of itself is the zero vector. Hence, T has neither eigenvalues nor eigenvectors.

7.1.5 Theorem. *Let X be a linear space over the field $\mathbb{K}$ and $T \in L(X)$. Let λ_1, λ_2, ..., λ_n be distinct eigenvalues of T. Then, the eigenvectors $x_1, x_2, ..., x_n$ of T associated with $\lambda_1, \lambda_2, ... , \lambda_n$, respectively, constitute a linearly independent set.*

Proof. Let, if possible, the set $\{x_1, x_2, ..., x_n\}$ be linearly dependent. Let x_m be the first vector which is a linear combination of the predecessors. Then

$$(1) \qquad x_m = \alpha_1 x_1 + \alpha_2 x_2 + ... + \alpha_{m-1} x_{m-1}$$

for certain scalars $\alpha_1, \alpha_2, ..., \alpha_{m-1}$. Clearly, the set $\{x_1, x_2, ..., x_{m-1}\}$ is linearly independent. Applying $T - \lambda_m I$ on both sides of (1), we obtain

$$(T - \lambda_m I)(x_m) = \sum_{i=1}^{m-1} \alpha_i \, (T - \lambda_m I)(x_i)$$

$$(2) \qquad\qquad = \sum_{i=1}^{m-1} \alpha_i \, (\lambda_i - \lambda_m) \, x_i \, .$$

Since x_m is an eigenvector corresponding to λ_m, the left hand side of (2) is zero. But the vectors $\{x_1, x_2, ..., x_{m-1}\}$ in the summation on the right hand side form a linearly independent set. We must have

$$\alpha_i \, (\lambda_i - \lambda_m) = 0$$

$$\Rightarrow \qquad\qquad \alpha_i = 0, \quad (i = 1, 2, ..., m-1).$$

$$\text{since} \quad \lambda_i \neq \lambda_m, \; 1 \leq i \leq m-1.$$

Thus in view of (1), $x_m = 0$. But x_m cannot be zero since x_m is an eigenvector. This is a contradiction. Hence, the theorem follows. $\blacksquare$

7.1.6 Definition. Let X be a linear space over the field $\mathbb{K}$ and $T \in L(X)$. Let λ be an eigenvalue of T. The null space of $T - \lambda I$ is called the *eigenspace* of T corresponding to λ and is denoted by $M_\lambda(T)$, or simply by M_λ, *i.e.,*

$$M_\lambda = \{x \in X : T x = \lambda x\}.$$

Briefly, the set M_λ is known as the *λ-space* of T.

The set M_λ consists of all eigenvectors of T corresponding to λ together with the zero vector. In fact, M_λ is different from $\{0\}$ if and only if λ is an eigenvalue of T.

7.1.7 Theorem. *Let X be a linear space over the field $\mathbb{K}$ and $T \in L(X)$. Then, $\lambda \in \mathbb{K}$ is an eigenvalue of T if and only if the operator $T - \lambda I$ is not one-to-one.*

Proof. By Definition 7.1.1, λ is an eigenvalue of T if and only if $\exists\, a\, 0 \neq x \in X$ such that

$$(T - \lambda I)(x) = 0.$$

Equivalently, λ is an eigenvalue of T if and only if $T - \lambda I$ has eigenvalue 0 (Note that $0 = 0\,x$). Since we also have

$$(T - \lambda I)\, 0 = 0,$$

it follows that $T - \lambda I$ is not one-to-one.

Conversely, we assume that $T - \lambda I$ is not one-to-one. Then, $\exists$ two different vectors $x_1, x_2 \in X$ such that

$$(T - \lambda I)(x_1) = (T - \lambda I)(x_2)$$

$$\Rightarrow \qquad (T - \lambda I)(x_1 - x_2) = 0$$

$$\Rightarrow \qquad T x = \lambda x, \text{ where } x = x_1 - x_2.$$

Since $x \neq 0$ and $x \in X$, it follows that λ is an eigenvalue of T with corresponding eigenvector $x = x_1 - x_2$. $\blacksquare$

Recall that if $T \in L(X)$, where X is a finite dimensional linear space, then we can define the matrix of T related to any ordered basis for X (Theorem 1.6.34).

7.1.8 Theorem. *Let X be a finite dimensional linear space and $T \in L(X)$. Then, the eigenvalues of T are precisely the distinct roots of the equation*

$$\det (T - \lambda I) = 0.$$

Proof. The scalar λ is an eigenvalue of T

$$\Leftrightarrow \quad \exists\, 0 \neq x \in X \text{ such that } T x = \lambda x$$

$$\Leftrightarrow \quad (T - \lambda I)(x) = 0$$

$\Leftrightarrow$ $T - \lambda\, I$ is not one-to-one (Theorem 7.1.7)

$\Leftrightarrow$ $T - \lambda\, I$ is a singular matrix

$\Leftrightarrow$ $\det (T - \lambda\, I) = 0.$ ∎

7.1.9 Definition. The equation

$$\det (T - \lambda\, I) = 0$$

is called the *characteristic equation* of T.

Remark. If the matrix $[T]$ of T relative to any basis for X is $[\alpha_{ij}]$, then the characteristic equation of T is given by

$$\begin{vmatrix} \alpha_{11} - \lambda & \alpha_{12} & \cdots\cdots & \alpha_{1n} \\ \alpha_{21} & \alpha_{22} - \lambda & \cdots\cdots & \alpha_{2n} \\ \vdots & & & \\ \alpha_{n1} & \alpha_{n2} & \cdots\cdots & \alpha_{nn-\lambda} \end{vmatrix} = 0$$

The roots of the characteristic equation of T give all the eigenvalues of T.

Problems

1. Let X be a linear space over the field $\mathbb{K}$ and $T \in L(X)$. If λ be an eigenvalue of T, prove that the eigenspace M_λ is a non-trivial subspace of X.

2. Let a square matrix A have eigenvalues λ_i, $(i = 1, 2, ..., n)$. Prove that:

(a) A is non-singular if and only if $\lambda_i \neq 0$, $(i = 1, 2, ..., n)$.

(b) If A^{-1} exists, it has eigenvalues λ_i^{-1}, $(i = 1, 2, ..., n)$.

(c) kA has eigenvalues $k\,\lambda_i$, $(i = 1, 2, ..., n)$.

(d) A^m has eigenvalues λ_i^m, $(i = 1, 2, ..., n)$, where m is a positive integer.

7.2 The Spectrum of a Bounded Linear Operator

It is well known that the complex field is algebraically closed while the real field is not, *i.e.,* every non-constant polynomial with complex coefficients has a complex zero, but there are non-constant polynomials with real coefficients that have no real zeros. Due to this fundamental difference between real linear spaces and complex linear spaces, the spectral theory becomes more interesting if we take complex scalars and, furthermore, we can utilize much of the machinery of the classical theory of functions of a complex variable.

Let $X \neq \{0\}$ be a complex normed space and $T \in B(X)$. In Example 7.1.4, we have seen that T may not have any eigenvalue. In fact, if $\lambda \in \mathbb{C}$ is not an eigenvalue of T, then the operator $T - \lambda I$ is one-to-one (Theorem 7.1.7) and, therefore, the inverse operator $(T - \lambda I)^{-1}$ is defined on the range set $(T - \lambda I)(X)$. There are still two possibilities that something goes wring with this inverse :

(i) $(T - \lambda I)(X) \neq X$.

(ii) $(T - \lambda I)^{-1}$ might not be bounded.

If everything goes well, *i.e.,* $T - \lambda I$ is one-to-one and $(T - \lambda I)^{-1}$ is a bounded linear operator defined on the entire space X, then λ is called a *regular value* for T. In all other cases λ is said to belong to the *spectrum of T.*

7.2.1 Definition. Let $X \neq \{0\}$ be a complex normed space and $T \in B(X)$. A number $\lambda \in \mathbb{C}$ is said to be *regular value* of T if

(i) $T - \lambda I$ is one-to-one,

(ii) $(T - \lambda I)^{-1}$ is defined on X, and

(iii) $(T - \lambda I)^{-1}$ is bounded.

(In other words, λ is a regular value of T if $(T - \lambda I)^{-1}$ exists and is bounded on X.)

In case, $(T - \lambda I)^{-1}$ is not bounded, the condition (*ii*) may be replaced by

(*ii*)′ $(T - \lambda I)^{-1}$ is defined on a set which is dense in X.

Note. When $(T - \lambda I)^{-1}$ is bounded, the conditions (*ii*) and (*ii*)′ are equivalent.

7.2.2 Definition. Let $X \neq \{0\}$ be a complex normed space and $T \in B(X)$.

(a) The set of all regular values of T, denoted by $\rho(T)$, is called the *resolvent set* of T, *i.e.,* $\rho(T) = \{\lambda \in \mathbb{C} : \lambda$ is a regular value of $T\}$.

(b) The complement of $\rho(T)$ in $\mathbb{C}$, denoted by $\sigma(T)$, is called the *spectrum* of T, *i.e.,* $\sigma(T) = \{\lambda \in \mathbb{C} : \lambda$ is not a regular value of $T\}$.

(c) A complex number $\lambda \in \sigma(T)$ is called a *spectral value* of T.

(d) Let $\lambda \in \rho(T)$. The function

$$\lambda \to (T - \lambda I)^{-1}$$

is called the *resolvent operator* (or, simply, *resolvent*) of T. It is denoted by $R_\lambda(T)$, *i.e.,*

$$R_\lambda (T) = (T - \lambda I)^{-1}.$$

We may write R_λ instead of $R_\lambda (T)$ if it is clear to what operator T we refer to.

Note. In the preceding definitions, we have assumed that the space X is complex. However, these definitions also hold when X is a real normed space. But, the concepts are not very useful in this case because $\sigma (T)$ may be an empty set even when X is finite dimensional (Example 7.1.4). On the other hand, we shall shortly see that $\sigma (T) \neq \phi$ when X is a complex normed space. Throughout, we shall assume X to be a complex normed space unless specified otherwise.

Although the regular values of $T \in B (X)$ behave nicely, they are not of much interest because they have the operator cold. On the other hand, if $\lambda \in \sigma (T)$, then with $R_\lambda (T)$ the operator reacts algebraically and thus gives rise to a beautiful spectral theory and such an operator recognizes λ as an integral part of its own. The spectrum $\sigma (T)$ is partitioned into three disjoint sets:

(a) The *point spectrum* (or *discrete spectrum*) of T, denoted by $\sigma_p (T)$, the set of all $\lambda \in \sigma (T)$ such that $R_\lambda (T)$ does not exist.

(b) The *continuous spectrum* of T, denoted by $\sigma_c (T)$, the set of all $\lambda \in \sigma (T)$ such that $R_\lambda (T)$ exists and satisfies $(ii)'$ but not (iii); in other words, $R_\lambda (T)$ exists, unbounded and the domain of it is dense in X.

(c) The *residual spectrum* of T, denoted by $\sigma_r (T)$, the set of all $\lambda \in \sigma (T)$ such that $R_\lambda (T)$ exists (may be bounded or unbounded) but does not satisfy $(ii)'$ (and hence not (ii)).

Remark. Some of the sets $\sigma_p (T)$, $\sigma_c (T)$, $\sigma_r (T)$ may be empty. For instance, in a finite dimensional case, we have

$$\sigma_c (T) = \phi = \sigma_r (T),$$

and in that case $\qquad \sigma_p(T) = \sigma (T).$

7.2.3 Observations

(i) The four sets $\rho (T)$, $\sigma_p (T)$, $\sigma_c (T)$ and $\sigma_r (T)$ are disjoint.

(ii) $\qquad \mathbb{C} = \rho (T) \cup \sigma (T)$

$$= \rho (T) \cup \sigma_p (T) \cup \sigma_c (T) \cup \sigma_r (T).$$

(iii) The resolvent operator $R_\lambda (T)$, if exists, is linear.

(iv) $R_\lambda(T) : \mathscr{R}(T_\lambda) \to \mathscr{D}(T_\lambda)$ exists if and only if

$$T_\lambda x = 0 \quad \Rightarrow \quad x = 0.$$

(Equivalently, $\mathscr{N}(T_\lambda) = \{0\}$). Here $T_\lambda = T - \lambda I$.

(v) If $T_\lambda x = (T - \lambda I)(x) = 0$ for some $x \neq 0$, then in view of (iv), $R_\lambda(T)$ does not exist and as such λ is not a regular value of T ; meaning thereby that $\lambda \notin \rho(T)$ and thus $\lambda \in \sigma(T)$. Furthermore, more precisely, we have $\lambda \in \sigma_p(T)$. Conversely, if $\lambda \in \sigma_p(T)$, then $T_\lambda x = 0$ for some $x \neq 0$. In summary, λ is an eigenvalue of T (Definition 7.1.1) if and only if $\lambda \in \sigma_p(T)$.

We give below a complete description of $\sigma(T)$ in case when X is finite dimensional.

7.2.4 Theorem. *Let dim $X = n$ and $T \in B(X)$. Then, $\sigma(T)$ is a non-empty finite subset of the complex plane $\mathbb{C}$ consisting of not more than n points, each of which is an eigenvalue of T.*

Proof. Since X is finite dimensional, T is bounded and

$$\sigma(T) = \{\lambda \in \mathbb{C} : T - \lambda I \text{ is not invertible}\}$$

$$= \{\lambda \in \mathbb{C} : T - \lambda I \text{ is not one-to-one}\}$$

$$= \{\lambda \in \mathbb{C} : (T - \lambda I)(x) = 0, \text{ for some } 0 \neq x \in X\}$$

$$= \sigma_p(T).$$

Therefore $\sigma(T)$ consists entirely of eigenvalues of T. By Theorem 1.6.34, we can represent T by $n \times n$ matrix. Let $[T] = [a_{ij}]$ be the matrix of order $n \times n$ of T relative to any basis for X. Now

$$\lambda \in \sigma(T) \Leftrightarrow \exists\, 0 \neq x \in X \text{ such that } (T - \lambda I)(x) = 0$$

$$\Leftrightarrow T - \lambda I \text{ is a singular matrix}$$

$$\Leftrightarrow \det(T - \lambda I) = 0.$$

In other words, the elements of $\sigma(T)$ are entirely those scalars λ which are the roots of the characteristic equation

$$\begin{vmatrix} a_{11} - \lambda & a_{12} & \cdots & a_{1n} \\ a_{21} & a_{22} - \lambda & \cdots & a_{2n} \\ \vdots & & & \\ a_{n1} & a_{n2} & \cdots & a_{nn-\lambda} \end{vmatrix} = 0.$$

The determinant here is a polynomial of degree n in complex variable λ with complex coefficients. By the fundamental theorem of algebra, the above polynomial equation must have exactly n complex roots. Hence, there will be at most n different elements in $\sigma(T)$ as some of the roots may be repeated. This completes the proof of the theorem. ∎

The spectrum of a bounded linear operator T on an infinite dimensional normed space is not so simple to describe. The following example shows that in infinite dimensional spaces $\sigma(T)$ may not be a finite set.

7.2.5 Example. Take $X = l^2$ and consider the linear operator $T : l^2 \to l^2$ defined by

$$T(\xi_1, \xi_2, ...) = (\xi_2, \xi_3, ...), \ \{\xi_i\} \in l^2. \qquad \text{(Example 6.2.5 (2))}$$

Then T is bounded on X. For any $\lambda \in \mathbb{C}$ with $|\lambda| < 1$, define the sequence $x_\lambda = \{\xi_n\}$, where

$$\xi_n = \begin{cases} 1, & n = 1 \\ \lambda^{n-1}, & n \neq 1. \end{cases}$$

Clearly, $x_\lambda \in l^2$, $x_\lambda \neq 0$ and $T x_\lambda = \lambda x_\lambda$. Thus, it follows that λ is an eigenvalue of T. Hence

$$\{\lambda \in \mathbb{C} : |\lambda| < 1\} \subset \sigma(T). \ ∎$$

It may be noted that in case of a linear operator on a finite dimensional space, every spectrum value is an eigenvalue. However, if X is infinite dimensional, then T can have spectral values which are not eigenvalues.

7.2.6 Example. Take $X = l^2$ and consider the linear operator $T : l^2 \to l^2$ defined by

$$T(\xi_1, \xi_2, ...) = (0, \xi_1, \xi_2, ...), \ \{\xi_i\} \in l^2. \quad \text{(Example 6.2.5 (2))}$$

Then:

(a) T is bounded.

(b) The operator

$$R_0(T) = T^{-1} : T(X) \to X$$

exists. Indeed, it is the left shift operator given by

$$T(\xi_1, \xi_2, ...) = (\xi_2, \xi_3, ...).$$

(c) Note that $T(X)$ is the subspace of X consisting of all $\{\xi_i\} \in l^2$ with $\xi_1 = 0$ and as such $T(X)$ is not dense in X. Consequently, $\lambda = 0$ is a spectral value of T.

(d) $\lambda = 0$ is not an eigenvalue of T since $R_0(T)$ exists.

Problems

3. Let I be the identity operator on a normed space X. Find:

(a) The eigenvalues of I.

(b) The eigenspaces of I.

(c) $\sigma(I)$.

(d) $R_\lambda(I)$.

(e) $\rho(I)$.

4. Let $T : l^1 \to l^1$ be defined by

$$T(\{x_n\}) = \left\{\frac{x_n}{n}\right\}, \quad \{x_n\} \in l^1.$$

Show that $\rho(T) \subset \mathbb{C} - \{0, 1, \frac{1}{2}, \frac{1}{3}, ..., \frac{1}{n}, ...\}$.

5. Determine $\sigma_p(T)$, $\sigma_c(T)$ and $\sigma_r(T)$ in each of the following cases.

(a) $T : l^2 \to l^2$ be defined by

$$T(\{\xi_i\}) = \{\alpha_i \xi_i\}, \quad \{\xi_i\} \in l^2$$

where $\{\alpha_i\} \subset \mathbb{C}$ is a abounded sequence.

(b) $T : C[0, 1] \to C[0, 1]$ be defined by

$$Tx(t) = t\,x(t), \quad t \in [0, 1], x \in C[0, 1].$$

(c) $T : C[0, 1] \to C[0, 1]$ be defined by

$$Tx(t) = \int_0^t x(\tau)\,d\tau, \quad t \in [0, 1], x \in C[0, 1].$$

6. Show that any compact set in $\mathbb{C}$ is the spectrum of an operator.

7. Let $T : X \to Y$ be a linear operator, where X is a finite dimensional inner product space. Prove that:

(a) If T is self-adjoint, then $\sigma(T) \subset \mathbb{R}$.

(b) If T is unitary, then its eigenvalues have absolute value 1.

7.3 Spectral Properties of Bounded Linear Operators

Recall that $B(X)$ denotes the Banach algebra of all bounded linear operators T on a Banach space X. Next result gives a series representation of $(I - T)^{-1}$ for $T \in B(X)$; this formula is being suggested by the geometric series $1 + r + r^2 + \ldots = (1 - r)^{-1}, 0 \le r < 1$.

7.3.1 Theorem (Neumann Series). *Let X be a complex Banach space and $T \in B(X)$. If $\| T \| < 1$, then $(I - T)^{-1} \in B(X)$ and*

$$(3) \qquad (I - T)^{-1} = \sum_{k=0}^{\infty} T^k = I + T + T^2 + \ldots.$$

where the series $\displaystyle\sum_{k=0}^{\infty} T^k$ converges in the norm of $B(X)$.

Proof. Consider the series $\displaystyle\sum_{k=0}^{\infty} T^k$. Since $B(X)$ is a Banach algebra, we have

$$\| T^k \| \le \| T \|^k, \quad (k = 1, 2, \ldots)$$

and also the series $\displaystyle\sum_{k=0}^{\infty} \| T \|^k$ converges for $\| T \| < 1$. Therefore, the series in

(3) converges absolutely under the hypothesis $\| T \| < 1$. Since X is complete, $B(X)$ is complete. Consequently, absolute convergence implies convergence (Theorem 2.1.10).

Let us denote the sum of the series in (3) by S. We now claim that $S = (I - T)^{-1}$.

We have

$$(I - T) \sum_{k=0}^{n} T^k = \left(\sum_{k=0}^{n} T^k \right) (I - T) = I - T^{n+1}.$$

But

$$\| T \| < 1 \quad \Rightarrow \quad T^{n+1} \to 0, \text{ as } n \to \infty.$$

Therefore, it follows that

$$(I - T)\, S = S\, (I - T) = I.$$

Hence, $S = (I - T)^{-1}$. ∎

7.3.2 Corollary. *Let $T,\, S \in B\,(X)$. If T is invertible and*

$$\| T - S \| < \frac{1}{\| T^{-1} \|},$$

then S is invertible.

Proof. By the hypothesis, we have

$$\| I - T^{-1}\, S \| = \| T^{-1}\, (T - S) \| \le \| T^{-1} \|\, \| T - S \| < 1.$$

By Theorem 7.3.1, the operator $A = T^{-1}\, S = I - (I - T^{-1}\, S)$ is invertible and so the operator $S = T\, A$ is also invertible since T is invertible. Furthermore,

$$S^{-1} = (T\, A)^{-1} = A^{-1}\, T^{-1}. \;\blacksquare$$

7.3.3 Corollary. *Let $T \in B\,(X)$. If $\lambda \in \rho\,(T)$ and*

$$|\mu - \lambda| < \frac{1}{\|(T - \lambda\, I)^{-1}\|},$$

then $\mu \in \rho\,(T)$.

Proof. Replace T and S by $T - \lambda\, I$ and $T - \mu\, I$, respectively, in Corollary 7.3.2. ∎

7.3.4 Theorem. *Let X be a complex Banach space and $T \in B\,(X)$. Then:*

(a) *The resolvent set $\rho\,(T)$ is an open set.*

(b) *The spectrum $\sigma\,(T)$ is a closed subset of the complex plane.*

Proof. (a) If $\rho\,(T) = \phi$, then there is nothing to prove. Assume that $\rho\,(T) \ne \phi$. Let $\lambda_0 \in \rho\,(T)$ be arbitrary but fixed. Then, $T - \lambda_0\, I$ is invertible and $(T - \lambda_0\, I)^{-1} \in B\,(X)$. Now, for any $\lambda \in \mathbb{C}$, we have

$$\begin{aligned}
T - \lambda\, I &= T - \lambda_0\, I - (\lambda - \lambda_0)\, I \\[6pt]
&= (T - \lambda_0\, I)\, [I - (\lambda - \lambda_0)\, (T - \lambda_0\, I)^{-1}] \\[6pt]
&= (T - \lambda_0\, I)\, (I - S),
\end{aligned}$$

where $S = (\lambda - \lambda_0)\, (T - \lambda_0\, I)^{-1}$. Since $(T - \lambda_0\, I)^{-1} \in B\,(X)$, it follows that $S \in B\,(X)$.

Now, suppose $\| S \| < 1$, *i.e.*,

$$(4) \qquad | \lambda - \lambda_0 | < \frac{1}{\| (T - \lambda_0 I)^{-1} \|}.$$

Then, by Theorem 7.3.1, $I - S$ is invertible, $(I - S)^{-1} \in B(X)$ and

$$(5) \qquad (I - S)^{-1} = \sum_{k=0}^{\infty} S^k = \sum_{k=0}^{\infty} (\lambda - \lambda_0)^k (T - \lambda_0 I)^{-k}.$$

Therefore

$$(6) \qquad (T - \lambda I)^{-1} = (I - S)^{-1} (T - \lambda_0 I)^{-1},$$

for all those scalars λ for which (4) holds. Thus, (4) represents a neighbourhood of λ_0 consisting of regular values λ of T. Hence, $\rho(T)$ is open.

(b) Since $\sigma(T) = \mathbb{C} - \rho(T)$ and $\rho(T)$ is an open set, it follows that the spectrum $\sigma(T)$ is a closed subset of the complex plane. ∎

Remark. If we replace T and S by $T - \lambda_0 I$ and $S - \lambda I$, respectively, in Corollary 7.3.2, the first part of the above theorem also follows directly.

7.3.5 Theorem. *Let X be a complex Banach space and $T \in B(X)$. Then, $\sigma(T)$ is compact and that*

$$\sigma(T) \subset \{ \lambda \in \mathbb{C} : | \lambda | \leq \| T \| \}.$$

Proof. Let $\lambda \neq 0$. Observe that $T - \lambda I = - \lambda (I - \lambda^{-1} T)$ and that $\lambda^{-1} T \in B(X)$. By Theorem 7.3.1, if $\| \lambda^{-1} T \| < 1$, then $(I - \lambda^{-1} T)^{-1}$ exists, $(I - \lambda^{-1} T)^{-1} \in B(X)$ and

$$(I - \lambda^{-1} T)^{-1} = \sum_{k=0}^{\infty} (\lambda^{-1} T)^k.$$

In other words, if $\| T \| < | \lambda |$, then

$$(7) \qquad (T - \lambda I)^{-1} = - \lambda^{-1} (I - \lambda^{-1} T)^{-1} = - \lambda^{-1} \sum_{k=0}^{\infty} \lambda^{-k} T^k.$$

Thus, we conclude that if $| \lambda | > \| T \|$, then $\lambda \in \rho(T)$.

Consequently, $\lambda \in \sigma(T) \;\Rightarrow\; | \lambda | \leq \| T \|$. This proves that $\sigma(T)$ is bounded and

$$\sigma(T) \subset \{ \lambda \in \mathbb{C} : |\lambda| \leq \|T\| \}.$$

Hence, $\sigma(T)$ being closed (Theorem 7.3.3), it follows that $\sigma(T)$ is a compact subset of $\mathbb{C}$. ∎

7.3.6 Corollary. *The resolvent set $\sigma(T)$ of T is not empty.*

We next give a basic representation of the resolvent of T by means of a power series in λ.

7.3.7 Theorem. *Let X be a complex Banach space and $T \in B(X)$. Then, for every $\lambda_0 \in \rho(T)$, the resolvent operator $R_\lambda(T)$ has the representation given by*

$$R_\lambda = \sum_{k=0}^{\infty} (\lambda - \lambda_0)^k R_{\lambda_0}^{k+1},$$

where the series on the right side is absolutely convergent for every λ in the open disc

$$\Delta = \{ \lambda \in \mathbb{C} : |\lambda - \lambda_0| < \frac{1}{\|R_{\lambda_0}\|} \}.$$

Furthermore, $\Delta \subset \rho(T)$.

Proof. From (5) and (6), it follows that

$$R_\lambda = (I - S)^{-1} (T - \lambda_0 I)^{-1}$$

$$= \sum_{k=0}^{\infty} (\lambda - \lambda_0)^k (T - \lambda_0 I)^{-k-1}$$

$$= \sum_{k=0}^{\infty} (\lambda - \lambda_0)^k (T - \lambda_0 I)^{-1[k+1]}$$

$$= \sum_{k=0}^{\infty} (\lambda - \lambda_0)^k R_{\lambda_0}^{k+1},$$

where λ (from (7)) lies in the open disc given by

$$|\lambda - \lambda_0| < \frac{1}{\|R_{\lambda_0}\|}.$$

The assertion that $\Delta \subset \rho(T)$ is available in the proof of Theorem 7.3.4. ∎

Problems

8. Let $X = C[0, 1]$ and $T : X \to X$ be defined by $Tx = \xi\, x$, where $\xi \in X$ is fixed. Find $\sigma(T)$.

9. Find a linear operator $T : C[0, 1] \to C[0, 1]$ whose spectrum is a given interval.

10. Let $T : l^p \to l^p$, $1 \le p < \infty$, be the left shift operator defined by
$$T(\alpha_1, \alpha_2, ...) = (\alpha_2, \alpha_3, ...).$$

(a) Find $\sigma(T)$.

(b) Find series representation for $R_\lambda(T)$ and hence define $R_\lambda(T)$.

11. Let X be a complex Banach space and let M_λ be the eigenspace corresponding to an eigenvalue λ of $T \in B(X)$. Find the spectrum of $T\big|_{M_\lambda}$. *[Ans:* $\{\lambda\}$ *]*

12. Let X be a complex Banach space and $T \in B(X)$. Show that
$$\lim_{\lambda \to \infty} \| R_\lambda(T) \| = 0.$$

13. Let $T : l^\infty \to l^\infty$ be defined by
$$T(\xi_1, \xi_2, ...) = (\xi_2, \xi_3, ...).$$

Prove that:

(a) If $|\lambda| > 1$, then $\lambda \in \rho(T)$.

(b) If $|\lambda| \le 1$, then λ is an eigenvalue of T and find the eigenspace M_λ.

14. Let X be a complex Banach space and $T \in B(X)$. Let $\lambda, \mu \in \rho(T)$. Prove that:

(a) $R_\mu - R_\lambda = (\mu - \lambda)\, R_\mu\, R_\lambda.$

   (This is known as *Hilbert relation* or *resolvent equation.*)

(b) For any $S \in B(X)$
$$R_\lambda\, S = S\, R_\lambda,\ \text{whenever}\ T\, S = S\, T.$$

(c) $R_\lambda\, R_\mu = R_\mu\, R_\lambda.$

15. Let X be a complex Banach space and $S, T \in B(X)$. If $\lambda \in \rho(S) \cap \rho(T)$, then show that
$$R_\lambda(S) - R_\lambda(T) = R_\lambda(S)\, (T - S)\, R_\lambda(T).$$

16. Let X be a complex Banach space and $T \in B(X)$ be idempotent. If $T \neq O$, $T \neq I$, then show that the spectrum of T is given by
$$\sigma(T) = \{0, 1\}.$$

17. Let $T : \mathbb{R}^3 \to \mathbb{R}^3$ be an idempotent operator and let the matrix representation of T with respect to an orthonormal basis B is given by

$$[T]_B = \begin{bmatrix} \dfrac{1}{2} & \dfrac{1}{2} & 0 \\ \dfrac{1}{2} & \dfrac{1}{2} & 0 \\ 0 & 0 & 1 \end{bmatrix}.$$

Then:

(a) Find $\sigma(T)$.

(b) Find eigenvectors and the eigenspaces of T.

7.4 Complex Analysis and Spectral Theory

It is of great importance and significance to note that the spectrum of a bounded linear operator T on a complex Banach space can never be the empty set (Theorem 7.2.4). Defining the spectral radius of T, a formula for the same has been derived, the credit of which goes to I. Gelfand (1941). Before we do all this, we need some important tools from complex analysis and connections between the two areas, namely, complex analysis and spectral theory.

Let $\Omega \subset \mathbb{C}$ be an open set. A complex valued function g of a complex variable is said to be *analytic* (or *holomorphic*) on Ω if g is defined and differentiable at points of Ω. By the derivative of g at a point $z_0 \in \Omega$, we mean

$$g'(z_0) = \lim_{\Delta z \to 0} \frac{g(z_0 + \Delta z) - g(z_0)}{\Delta z}$$

exists. The function g is said to be analytic at a point $z_0 \in \Omega$ if it is analytic in a neighbourhood of z_0 contained in Ω. It is known that g is analytic on Ω if and only if for every point $z_0 \in \Omega$, it has a power series representation of the form

$$g(z) = \sum_{n=0}^{\infty} a_n (z - z_0)^n,$$

with non-zero radius of convergence.

By a vector valued (or operator valued) function defined on $\Omega \subset \mathbb{C}$, we mean a mapping

(8) $S : \Omega \to B (X)$

$$z \to S_z.$$

We write S_z instead of $S (z)$ to have a notation similar to R_λ. Now, S being given, consider an $x \in X$ so that we get a mapping

(9) $\Omega \to X$

$$z \to S_z x.$$

We may consider similarly $x \in X$ and $f \in X^*$ to get a mapping

(10) $\Omega \to \mathbb{C}$

$$z \to f (S_z x).$$

Thus mapping (10) gives a complex valued function of a complex variable defined on Ω which is obtained as composition of the mappings:

$$\Omega \to B (X) \to X \to \mathbb{C}.$$

7.4.1 Definition. Let $\Omega \subset \mathbb{C}$ be an open set and X a complex Banach space. The mapping S in (8) is said to be *locally analytic* (or *locally holomorphic*) on Ω if for every $x \in X$ and $f \in X^*$, the function $g : \Omega \to \mathbb{C}$ defined by

$$g (z) = f (S_z x),$$

is analytic at every $z_0 \in \Omega$ in the usual sense. It is said to be analytic on Ω if it is locally analytic on Ω and Ω is a domain (connected and open set in $\mathbb{C}$), and analytic at a point $z_0 \in \mathbb{C}$ if it is analytic in some neighbourhood of z_0.

7.4.2 Theorem. *Let X be a complex Banach space and $T \in B (X)$. Then, $R_\lambda (T)$ is analytic at every point $\lambda_0 \in \rho (T)$ and hence it is locally analytic on $\rho (T)$.*

Proof. By Theorem 7.3.7, $R_\lambda (T)$ has a power series representation

(11) $$R_\lambda (T) = \sum_{k = 0}^{\infty} R_{\lambda_0}^{k + 1} (T) (\lambda - \lambda_0)^k ,$$

which converges absolutely for each λ in the disc

(12) $$| \lambda - \lambda_0 | < \frac{1}{\| R_{\lambda_0} \|} .$$

Taking any $x \in X$ and $f \in X^*$, define g by

$$g(\lambda) = f(R_\lambda(T) x).$$

Then, from (8), we obtain the power series representation

$$g(\lambda) = f\left(\left(\sum_{k=0}^{\infty} R_{\lambda_0}^{k+1}(T)(\lambda - \lambda_0)^k\right) x\right)$$

$$= \sum_{k=0}^{\infty} f(R_{\lambda_0}^{k+1}(T) x)(\lambda - \lambda_0)^k$$

$$= \sum_{k=0}^{\infty} C_k (\lambda - \lambda_0)^k,$$

where $C_k = f(R_{\lambda_0}^{k+1}(T)x)$, and this series converges absolutely on the disc given by (9). This verifies the result. ∎

7.4.3 Theorem (Spectrum). *Let $X \neq \{0\}$ be a complex Banach space and $T \in B(X)$. Then $\sigma(T) \neq \phi$.*

Proof. If $T = 0$, then $\sigma(T) = \{0\} \neq \phi$. Assume that $T \neq O$. Then

$$\| T \| \neq 0.$$

For $\lambda \neq 0$, we have the following series

$$R_\lambda = (T - \lambda I)^{-1} = -\lambda^{-1}(I - \lambda^{-1} T)^{-1}$$

(13)
$$= -\lambda^{-1} \sum_{k=0}^{\infty} (\lambda^{-1} T)^k, \quad |\lambda| > \| T \|.$$

Since the series converges for $|\lambda|^{-1} < \| T \|^{-1}$, it converges absolutely for

$$|\lambda|^{-1} < \frac{1}{2} \| T \|^{-1}, \ i.e., \ |\lambda| > 2 \| T \|.$$

Therefore, from (13), we get

$$(14) \qquad \| R_\lambda \| \leq | \lambda |^{-1} \sum_{k=0}^{\infty} \| \lambda^{-1} T \|^k$$

$$= \frac{1}{| \lambda | - \| T \|} \leq \| T \|^{-1}.$$

Let, if possible, $\sigma (T) = \phi$. Then, $\rho (T) = \mathbb{C}$. Therefore R_λ is analytic on the entire complex plane $\mathbb{C}$ (Theorem 7.4.2). Thus, for fixed $x \in X$ and a fixed $f \in X^*$, the function g defined by

$$g (\lambda) = f (R_\lambda x)$$

is analytic on $\mathbb{C}$. In other words, g is an entire function and hence continuous. Thus g is bounded on the compact disc $| \lambda | < 2 \| T \|$. But

$$| g (\lambda) | = | f (R_\lambda x) |$$

$$\leq \| f \| \| R_\lambda x \|$$

$$\leq \| f \| \| R_\lambda \| \| x \|$$

$$\leq \| f \| \| T \|^{-1} \| x \|, \ \text{for} \ | \lambda | \geq 2 \| T \| \qquad \text{(by (14))}$$

$\Rightarrow \quad g$ is bounded for $| \lambda | \geq 2 \| T \|$.

Hence, g is bounded on $\mathbb{C}$. Applying Liouville's Theorem, it follows that g is a constant function.

Since $x \in X$ and $f \in X^*$ are arbitrary and $g = $ constant, it follows that R_λ is independent of λ and thus

$$R_\lambda^{-1} = T - \lambda I$$

is independent of λ. But this is not possible. Hence, the theorem follows. ∎

7.4.4 Theorem. *Let X be a complex Banach space, $T \in B (X)$ and $\lambda \in \rho (T)$. Then*

$$\| R_\lambda (T) \| \geq \frac{1}{\delta (\lambda)},$$

where $\delta (\lambda) = d (\lambda, \sigma (T)) = \inf_{S \in \sigma(T)} | \lambda - S |$. Hence

$$\| R_\lambda (T) \| \to \infty, \ as \ \delta (\lambda) \to 0.$$

Proof. Note that, for every $\lambda_0 \in \rho(T)$, the disc $|\lambda - \lambda_0| < \dfrac{1}{\| R_{\lambda_0} \|}$ is a subset of

$\rho(T)$. But, by Theorem 7.4.3, $\sigma(T) \neq \phi$. Thus we note that the distance from λ_0 to $\sigma(T)$ must at least be equal to the radius of the disc and so

$$\delta(\lambda_0) \geq \frac{1}{\| R_{\lambda_0} \|}.$$

This proves the result. ∎

In Theorem 7.3.5, we have shown an important fact that the spectrum $\sigma(T)$ of a bounded linear operator T on a complex Banach space is a bounded set in the complex plane. Therefore, it is now natural to look for the smallest disc about the origin which contains $\sigma(T)$. This motivates the following concept.

7.4.5 Definition. Let X be a complex Banach space and $T \in B(X)$. The *spectral radius* of T, denoted by $r_\sigma(T)$, is the number defined by

$$r_\sigma(T) = \sup \{ |\lambda| : \lambda \in \sigma(T) \}.$$

Clearly, $0 \leq r_\sigma(T) \leq \| T \|$ (Theorem 7.3.5). We now give a precise formula for $r_\sigma(T)$ as ascertained by I. Gelfand [13].

7.4.6 Theorem. *Let X be a complex Banach space and $T \in B(X)$. Then*

$$r_\sigma(T) = \lim_{n \to \infty} \| T^n \|^{\frac{1}{n}}.$$

Proof. Note that for any $0 \neq \lambda \in \mathbb{C}$, we have the factorization

$$T^n - \lambda^n I = (T - \lambda I) p(T) = p(T)(T - \lambda I),$$

where $p(T)$ is a polynomial in T. This statement shows that if $T^n - \lambda^n I$ has an inverse in $B(X)$, then $T - \lambda I$ also has an inverse in $B(X)$. Therefore

$$\lambda^n \in \rho(T^n) \implies \lambda \in \rho(T)$$

and so

(15) $\qquad \lambda \in \sigma(T) \implies \lambda^n \in \sigma(T^n).$

Hence, if $\lambda \in \sigma(T)$, then

$$|\lambda|^n \leq \| T^n \| \quad \text{(by (15) and Theorem 7.3.5)}$$

$$\implies \qquad |\lambda| \leq \| T^n \|^{\frac{1}{n}}, \quad \text{for } \lambda \in \sigma(T).$$

Hence

$$r_\sigma (T) = \sup \{ \, |\lambda| : \lambda \in \sigma (T) \, \} \le \| T^n \|^{\frac{1}{n}}, \ \forall \, n \in \mathbb{N}.$$

This gives

$$(16) \qquad r_\sigma (T) \le \lim_{n \to \infty} \inf \| T^n \|^{\frac{1}{n}}.$$

Further, in view of (7) the resolvent operator is represented by

$$R_\lambda (T) = -\lambda \sum_{k=0}^{\infty} \lambda^{-k} T^k, \quad |\lambda| \ge \| T \|.$$

Also, by Definition 7.4.5, $r_\sigma (T) \le \| T \|$ and by Theorem 7.4.2, $R_\lambda (T)$ is analytic at every point $\lambda \in \rho (T)$. Let $x \in X$ and $f \in X^*$. Then, the function g defined by

$$g (\lambda) = f (R_\lambda (T) \, x) = -\lambda^{-1} \sum_{k=0}^{\infty} f (\lambda^{-k} T^k x)$$

is analytic for $|\lambda| > r_\sigma (T)$. Hence, the singularities of the function g all lie in

the disc $\{\lambda : |\lambda| \le r_\sigma (T)\}$. Therefore, the series $\sum_{k=1}^{\infty} f (\lambda^{-k} T^k x)$ converges for

$|\lambda| > r_\sigma (T)$. This implies that the terms of $\sum_{k=1}^{\infty} f (\lambda^{-k} T^k x)$ form a bounded

sequence. Since this is true for every $f \in X^*$, an application of Uniform Boundedness Principle (Theorem 3.7.3) shows that the elements $\lambda^{-n} T^n$ form a bounded sequence in $B (X)$. Thus

$$\| \lambda^{-n} T^n \| \le K < \infty,$$

for some positive constant K (depending on λ). Hence

$$\| T^n \|^{\frac{1}{n}} \le K^{\frac{1}{n}} |\lambda|$$

$$\Rightarrow \lim_{n \to \infty} \sup \| T^n \|^{\frac{1}{n}} \le |\lambda|.$$

Since λ is arbitrary with $|\lambda| > r_\sigma (T)$, it follows that

$$(17) \qquad \lim_{n \to \infty} \sup \| T^n \|^{\frac{1}{n}} \le r_\sigma (T).$$

The result now follows from (16) and (17). ∎

Problems

18. Let $X \neq \{0\}$ be a complex Banach space and $T \in B(X)$ be nilpotent. Find $\sigma(T)$.

[A linear operator $T : X \to X$ is said to be *nilpotent* if $\exists$ a positive integer n such that $T^n = O$.]

19. Let

$$A = \begin{bmatrix} 0 & 1 & 0 \\ 1 & 0 & 0 \\ 0 & 0 & 1 \end{bmatrix}.$$

Find the resolvent $R_\lambda(A)$.

[*Hint:* Note that A is idempotent. Use the series

$$R_\lambda(A) = -\lambda^{-1} \sum_{k=0}^{\infty} (\lambda^{-1} A)^k, \quad (|\lambda| > \|A\|)$$

and prove that $R_\lambda(A) = (1 - \lambda^2)^{-1}(A + \lambda I).$]

20. Let X be a complex Banach space and $S, T \in B(X)$. If $ST = TS$, then show that

$$r_\sigma(ST) \leq r_\sigma(S)\, r_\sigma(T).$$

21. Let $T : l^1 \to l^1$ be defined by

$$T(\xi_1, \xi_2, ...) = (0, \xi_1, 2\xi_2, \xi_3, 2\xi_4, \xi_5, ...).$$

Show that the sequence $\{\| T^n \|^{\frac{1}{n}}\}$ in the spectral radius formula for T is not monotone.

[*Hint:* $\| T \| = 2$, $\| T^2 \|^{\frac{1}{2}} = \sqrt{2}$, $\| T^3 \|^{\frac{1}{3}} = (4)^{\frac{1}{2}}$, and so on.]

Appendix
SCHAUDER BASES

In general, it is not always possible to find out the exact form of a Hamel basis in an arbitrary infinite dimensional linear space, although it does exist (Theorem 1.6.7). Thus, the Hamel basis theory is not as helpful as one might expect for all practical purposes. To overcome this difficulty, at least in many important spaces of interest which have natural linear topologies, in particular, in the Banach spaces, we may introduce another notion of a basis, namely, Schauder basis.

Definition 1. Let X be a Banach space over the field $\mathbb{K}$. A sequence $\{x_n\} \subset X$ is called a *basis* in X if for every $x \in X$, $\exists$ a unique sequence $\{\alpha_n\} \subset \mathbb{K}$ such that

$$x = \sum_{i=1}^{\infty} \alpha_i x_i, \qquad \qquad ...(1)$$

the convergence of the series on the right of (1) being in the norm topology of X, *i.e.*,

$$\lim_{n \to \infty} \left\| x - \sum_{i=1}^{\infty} \alpha_i x_i \right\| = 0.$$

The elements in the sequence $\{\alpha_n\}$ naturally depend upon x and as such we may define the linear functionals $\{f_n\}$ on X :

$$f_n(x) = \alpha_n \equiv \alpha_n(x), \quad (n = 1, 2, ...).$$

The sequence $\{f_n\}$ is called the *associated sequence of coefficient functionals* (a.s.c.f.) to the basis $\{x_n\}$.

Each α_n being a unique function of x implies that every element in the sequence $\{x_n\}$ is non-zero. If $\{x_n\}$ is a basis in X and $\{f_n\}$ is the a.s.c.f., then every $x \in X$ has the unique expansion of the form

---

*For details of the concepts presented in the Appendix cf [38] [39].

$$x = \sum_{i=1}^{\infty} f_i(x)\, x_i. \qquad\qquad\qquad ...(i)$$

If each f_n is continuous, the basis $\{x_n\}$ is said to be a *Schauder basis.*

The notion of basis in a Banach space (Definition 1) was introduced by J. Schauder in 1927. In his definition of basis, he separately required the continuity of the coefficient functionals f_n. However, S. Banach proved in 1932 that the condition of continuity on the coefficient functionals is satisfied for all bases in a Banach space, *i.e., every basis in a Banach space is a Schauder basis**. Thus, in a Banach space, whether we talk of a basis or a Schauder basis, they are one and the same.

Theorem 1. *A Banach space with a basis is separable.*

Proof. Let X be a Banach space over the field $\mathbb{K}$ ($\mathbb{R}$ or $\mathbb{C}$) with a basis $\{x_n\}$. Consider the set

$$A = \left\{ \sum_{i=1}^{n} r_i x_i : r_i \text{'s are rationals in } \mathbb{K} \text{ and } n \in \mathbb{N} \right\}.$$

Clearly, A is a countable subset of X.

Claim. A is dense in X.

Let $\varepsilon > 0$ be given. Since $\{x_n\}$ is a basis in X, $\overline{\text{span}\,\{x_n\}} = X$. Then, for an $x \in X, \exists\, a\; y \in \text{span}\,\{x_n\}$ such that

$$\|x - y\| < \frac{\varepsilon}{2}.$$

We can write

$$y = \sum_{i=1}^{k} \alpha_i x_{n_i},$$

for some finite scalars $\alpha_1, \alpha_2, ..., \alpha_k$ in $\mathbb{K}$ and also we can find rationals $r_i\,(1 \le i \le k)$ such that

$$|\alpha_i - r_i| < \frac{\varepsilon}{2k\,\|x_{n_i}\|}.$$

---

* The details of the proof of this result is beyond the present scope and hence omitted.

Now

$$\left\| x - \sum_{i=1}^{k} r_i\, x_{n_i} \right\| \leq \left\| x - \sum_{i=1}^{k} \alpha_i x_{n_i} \right\| + \sum_{i=1}^{k} |\alpha_i - r_i|\, \left\| x_{n_i} \right\|$$

$$< \frac{\varepsilon}{2} + \frac{\varepsilon}{2} = \varepsilon.$$

This means that for a given $x \in X$, $\exists$ an element $z \left(= \sum_{i=1}^{n} r_i x_{n_i} \right)$ in A such that

$$\| x - z \| < \varepsilon.$$

Hence A is dense in X. ∎

Remark. In 1932, S. Banach raised the problem: Is the converse of Theorem 1 true, *i.e.*, *Does every separable Banach space possess a basis ?* This is known as the *Basis Problem* which is settled in 1973 by P. Enflo. In fact, Enflo constructed a separable Banach space without a basis.

Definition 2. Let X be a Banach space. A system (x_n, f_n), where $\{x_n\} \subset X$ and $\{f_n\} \subset X^*$, is said to be a *biorthogonal system* in X if

$$f_n(x_m) = \delta_{nm}, \ (n, m \in \mathbb{N})$$

where δ_{nm} is the Kronecker delta.

Note that, if $\{x_n\}$ is a basis in X and $\{f_n\} \subset X^*$, the a.s.c.f., then:

(i) $\{x_n, f_n\}$ is a biorthogonal system.

(ii) $\{x_n\}$ is X-complete, *i.e.*, $\overline{\text{span}\, \{x_n\}} = X$.

(iii) $\{f_n\}$ is total over X, *i.e.*,

$$\{x \in X : f_n(x) = 0, \ \forall\, n \in \mathbb{N}\} = \{0\}.$$

(iv) $\lim_{n \to \infty} S_n(x) = x, \ \forall\, x \in X$, where $S_n(x) = \sum_{i=1}^{n} f_i(x)x_i$.

the convergence being in the norm topology of X. (The sequence $\{S_n\}$ is called the *sequence of partial sum operators* to the basis $\{x_n\}$).

Examples

1. The unit vectors $\{e_i\}$ form a basis for c_0.

Let $x = \{\alpha_i\} \subset \mathbb{K}$ be an arbitrary element in c_0. Then, for $n \in \mathbb{N}$, we have

$$\left\| x - \sum_{i=1}^{n} \alpha_i e_i \right\|_{c_0} = \left\| (0, \ldots\ldots, 0, \alpha_{n+1}, \alpha_{n+2}, \ldots\ldots) \right\|_{c_0}$$

$$= \sup_{n+1 \le i < \infty} |\alpha_i|$$

$$\to 0, \text{ as } n \to \infty \quad \left(\because \lim_{i \to \infty} \alpha_i = 0 \right)$$

$$\Rightarrow \qquad x = \lim_{n \to \infty} \sum_{i=1}^{n} \alpha_i e_i = \sum_{i=1}^{\infty} \alpha_i e_i \qquad \ldots(2)$$

Note that the sequence $\{\alpha_i\} \subset \mathbb{K}$ satisfying (2) is unique; for if possible, let $\{\beta_i\} \subset \mathbb{K}$ be the another sequence satisfying (2). Then

$$\sum_{i=1}^{\infty} \alpha_i e_i = \sum_{i=1}^{\infty} \beta_i e_i$$

$$\Rightarrow \qquad \lim_{n \to \infty} \sum_{i=1}^{\infty} (\alpha_i - \beta_i) e_i = 0$$

$$\Rightarrow \qquad \lim_{n \to \infty} \left(\sup_{1 \le i \le n} |\alpha_i - \beta_i| \right) = 0.$$

Therefore

$$|\alpha_i - \beta_i| \le \sup_{1 \le i \le n} |\alpha_i - \beta_i| \to 0, \text{ as } n \to \infty$$

$$\Rightarrow \quad \alpha_i = \beta_i, \quad \forall \, i \in \mathbb{N}.$$

Hence $\{e_i\}$ form a basis for c_0. ∎

2. The unit vectors $\{e_i\}$ form a basis for l^1.

Let $x = \{\alpha_i\} \subset \mathbb{K}$ be an arbitrary element in l^1. Then

$$\sum_{i=1}^{\infty} |\alpha_i| < \infty$$

For $n \in \mathbb{N}$, we have

$$\left\| x - \sum_{i=1}^{n} \alpha_i e_i \right\|_{l^1} = \left\| (0, \ldots\ldots, 0, \alpha_{n+1}, \alpha_{n+2}, \ldots\ldots) \right\|_{l^1}$$

$$= \sum_{i=n+1}^{\infty} |\alpha_i| \to 0, \text{ as } n \to \infty$$

$$\Rightarrow \qquad x = \lim_{n \to \infty} \sum_{i=1}^{n} \alpha_i e_i. \qquad\qquad ...(3)$$

Note that the sequence $\{\alpha_i\} \subset \mathbb{K}$ satisfying (3) is unique; for if possible, let $\{\beta_i\} \subset \mathbb{K}$ be another sequence satisfying (3). Then

$$\sum_{i=1}^{n} \alpha_i e_i = \sum_{i=1}^{\infty} \beta_i e_i$$

$$\Rightarrow \qquad \lim_{n \to \infty} \sum_{i=1}^{n} (\alpha_i - \beta_i) e_i = 0$$

$$\Rightarrow \qquad \lim_{n \to \infty} \sum_{i=1}^{n} |\alpha_i - \beta_i| = 0$$

$$\Rightarrow \qquad \alpha_i = \beta_i, \quad \forall\, i \in \mathbb{N}.$$

Hence $\{e_i\}$ forms a basis for l^1.

3. The unit vectors $\{e_i\}$ form a basis for l^p, $1 \le p < \infty$.

4. The unit vectors $\{e_i\}$ do not form a basis for l^∞; rather, the space l^∞, being non-separable, does not possess any basis, in view of Theorem 1.

5. The sequence $\{x_n\} \subset C\,[0, 1]$, where

$$x_0(t) \equiv 1$$

$$x_1(t) = t$$

$$x_{2^n + m}(t) = \begin{cases} 0, & t \notin \left]\dfrac{2m-2}{2^{n+1}}, \dfrac{2m}{2^{n+1}}\right[\\[2ex] 1, & t = \dfrac{2m-1}{2^{n+1}} \\[2ex] \text{linear in } \left[\dfrac{2m-2}{2^{n+1}}, \dfrac{2m-1}{2^{n+1}}\right] \text{and} \left[\dfrac{2m-1}{2^{n+1}}, \dfrac{2m}{2^{n+1}}\right] \end{cases}$$

$$(m = 1, 2, ..., 2^n\,; n = 0, 1, 2, ...)$$

constitutes a basis for the Banach space $C\,[0, 1]$.

(The sequence $\{x_n\}$ is called the *Schauder system* for $C\,[0, 1]$).

6. The sequence of equivalence classes $\{\tilde{y}_n\}$, where y_n are the functions defined on $[0, 1]$ by

$$y_1(t) \equiv 1$$

$$y_{2^n + m}(t) = \begin{cases} \sqrt{2^n}, & t \in \left[\dfrac{2m-2}{2^{n+1}}, \dfrac{2m-1}{2^{n+1}}\right[\\[2ex] \sqrt{2^n}, & t \in \left[\dfrac{2m-1}{2^{n+1}}, \dfrac{2m}{2^{n+1}}\right[\\[2ex] 0, & \text{for other } t \end{cases}$$

$$(m = 1, 2, \,..... \, 2^n \, ; n = 0, 1, 2, \,.... \,)$$

constitutes a basis for the space $L^p[0, 1]$, $1 \le p < \infty$.

(The sequence $\{y_n\}$ is called the *Haar System* for $L^p[0, 1]$).

We now give below the comparison between the two types of bases; namely, Hamel basis and basis (Schauder basis).

Case I. When dim $X < \infty$.

In this case, every Hamel basis of X is a basis; and vice versa. In fact, when X is finite dimensional linear space, any two norms defined on X are equivalent and hence both generate the same topology on X in which X is complete.

Case II. When dim $X = \aleph_0$.

In this case a Hamel basis is also a basis. Indeed, let

$$S = \{x_n : n \in \mathbb{N}\}$$

be a Hamel basis for X. Then, every, $x \in X$ has the unique representation of the form

$$x = \sum_{i=1}^{k} \alpha_{n_i} x_{n_i},$$

where α_{n_i}'s are scalars and k is a positive integer. Define

$$\alpha_n = \begin{cases} \alpha_{n_i}, & n = n_i \\ 0, & n \ne n_i \end{cases}.$$

Then

$$x = \sum_{n=1}^{\infty} \alpha_n x_n.$$

However, the converse need not be true, *i.e.*, a basis in X need not be a Hamel basis for X (of course, the cardinality of the Hamel basis and the basis is the same, *i.e.*, $\aleph_0$). Indeed, consider the linear space

$$\Phi = \{ \{\alpha_i\} \subset \mathbb{K} : \alpha_i \neq 0 \text{ for at most finite } i\text{'s}\}$$

equipped with the norm given by

$$\| \{\alpha_i\} \|_\infty = \sup_{1 \leq i < \infty} |\alpha_i|.$$

Then $(\Phi, \| \cdot \|_\infty)$ is a (incomplete) normed linear space. In fact, a normed space of dimension $\aleph_0$ is always incomplete (Problem 3.13). Clearly, $\{e_n\}$ is a sequence of vectors in Φ such that

$$\Phi = \text{span } \{e_n : n \geq 1\}.$$

Here $\{e_n\}$ is a Hamel basis for Φ which is also a basis for $(\Phi, \| \cdot \|_\infty)$.

We now construct a basis for $(\Phi, \| \cdot \|_\infty)$ which is not a Hamel basis for Φ as below:

Let $\{x_n\}$ be a sequence of vectors in Φ, where

$$x_1 = (1, \frac{1}{2}, 0, 0, \dots)$$

$$x_2 = (0, \frac{1}{2}, \frac{1}{3}, 0, \dots)$$

$$x_3 = (0, 0, \frac{1}{3}, \frac{1}{4}, 0, \dots)$$

$$\dots\dots\dots\dots\dots\dots\dots$$

$$\dots\dots\dots\dots\dots\dots\dots$$

$$\dots\dots\dots\dots\dots\dots\dots$$

$$x_n = (0, 0, \dots , 0, \frac{1}{n}, \frac{1}{n+1}, 0, \dots)$$

$$\dots\dots\dots\dots\dots\dots\dots\dots\dots$$

$$\dots\dots\dots\dots\dots\dots\dots\dots\dots$$

$$\dots\dots\dots\dots\dots\dots\dots\dots\dots$$

Claim. $\{x_n\}$ is a basis but not a Hamel basis for $(\Phi, \| \cdot \|_\infty)$.

Let $x = (\alpha_1, \alpha_2, \dots, \alpha_m, 0, 0, \dots) \in \Phi$. Then, we can find a sequence $\{\beta_i\}_{i=1}^m \subset \mathbb{K}$ such that

$$\alpha_1 = \beta_1$$

$$\alpha_2 = \frac{1}{2}\beta_1 + \frac{1}{2}\beta_2$$

$$\alpha_3 = \frac{1}{3}\beta_2 + \frac{1}{3}\beta_3$$

$$\cdots\cdots\cdots\cdots\cdots\cdots\cdots\cdots$$

$$\cdots\cdots\cdots\cdots\cdots\cdots\cdots\cdots$$

$$\cdots\cdots\cdots\cdots\cdots\cdots\cdots\cdots$$

$$\alpha_m = \frac{1}{m}\beta_{m-1} + \frac{1}{m}\beta_m.$$

Then

$$x = \sum_{i=1}^{m} \beta_i x_i + \sum_{i=m+1}^{\infty} (-1)^j \beta_m x_i,$$

where $j = i + 1$, if m is odd and $j = i$ if m is even, and the series converges in the norm topology of Φ. Thus, $\{x_n\}$ is a basis for $(\Phi, \|\cdot\|_\infty)$. But it is not a Hamel basis for Φ since one may easily note that

$$e_1 = \sum_{i=1}^{\infty} (-1)^{i+1} x_i$$

is the unique representation for e_1 which, in fact, is an infinite sum.

Case III. When $\dim X > \aleph_0$.

In this case, the cardinality of any Hamel basis for X is greater than $\aleph_0$ while that of a basis, in case it exists, is $\aleph_0$. Hence, neither a Hamel basis can be a basis nor vice versa. As an illustration, consider the Banach space $(c_0, \|\cdot\|_\infty)$ (Problem 2.11). Here $\dim c_0 = 2^{\aleph_0} > \aleph_0$. Thus, any Hamel basis for c_0, as a linear space, has cardinal number $2^{\aleph_0}$ ($= c$). On the other hand, $\{e_n\}$ forms a basis for $(c_0, \|\cdot\|_\infty)$ (Example 1). ∎

Problems

1. Let $\{x_n\}$ be a basis in a Banach space X and $\{S_n\}$ the sequence of partial sum operators. Prove that:

(a) $S_n \in B(X)$, $(n = 1, 2, ...)$

(b) If $\| x \|' = \sup_{1 \leq n < \infty} \| S_n (x) \|$, then $(X, \| \cdot \|')$ is a complete normed space.

(c) $\sup_{1 \leq n < \infty} \| S_n \| < \infty$.

(The number $\sup_{1 \leq n < \infty} \| S_n \|$ is called the *basis constant* of the basis $\{x_n\}$.)

2. A sequence $\{x_n\}$ in a Banach space X is said to be a *basic sequence* in X if it is a basis for $\overline{\text{span} \{x_n\}}$. Prove that:

(a) The unit vectors $\{e_i\}$ form a basic sequence in l^∞.

(b) Every subsequence of a basis in X is a basic sequence in X.

(c) If dim $X = \infty$, then X has a basic sequence.

3. Prove that a sequence $\{x_n\}$ in a Banach space X is a basic sequence in X if and only if $\exists\, M \geq 1$ such that

$$\left\| \sum_{i=1}^{m} \alpha_i x_i \right\| \leq M \left\| \sum_{i=1}^{n} \alpha_i x_i \right\|$$

for all m, n with $m \leq n$ and arbitrary scalars $\alpha_1, \alpha_2, ..., \alpha_n$.

(This inequality is known as *Nikolskii inequality.*)

4. Let $\{x_n\}$ be a basis in a Banach space X and $\{f_n\} \subset X^*$ be the a.s.c.f. Prove that:

(a) $\{f_n\}$ is a basic sequence in X^* and that

$$f = \sum_{i=1}^{\infty} f(x_i)\, f_i, \quad f \in \overline{\text{span} \{f_n\}}.$$

(b) Give an example to show that $\{f_n\}$, in general, does not form a basis for X^*.

(c) In case X is reflexive, then $\{f_n\}$ is a basis for X^*.

5. With the notations of Problem 4, if $\overline{\text{span} \{f_n\}} = X^*$, the basis $\{x_n\}$ of X is said to be *shrinking*. Prove that:

(a) The unit vector basis $\{e_n\}$ of c_o is shrinking.

(b) The unit vector basis $\{e_n\}$ of l^1 is not shrinking.

(c) The unit vector basis $\{e_n\}$ in l^p, $1 < p < \infty$, is shrinking.

(d) If X is a Banach space with a basis $\{x_n\}$ whose conjugate space X^* is non-separable, then X has no shrinking basis.

(e) If X is a reflexive Banach space with a basis, then all bases of X are shrinking.

6. With the notations of Problem 4, prove that the following statements are equivalent:

(a) $\{x_n\}$ is shrinking.

(b) $\lim\limits_{n \to \infty} \|f\|_n = 0, \quad \forall f \in X^*$

  where $\|f\|_n = \|\, f\,|_{\overline{\mathrm{span}}\{x_{n+1},\, x_{n+2},\, \ldots\}}\|$.

(c) $S_n(x) \xrightarrow{w} x$ uniformly on the unit ball of X.

(d) A bounded sequence $\{y_k\} \subset X$ converges weakly to 0 whenever

$$\lim_{k \to \infty} f_n(y_k) = 0, \ (n = 1, 2, \ldots).$$

7. A basis $\{x_n\}$ in a Banach space X is said to be *boundedly complete* if for each sequence $\{\alpha_n\} \subset \mathbb{K}$ for which

$$\sup_{1 \le n < \infty} \ \Big\| \sum_{i=1}^{n} \alpha_i x_i \Big\| < \infty,$$

there exists an $x \in X$ such that

$$x = \sum_{i=1}^{\infty} \alpha_i x_i.$$

Prove that:

(a) The unit vector basis $\{e_n\}$ in c_0 is not boundedly complete.

(b) The unit vector basis $\{e_n\}$ in l^1 is boundedly complete.

(c) The unit vector basis $\{e_n\}$ in l^p, $1 < p < \infty$, is boundedly complete.

8. Let $\{x_n\}$ be a basis in a Banach space X and $\{f_n\} \subset X^*$ be the a.s.c.f. Prove that the following statements are equivalent:

(a) $\{x_n\}$ is boundedly complete in X.

(b) $\displaystyle\sum_{i=1}^{\infty} \varphi(f_i)\, x_i$ converges for each $\varphi \in X^{**}$.

(c) $X^{**} = \pi(X) \oplus \left[\overline{\text{span} \{f_n\}}\right]^{\perp}$, where $\pi : X \to X^{**}$ is the canonical mapping.

9. If $\{x_n\}$ is a shrinking basis in a Banach space X, then prove that its a.s.c.f. $\{f_n\} \subset X^*$ is a boundedly complete basis for X^*.

10. Let X be a Banach space with a basis. Prove that X is reflexive if and only if the basis is both shrinking and boundedly complete (*James Theorem*).

11. A basis $\{x_n\}$ in a Banach space X is said to be *unconditional* if every

convergent series of the form $\sum\limits_{i=1}^{\infty} \alpha_i x_i$ is unconditionally convergent, *i.e.*,

for every $x \in X$, the series $\sum\limits_{i=1}^{\infty} f_i(x) x_i$ converges unconditionally, where

$\{f_n\} \subset X^*$ is the a.s.c.f. to the basis $\{x_n\}$. A basis $\{x_n\}$ is said to be *conditional* it if is not unconditional. Prove that:

(a) The natural basis $\{e_n\}$ in c_0 is unconditional.

(b) The sequence $\{x_n\} \subset X$, where

$$x_n = \sum_{j=1}^{n} e_j, \quad (n = 1, 2, \ldots.)$$

defines a conditional basis in c_0.

(c) The Schauder system for $C[0, 1]$ is not an unconditional basis.

(d) The space $C[0, 1]$ has no unconditional basis.

(e) The Haar system for $L^1[0, 1]$ is not an unconditional basis.

(f) The space $L^1[0, 1]$ has no unconditional basis.

(g) In an infinite dimensional Banach space with a basis, there exists a conditional basis.

(h) All the bases in a Banach space are unconditional if and only if it is finite dimensional.

BIBLIOGRAPHY

1. Ahlfors, L.V. : *Complex Analysis,* 2nd ed., McGraw-Hill, New York (1966).

2. Bachman, G. and Narici, L. : *Functional Analysis,* Academic Press, New York (1966).

3. Banach, S. : *Theorie des Operations Lineares,* Monografje Matematyczne, Warsaw (1932).

4. Berberian, S. : *Introduction to Hilbert Space,* Oxford University Press, New York (1961).

5. Bohenblust, H.F. and Sobczyk, A. : *Extensions of Functionals on Complex Linear Spaces,* Bull. Amer. Math. Soc. 44 (1938), 91-93.

6. Bollobas, B. : *Linear Analysis,* Cambridge University Press, New York (1990).

7. Conway, J.B. : *A Cource in Functional Analysis,* Springer-Verlag, New York Inc. (1985).

8. Curtain, R. F. and Pritchard, A. J. : *Functional Analysis in Modern Applied Mathematics,* Academic Press Inc. (London) Ltd., (1977).

9. Day, M.M. : *Normed Linear Spaces,* Springer-Verlag, New York (1973).

10. Debnath, L. and Mikusiński, P. : *Introduction to Hilbert Spaces with Applications,* Academic Press (1990).

11. Dieudonnè, J. : *Foundations of Modern Analysis,* Academic Press, New York (1960).

12. Dixmier, J. : *Sur les Bases Orthonormales dans les Espaces Pre ′hilbertiens,* Acta Math. (Szeged), 15 (1953), 29-30.

13. Dunford, N. and Schwartz, J.T. : *Linear Operators,* Part I, Interscience, New York (1958).

14. Edwards, R.E. : *Functional Analysis,* Holt Rinehart and Winston, New York (1965).

15. Enflo, P. : *A Counterexample to the Approximation Property,* Acta Math. 130 (1973), 309-317.

16. Flaherty, R.E. : *Functional Analysis,* (English edition edited by G.F. Votruba), Walters-Noordhoff Publishing Groningen, The Netherlands (1972).

17. Gelfand, I. : Normierte Ringe, Mat. Sbornik N S. 9 (51) (1941), 3-24.

18. Goffman, C. and Pedrick, G. : *First Course in Functional Analysis,* Prentice-Hall of India Pvt. Ltd., New Delhi (1987).

19. Halmos, P.R. : *Finite Dimensional Vector Spaces,* 2nd ed., Van Nostrand Rienhold, New York (1958).

20. Halmos, P.R. : *A Hilbert Space Problem Book,* Springer-Verlag, New York (1974).

21. Helmberg, G. : *Introduction to Spectral Theory in Hilbert Space,* North-Holland, Amsterdam (1969).

22. Hille, E. and Phillips, R.S. : *Functional Analysis and Semi-groups,* Amer. Math. Soc. Coll. Publ. 31 (1957).

23. Holmes, R.B. : *Geometric Functional Analysis and its Applications,* Springer-Verlag, New York (1975).

24. Jain, P.K. and Gupta, V.P. : *Lebesgue Measure and Integration,* New Age International (P) Ltd., Publishers, New Delhi (1986), reprint (2007).

25. Jain, P.K. and Ahmad, K. : *Metric Spaces,* Narosa Publishing House, New Delhi, Second Ed., Second Reprint (2009).

26. James, R.C. : *Bases and Reflexivity of Banach Spaces,* Ann. Math. 52 (1950), 518-527.

27. James, R.C. : *A Non-reflexive Banach Space Isometric with its Conjugate Space,* Proc. Nat. Acad. Sci. U.S.A. 37 (1951), 174-177.

28. Kreyszig, E. : *Introductory Functional Analysis with Applications,* John Wiley & Sons, New York (1978).

29. Limaye, B.V. : *Functional Analysis,* New Age International (P) Ltd., Publishers, New Delhi (1984).

30. Lorch, E.R. : *Spectral Theory,* Oxford University Press, New York (1962).

31. Lusternik, L.A. and Sobolev, V.J. : *Elements of Functional Analysis,* Hindustan Publishing Corporation, New Delhi (1971).

32. Marti, J.T. : *Introduction to the Theory of Bases,* Springer-Verlag, New York (1969).

33. Riesz, F. : *Zur Theorie des Hilbertschen Raumes,* Acta Sci. Math. (Szeged), 7 (1934), 34-38.

34. Riesz, F. and Sz-Nagy, B. : *Functional Analysis,* Ungar, New York (1955).

35. Rynne, B.P. and Youngson, M.A.: *Linear Functional Analysis,* Springer-Verlag, London Limited, (2000).

36. Saxe, K. : *Beginning Functional Analysis,* Springer-Verlag, New York, Inc. (2002).

37. Simmons, G.F. : *Introduction to Topology and Modern Analysis,* McGraw-Hill Book Company, New York (1963).

38. Singer, I. : *Bases in Banach Spaces I,* Springer-Verlag, New York (1970).

39. Singer, I. : *Bases in Banach Spaces II,* Springer-Verlag, New York (1981).

40. Taylor, A.E. : *Introduction to Functional Analysis,* John Wiley & Sons, New York (1958).

41. Yosida, K. : *Functional Analysis,* 3rd ed., Springer-Verlag, New York (1971).

INDEX

Symbols

(Banach) inverse mapping theorem 136
3-dimensional space 40

A

Absolutely
    convergent 10
    summable 10, 66
Adjoint (or conjugate) 205
Affine hyperplane or simply a hyperplane 27
Algebra 115
    with identity 115
Algebraic dual 33
Algebraically reflexive space 34
Analytic 191
    (or holomorphic) 349
Associated sequence of coefficient
    functionals 356
Axiom of choice 11
Axioms of the
    metric space 38
    norm 62

B

B*-algebra 307
Baire's category theorem 52
Banach
    algebra 116, 307
    space 65
    Mazur distance 131
Basic sequence 364
Basis 356
    constant 364
    problem 358
Bessel
    generalised inequality 268
    inequality 266

Bicontinuous 58
Bijective 6
Biorthogonal system 358
Bilinear functional (or bilinear form) 34
Bolzano-Weierstrass Property (BWP) 59
Bounded 5, 43, 101, 152, 192, 293
    above 5
    below 5
    operator 103
Boundedly complete 365

C

Canonical mapping
    (or canonical embedding) 34
    (or, quotient mapping) 26
Canonical norm 246
Cantor's continuum hypothesis 14
Cardinal number (or the power) 13
Cartesian product 2
Cauchy
    sequence 9, 46, 63
    Schwartz inequality (Finite form) 15
    Schwartz inequality (Infinite form) 16
Characteristic equation 338
Closed
    graph theorem 142
    interval 6
    linear hull (or closed linear span) 92
    linear operator 139
    set 43
    sphere (or closed ball) 63
    subspace 89
    subspace 247
Co-dimension 26
Co-domain 4, 6
Commuting operators 319
Compact 58
Complement 3

Complementary pair 25
Complete 63
      (or maximal) orthonormal set 279
      metric space 47
Completion 56
      (or complete enclosure) 98
Complex
      Euclidean metric 40
      linear space 17
Composite function 7
Conjugate space, or adjoint space 160
Constant function 7
Continuous
      at a point 57
      at a point x_0 101
      on X 57, 101
      spectrum 340
Contraction 318
Convergent 8, 46, 63
Converges weakly 212
Convex hull 27
Countable 12
Countably infinite 12
Cover of X 58

D

Decreasing 8
Denumerable 12
Derived set 43
Diameter of A 45
Difference 2
Differentiation operator 107
Dimension 24
Dirac
      distribution 155
      δ-function 155
Direct
      sum 25
Discrete metric 38
      space 38
Disjoint 3
Distance 45
      between the sets 45
      function 38
Domain 4, 6
Dual space 160

E

Eigenspace 336
Eigenvalue (characteristic value or proper
      value) 335
Eigenvector (characteristic vector or proper
      vector 335
Empty (or void or null) set 2
Equivalence
      class 4
      relation 4
Equivalent
      function 81
      norms 119
Essential supremum 20
Essentially bounded 20
Euclidean Space $\mathbb{R}^n$ 69
Everywhere dense (or, simply dense) 48
Extended real number system 6
Extension of f 7

F

f-image 6
f-inverse 7
Family of indexed sets 3
Finite
      cardinal number 14
      dimensional 22
      subcover 58
Fourier
      coefficients 273
      series expansion 274
Function (or mapping) 6
Function space C [a, b] 77
Fundamental (or total) 217

G

Grahm-Schmidt process 275
Graph of T 139
Greatest lower bound (or infimum) 5

H

Haar system 361
Hamel basis 21
Hermitian sesquilinear form (or simply,
      Hermitian form) 35

Hilbert
  relation or resolvent equation 348
  space 239
  adjoint operator 298
Hölder
  inequality (Finite form) 15
  inequality (Infinite form) 16
  space 88
Homeomorphic 58
Homeomorphis 57

I

Idempotent 32
Identity
  function 6
  operator 28
Imbed 96
Increasing sequence 8
Index set 3
Indexed sets 3
Induced functionals 197
Infinite dimensional 22
  Euclidean and unitary spaces 75
Inner product 234
  space C $[0, 2\pi]$ 263
  space (or pre-Hilbert space) 235
Integral function 192
Interior of G 43
Interior point of G 43
Intersection 2
Invariant 145, 332
Inverse (algebraic) 29
Invertible 29
Involution 307
Isometric 96
  isomorphism 118
  operator 326
  spaces 46, 96
Isometrically isomorphic 118
Isometry 46, 96, 116
Isomorphic 31, 249
Isomorphism 31, 249

J

James theorem 366

K

Kernel of T 29

L

l^1–norm 70
Large 52
Least upper bound (or supremum) 5
Lebesgue norm 79
Left inverse of T 30
Left shift operator 298
Limit 8
  inferior 9
  inferior or the lower limit 10
  superior 9
  superior (or the upper limit) 9
Linear
  combination 21
  functional 152
  map 27
  projection 32
  space 17
  subspace (or, simply, a subspace) 20
  topological invariant property 121
  transformations, linear operators or
    homomorphism 28
  variety 27
Linearly independent 21
Locally
  analytic (or locally holomorphic) 350
  compact 60
Lower bound 5
l^p-norm 76
L^p-norm (or simply a p-norm on L^p $[a, b]$ 79

M

Maximal 27
Meagre sets, sets of Baire's first category or
    simply of first category 52
Metric 38
  induced by the norm 62
  space 38
  subspace 42
Minkowski inequality
  (Finite form) 15
  (Infinite form) 16
Minute 52
Monotone (monotonic) sequences 8
Multiplier 111

N

n-dimensional Euclidean space 40, 70

n-dimensional unitary space 40, 70
Natural embedding (or, the conical mapping) 197
Neighbourhood 43
    of zero 131
Neumann series 344
Nikolskii inequality 364
Nilpotent 355
Non-meagre sets or the sets of second (Baire) category 52
Norm 61
    (or bound) of T 103
    induced by the inner product 239
    topology 119
Normal operator 309
Normed
    algebra 116
    linear space 62
    space 62
Nowhere dense 51
Null space of T 29

O

One-to-one (injective) 6
Onto (surjective) 6
Open 63
Open cover 58
    interval 6
    map 131
    set 43
    sphere (or open ball) 42, 63
Order relation 4
Ordered basis 35
Orthogonal 331
    (or Hilbert) dimension 283
    (or perpendicular) projection operator 329
    complement 230
    projection 259
    projection operator (briefly, orthogonal projection) 259
    set 262
    vectors 250
Orthonormal
    basis 281
    set 262

P

Parallelogram equality 239

Parseval relation 274
Part of a projection 333
Partial order 11
Partially ordered set 11
Partition 4
Perpendicular projection operator 259
Point spectrum (or discrete spectrum) 340
Pointwise bounded 145
Polarization identity 243
Positive
    operators 316
    square root 319
Power set 2
Pre-image 6
Pythagorean
    identity 253
    theorem 251

Q

Quotient
    norm 95
    set 4
    space (or factor space) 25

R

Range 4, 6
Rare sets 52
Real
    linear space 17
    or complex Hilbert space 239
    valued function 6
Reduce 145
    T (or T is reduced by M) 333
Reflexive 197
Reflexivity 4
Regular value 339
Relation 4
Relative metric induced (or simply the metric induced 42
Relatively compact (or, precompact) 58
Residual spectrum 340
Resolvent
    operator (or, simply, resolvent) 339
    set 339
Restriction of f 7
Riesz
    Fisher theorem 82
    Fréchet representation theorem 290
    Hölder inequality 79

Minkowski inequality 80
Right
    inverse of T 30
    shift operator 298

S

Schauder basis 357
Schwartz inequality 237
Second algebraic dual space of V 33
Self-adjoint
    (or Hermitian) 309
    operators 309
Semi-open interval,
    closed at left 6
    closed at right 6
Separable 49, 92
Sequence 8
    of partial sum operators 358
    of partial sums 10
    space l^∞ 47, 75
    space l^p 47, 73
    space ω 86
Sequentially compact 59
Series 10
Sesquilinear functional (or sesquilinear form) 34
Set 1
Shrinking 364
Simple order or a linear order 4
Skew-adjoint 318
Small 52
Sobolev norm 88
Space 62
    $(C^n, <.,.>)$ 240
    $(\Phi, <.,.>)$ 240
    $(l^2, <.,.>)$ 241
    $(\mathbb{R}^n, <.,.>)$ 240
    B [a, b] 87
    bv 87
    bts 87
    BV [a, b] 88
    C 47
    c 87
    C [a, b] 47, 236, 241, 242
    $C_0(\mathbb{R})$ 88
    C^1 [0, 1] 85
    C_λ [0, 1] 88
    C^n [0,1] 86
    C^n [a, b] 87, 88
    c_0 87
    Φ 90, 236
    kr 87
    L^∞ [a, b] 242
    l^2 236
    $L^2 [-\pi, \pi]$ 264
    L^2 [a, b] 242
    l^p 241
    $l^p(n)$ 71
    P [a, b] 83
    R 47
Spaces $\mathbb{R}^n$ and $\mathbb{C}^n$ 47
Span S 21
Spectral
    radius 353
    value 339
Spectrum of T 339
Sphere
    (or closed ball) 43
    (or open ball) 63
Strictly
    decreasing sequences 8
    increasing 8
Strong
    limit 214
    operator limit 230
Strongly convergent (or convergent in the norm) 214
Strongly operator convergent 230
Subcover 58
Sublinear functional 180
Subsequence 8
Subset 2
Subspace 89, 246
Summable 10, 66
Sup norm
    (or the uniform norm) 70
    (or uniform norm) on C [a, b] 79
    on l^∞ 76
Superset 2
Symmetry 4

T

Topological
    inverse (or, simply, inverse) 109
    isomorphism (or linear homeomor-phism) 118
    linear space 65

Total
order relation 11
sets 280
Totally
bounded 59
ordered set (or chain or a linearly
ordered set 11
Transfinite cardinal number 14
Transitivity 4
Triangle inequality 14
Two-out of-three result 96

U

Unbounded 5, 101
interval 6
linear functionals 156
Unconditional 366
Uniform
boundedness principle 146
operator limit 230
Uniformly
bounded 145
operator convergent 230
Unilateral shift operator 117
Union 2
Unit open sphere 64
Unitarily equivalent 327
Unitary
operators 309
space $\mathbb{C}^n$ 70

Usual matrix 38

V

Vector space 17
Vectors 17
Voltra integral operator 108

W

Weak
Cauchy sequence 219
convergent 223
limit 212, 223
Weakly
closed 219
complete 219
convergent 212
operator convergent 230
sequentially compact 219
Well ordering theorem 12

Z

Zermelo's postulate 11
Zero
operator 28
space 18
Zorn's Lemma 12

The Concert Composer's
Business Handbook